Fundamentals of Geometric Dimensioning and Tolerancing

Second Edition

by

Alex Krulikowski

Delmar Publishers

an International Thomson Publishing company I(T)P®

Albany • Bonn • Boston • Cincinnati • Detroit • London • Madrid
Melbourne • Mexico City • New York • Pacific Grove • Paris • San Francisco
Singapore • Tokyo • Toronto • Washington

NOTICE TO THE READER

Cover Design by Susan Mathews,
Stillwater Studio

Delmar Staff
Acquisitions Editor: Sandy Clark
Senior Project Editor: Christopher Chien
Production Coordinator: Jennifer Gaines
Art and Design Coordinator: Mary Beth Vought
Editorial Assistant: Christopher Leonard

Printed in the United States of America

For more information, contact:

Effective Training Inc.
PO Box 756
Wayne, Michigan 48184
Tel. (734) 728-0909
FAX (734) 728-1260

7 8 9 10 XXX 03 02 01 00 99

Library of Congress Cataloging-in-Publication Data

Krulikowski, Alex.
Fundamentals of geometric dimensioning and tolerancing / by Alex Krulikowski.--2nd ed.
p. cm.
Includes index.
ISBN 0-8273-7995-1
1. Engineering drawings--Dimensioning. 2. Tolerance (Engineering)
I. Title.
T357.K78 1997
604.2 ' 43--dc21
97-21558
CIP

To: Donna, Jamy, and Mark

Thanks for all your love, patience, and understanding.

TABLE OF CONTENTS

ACKNOWLEDGMENTS

Ever hear the joke about how many ________________________ it takes to screw in a lightbulb? (You fill in the blank: engineers, managers, computer programmers, etc.) Seriously, have you ever wondered about how many people it takes to write a book? Many people think an author writes a book single-handedly.

Writing a book is a team effort. Yes, I (the author) have the skills necessary to write the technical content of this book. After all, I have already written several other books on this topic. I am able to write the first draft, but as the job progresses, it takes many different skills to write the finished book.

I am very fortunate to have a staff of talented people working for me: a technical writer/editor, Katherine Billings; a graphics expert, Jamy Krulikowski (yes, he's my son); and several more bright young men on my staff—Matthew Pride, Paul Moore, and Brandon Billings. This talented group was successful in overcoming several hardships during the writing of this book, from interpreting my markups or drawings to reconstructing (and retyping) lost data files when our office was plagued with the virus from hell. THEY did it! They worked days and evenings and late nights and weekends. I thank each and every one of you for your assistance during this project.

I want to tell the world about a very special person, my son Jamy. Simply put, this young man—at an age when most young men are pursuing their own personal ambitions—has taken the time to help his father realize his. Jamy, you are one of a kind; I thank you.

There is another part of my team who deserves credit. While I was spending much of my time working on this book for the last twelve months, several people on my staff have kept this company functioning. The people I am referring to are Donna Pokrywki, Kathy Darfler, and Tina White. These hardworking individuals have literally run Effective Training while I was working on the book. I thank you for your help.

And that's not all; I also had a crack team of technical proofreaders who were willing to pitch in and use their valuable time to read the drafts and offer suggestions on the technical content of this book. The technical proofreaders have many areas of expertise. The following list shows the individuals and their affiliations to colleges or industries.

The technical proofreaders who contributed to the development of this book:

Bourland, Robert - General Motors Powertrain
Burns, William - Macomb Community College
Davis, Brent - Ford Motor Co.
Day, Donald - Monroe Community College
Cavanaugh, Gerry - Schoolcraft College
Ferguson, Chuck - Steelcase Inc.
Keith, James - Boeing
Murphy, Michael - General Motors
Griess, Dr. Gerald - Eastern Michigan University
Honsinger, David - Watervliet Arsenal
Huebner, Glen - Waukesha Community College
Nirva, Raymond - Ford Motor Co.
Smith, Larry - St. Clair College
Smith, Nick - Boeing
Young, Roger - Storage Technology Corp.

I salute the proofreaders. What an outstanding job they did. They challenged the concepts, illustrations, and sometimes even my interpretation of Y14.5. They provided an invaluable service to the quality of the book. Several of the proofreaders are very close friends, and I sincerely appreciate the honesty in their comments.

Credit is gratefully given and acknowledgment made for the use of definitions and terms from the ASME Y14.5M-1994 Dimensioning and Tolerancing Standard. This standard is published by the American Society of Mechanical Engineers (ASME), New York, NY.

One more group whom I want to thank is all the students who used draft versions of this text and provided comments from a user's perspective. Your comments and suggestions have improved the usability of this text. Thank you.

Sincerely,

Alex Krulikowski
January, 1997

SPECIAL NOTE TO THE STUDENT

This section contains very important information about how to increase your success in learning GD&T.

Dear Student,

Welcome to the world of geometric dimensioning and tolerancing. This textbook is designed to introduce you to the fundamentals of GD&T; however, only *you* can ensure a successful understanding of the topic through proper goal-setting and initiative. I can lead you into the world of GD&T, but you must make a conscious effort to do the work required to successfully master the subject.

Because of my firm belief that success can be achieved through proper goal-setting and discipline, I have provided you with a list of goals and objectives at the beginning of each chapter. Acquire an understanding of these goals, step by step, and you will soon have mastered the world of GD&T. However, *you* must also set some goals: you must set aside a specific time each day to read, study, and practice the problems at the end of each chapter. The results of your hard work and discipline will be apparent when you successfully grasp each concept and topic. Best of all, you will soon realize the larger goal—an understanding of GD&T as a whole.

The Levels of Learning

Psychological studies on learning have discovered that there are several levels involved in the learning process. Many students try to learn by simply memorizing facts, without a thorough understanding of the topic at hand; however, these studies indicate that memorization alone will not properly prepare the student for tests or the application of any topic. The following list illustrates the levels of understanding:

1. **Knowledge** involves *remembering* (memorizing) factual material.
2. **Comprehension** involves *interpreting* information, changing it from one form to another, and/or making predictions.
3. **Application** involves *using* facts and fundamental principles when solving problems.
4. **Analysis** involves *identifying* and *sorting out* relevant and irrelevant facts to make comparisons.
5. **Synthesis** involves *combining* information and *developing* a plan or using original ideas.
6. **Evaluation** involves *judging* the value of observations and calculated results in order to reach a meaningful conclusion.

As you can see, memorization alone does not help with problem-solving or the application of concepts. It must be combined with other methods of learning before a real understanding of concepts can be accomplished.

The "Ten Principles to Productive Study"

Educators utilize many different study methods to assist students in the learning process. In order to assist you in this course, I have adapted ten commonly known principles of productive study. These "Ten Principles" can assist you in successful goal-setting and can lead the way to a proper understanding of GD&T. Before beginning your GD&T course, take time to read these concepts and make an effort to apply them to your study plan.

1. Learning occurs in *small steps*. Begin here and now—not tomorrow—to study and to solve problems.
2. Study *daily*. Don't expect to learn a lot the night before a test.
3. First *scan* the performance objectives, then carefully read the material and ask yourself relevant questions. Write down questions you cannot answer.
4. *Read* the material again, take notes, and *list* key points. Learning is aided by repetition.
5. *Think* about interconnections with what you know, including on-the-job applications.
6. *Visualize* GD&T applications, formulas, and key points until you can "see" them with your eyes shut.
7. *Write* down key points. You really don't know it if you can't write it.
8. *Think* about each key point. *Say* it! *Write* it! *Review* it! *Relate* key points to each other and compare their similarities and differences.
9. *Study sample problems* in the text. Consider the strategies used to solve these problems and how you would recognize and approach similar problems presented in the text or on a test or on a job.
10. *Solve problems* included in the exercise workbook. Work problems daily. Become familiar with different types of problems.

Your Commitment to Success

Although these principles can work for you, you alone can decide to commit the time and effort it will take to apply them. You must first commit yourself to attending the lectures. You must prepare for lectures by reading the list of performance objectives beforehand. You should also read the text before lectures, note key points and ask questions during the lecture, and review your notes and the text afterwards. You must do the assigned problems and study daily.

Remember, major geometric tolerancing topics are interrelated and build on one another, so after studying a chapter, review the performance objectives from the beginning of the chapter to be sure you know the major points and the terminology involved. Can you explain these terms and concepts to someone else? Try it! A person who understands a topic can use the vocabulary needed to discuss that topic.

Exams

Taking tests can be stressful for students; however, the more prepared you can help your students to be, the less stress they will feel. Have students work the problems and answer the questions at the end of each chapter. They should not waste time trying to guess what topics will or will not be included on exams. The exams are based on the performance objectives; if the students study them, they should do well. As the "Ten Principles" suggest, daily study, reading and rereading the topics, and trying practice problems assure successful study—and less stressful test-taking! Just as a marathon runner trains daily for an event, students should study daily for their "events." With enough preparation, a short review before the exam should be enough to guarantee successful results.

I hope that you can work together with your students to encourage them enough to begin the hard work needed to successfully accomplish the task at hand—if so, everyone involved will be well rewarded. Geometric dimensioning and tolerancing is a complex and exciting subject; understanding its principles can be rewarding and profitable. In writing this text, I have done everything possible to assist students with their studies—the rest of the work is up to you!

Sincerely yours,

Alex Krulikowski

Effective Training Inc.
P.O. Box 756
Wayne, Michigan 48184
(734) 728-0909

NOTE TO THE INSTRUCTOR

This second edition of *Fundamentals of Geometric Dimensioning and Tolerancing* preserves the best qualities of the first edition while adding new material that reflects the dynamic changes in the field of geometric tolerancing. We have adhered to two ideas from the first edition: first, the use of the Y14.5 dimensioning standard as a basis for the text; and second, the building block approach to learning geometric tolerancing.

New to This Edition

The second edition of *Fundamentals of Geometric Dimensioning and Tolerancing* continues to be the most practical and up-to-date text on the market. We discuss each dimensioning topic in a manner that is understandable and useful to the reader. New to the second edition are:

- A list of abbreviations and acronyms on the inside front cover
- A special note to the student explaining how to maximize learning from this course
- A new format that is easier to read and allows room for notes in the page margin
- Icons for study tips, author's comments, design tips, and "for more info" tips; the use of these icons helps the reader study the material and understand how it can be used in industry
- A list of text conventions that explain drawing conventions used in the text
- The text is divided into twelve chapters to allow for shorter, more specific lessons
- The integration of goals and objectives to aid the learning process; this is a major step to help the student understand what the important concepts are in each chapter, on what they will be tested, and on what they should focus their efforts
- Isometric drawings added to many of the figures to aid in the visualization of the part
- Information on how to inspect each geometric tolerance
- Information on when each geometric control is used in a part design
- Technical content updated to include the latest information from ASME Y14.5M-1994
- Specific references to the Y14.5 standard are included to allow the student to find additional information on a particular concept
- Problems that involve dimensioning drawings are included to allow the students to apply geometric tolerances to drawings
- Numerous comparison charts to understand trade-offs between using various symbols in a design application
- Appendices contain selected answers, a bibliography, comparison charts, ISO references

Organization

Fundamentals of Geometric Dimensioning and Tolerancing is divided into twelve chapters. Each chapter has several goals described at the beginning of the chapter. These thirty goals are the major topics that must be mastered to be fluent in the fundamentals of geometric tolerancing. The course design is based on this set of goals. Each chapter's goals are further defined and supported with a set of performance objectives.

The performance objectives describe specific, observable, measurable actions that the student must accomplish to demonstrate mastery of each goal. There are over 240 performance objectives in this text. These performance objectives are the key to success for both the student and the instructor. The text, problems, exercises, quizzes, and teaching materials are all based on the performance objectives. Using the performance objectives will make conducting the class easier for the instructor and will make attending the class more meaningful for the students.

Suggestions for Course Planning

Fundamentals of Geometric Dimensioning and Tolerancing is intended for a one semester college course in geometric tolerancing. The material can be divided in a variety of ways. It is designed to be segmented by chapter goals. The text contains thirty goals; the goals could easily be divided into two or three per class session.

Supplements

This book has a complete package of supplements, including an *Instructor Answer Guide*, a *Performance Based Instructor Kit*, and a student *Study Guide*.

The *Instructor Answer Guide* is an answer guide for the problems at the end of each chapter of the text. It is available from Delmar Publishers.

The *Performance Based Instructor Kit* is a complete set of teaching materials for this course. It contains over 275 overhead transparency masters; detailed lesson plans; several suggested course outlines; a complete set of quizzes and tests with answers; and answers to all the text and study guide problems. The *Performance Based Instructor Kit* is only available from Effective Training Inc.

The *Fundamentals of GD&T Exercise Workbook* is designed to reinforce the performance objectives for each goal with a set of activities and problems that engage the student in using the skills associated with those goals. The workbook contains over 200 additional problems for the student to solve. The workbook can be used in the class as an activity to reinforce concepts and create discussion of topics being taught. It is only available from Effective Training Inc.

The Ultimate Pocket Guide on Geometric Dimensioning and Tolerancing is a 77-page mini-book that is a great reference for GD&T. This pocket guide covers the definitions, rules, and major concepts—and also explains each symbol. *The Ultimate Pocket Guide on GD&T* is only available from Effective Training, Inc.

The Geometric Dimensioning and Tolerancing Self-Study Workbook, 2nd Edition, is a start-to-finish self-training course in Geometric Dimensioning and Tolerancing that has become a classic in its field. The self-study workbook is only available from Effective Training, Inc.

A Few Comments from the Author

I hope you enjoy teaching this course and using the materials I have designed. If you would like to contact me with a comment or suggestion, I can be reached at Effective Training Inc., PO Box 756., Wayne, MI 48184. I can also be reached at my E-mail address: GDT MAN@AOL.COM

Over twenty proofreaders have reviewed this textbook prior to publication. We have made numerous improvements and corrections as a result of their efforts. However, a few errors may have slipped through in the final stages of the book production. I apologize for any inconvenience this may cause. If you find an error, please send it to me. I will maintain an errata sheet and send it upon request.

As a parting thought, I want to share a quotation with all of you:

"He who dares to teach must never cease to learn."

John Cotton Dana

TEXT CONVENTIONS

Drawing Conventions

There are many engineering drawings used in this book. In order to focus on the dimensioning topic being discussed, many of the drawings are partial drawings. In some instances, figures show added detail for emphasis; in some instances, figures are incomplete by intent. Numerical values for dimensions and tolerances are illustrative only.

Notes shown in capital letters on drawings are intended to appear on actual industry drawings.

Notes shown in lowercase letters are for explanatory purposes only and are not intended to appear on industry drawings.

All drawings are in accordance with ASME Y14.5M-1994.

Unless otherwise specified, all angles ± 5°.

All units are metric.

The name of the dimensioning and tolerancing standard is ASME Y14.5M-1994. It is referred to in the text as Y14.5.

Gage Tolerances

The gages used in this text are described with basic dimensions; no tolerances are shown. In the product design field, gages are considered to have no tolerances; however, in industry, gages do have tolerances. The gage tolerances are usually quite small compared to part tolerances. A rule of thumb is that gage tolerances are 10% of the part tolerances. Gage tolerances are usually arranged so that a (marginally) good part may be rejected, but a bad part will never be accepted. (From paragraph 2.5.4.1 MIL-HD8K-204A[AR] Design of Inspection Equipment for Dimensional Characteristics.)

Technotes

Throughout the book you will find "Technotes." These are important facts that should be noted and remembered for better understanding of the text. These notes contain technical definitions and specific rules that are applied to information within the lessons.

Technotes are easy to locate in the chapters because they are highlighted by shadowed boxes. Each note is clearly labeled with a technote number that corresponds to the chapter where it is found and near the information where it will be of the most help.

ICON DEFINITIONS

Study Tip

Study Tip. When this symbol appears on a page, it is accompanied by a tip. This tip provides advice on how to maximize your learning while using the text.

Author's Comment

Author's Comment. In various places throughout this book, the author provides comments. Author's comments are strictly advisory and are not part of the Y14.5M-1994 dimensioning standard. When an author's comment is made, a symbol like the one shown here is shown adjacent to the text material to which the comment applies. The comments usually fall into one of two categories:

- They discuss a dimensioning situation that is not covered in the Y14.5M-1994 dimensioning standard.
- They offer the reader opinions, insights, or tips about the topic being discussed.

For More Info. . .

For More Info. . . When this symbol appears on a page, it is accompanied by page references in this text which contain information related to the topic. On a few occasions, this icon may refer to other sources for additional information.

Design Tip

Design Tip. In various places throughout the book, the author provides design tips. These tips are strictly advisory and are not part of the Y14.5M-1994 dimensioning standard. The design tips help designers to apply this tolerancing information in a cost-effective manner.

Chapter 1

Engineering Drawings and Tolerancing

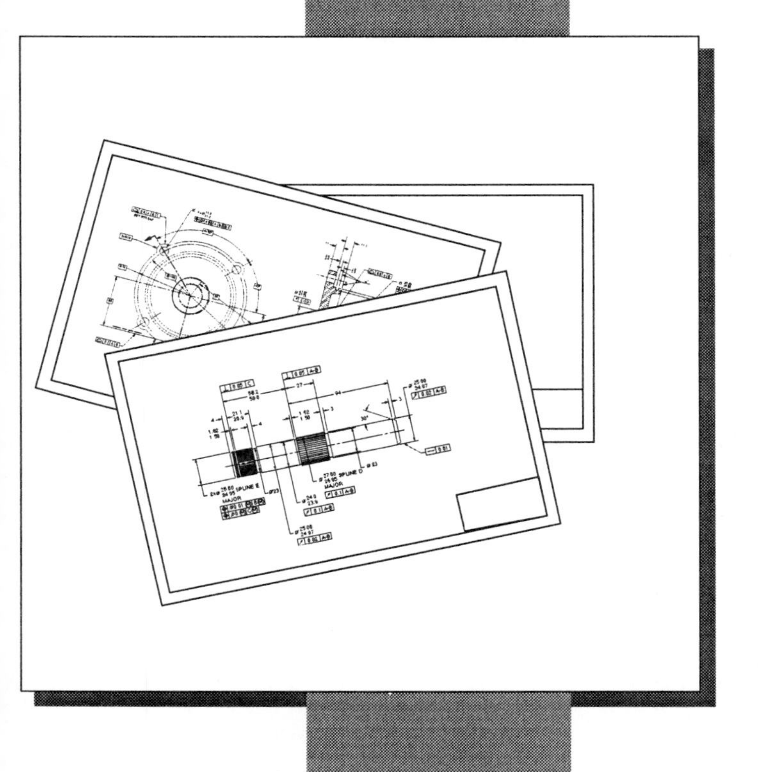

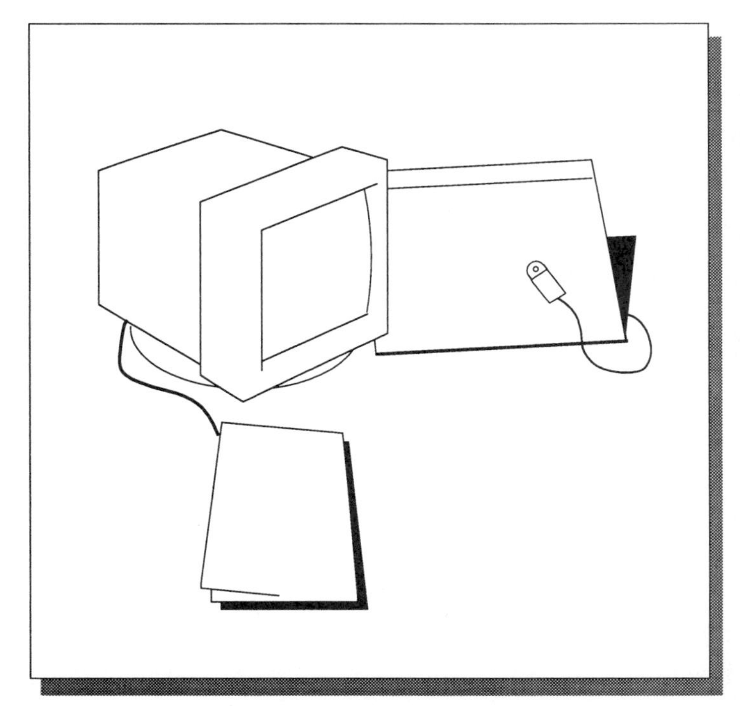

INTRODUCTION

An engineering drawing is a tool that is used to communicate the design and manufacturing information for a part. Important elements of an engineering drawing are the dimensions and tolerances. This chapter introduces engineering drawings, dimensions, geometric tolerances, and coordinate tolerances.

CHAPTER GOALS AND OBJECTIVES

There are Two Goals in this Chapter:

1-1. Understand what an engineering drawing is.

1-2. Understand why geometric tolerancing is superior to coordinate tolerancing.

Performance Objectives that Demonstrate Mastery of These Goals

Upon completion of this chapter, each student should be able to:

Study Tip
Take a few minutes to fully understand these objectives. When reading this chapter, look for information to help you master these objectives.

Goal 1-1 (pp. 3-12)

- Explain what an engineering drawing is.
- Describe how precisely drawings should communicate.
- List at least four consequences of drawing errors.
- Describe what a dimension is.
- Describe what a tolerance is.
- Describe what a limit tolerance is.
- Describe what a plus-minus tolerance is.
- Explain three conventions in the specification of metric unit dimensions on drawings.
- Explain how dimensional limits are interpreted.
- Explain ASME Y14.5M-1994.
- Describe seven of the ten fundamental dimensioning rules.

Goal 1-2 (pp. 13-22)

- Explain what coordinate tolerancing is.
- Explain the three major shortcomings of coordinate tolerancing.
- Explain three appropriate uses for coordinate tolerancing.
- Explain what the geometric tolerancing dimensioning system is.
- List three major benefits of geometric tolerancing.
- Explain how geometric tolerancing eliminates the shortcomings of coordinate tolerancing.
- Explain why the "great myth" about geometric tolerancing is untrue.

ENGINEERING DRAWINGS

What is an Engineering Drawing?

An ***engineering drawing*** is a document that communicates a precise description of a part. This description consists of pictures, words, numbers, and symbols. Together, these elements communicate part information to all drawing users. Engineering drawing information includes:

- Geometry (shape, size, and form of the part)
- Critical functional relationships
- Tolerances allowed for proper function
- Material, heat treat, surface coatings
- Part documentation information (part number, revision level)

For the last hundred years, most engineering drawings have been created by manual methods. The designer used tools like drafting boards, T-squares, compasses, triangles, etc. The drawing original was created on paper, linen, mylar, or other materials that could be used for making reproductions. The reproductions were generally referred to as "prints."

Today, many engineering drawings are created electronically. The designer uses a computer to create an electronic version of a drawing. Often, no physical original drawing exists; the original is a computer file. Copies (prints) are made through the use of a printer/plotter. Whether engineering drawings are manual drawings or electronic computer files, their basic purpose remains essentially the same: to record and communicate important part information.

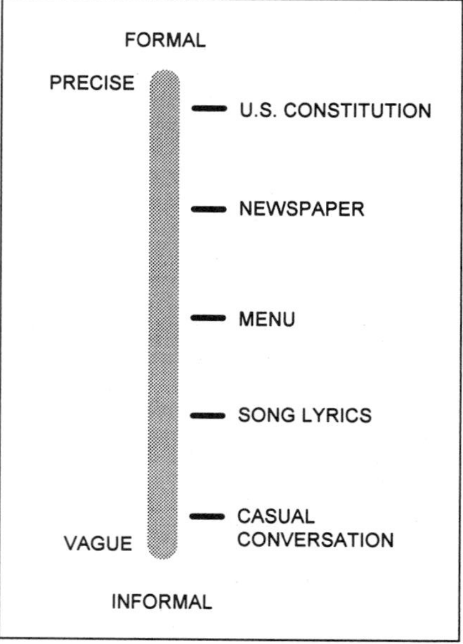

FIGURE 1-1 Communications Model

The Need for Precise Communications

There are many kinds of communications; some are formal and some are informal. Figure 1-1 shows a scale with different types of communications.

A casual conversation and song lyrics are examples of informal communications. They do not need to be very precise. However, other communications may be very formal and precise. The United States Constitution is an example of a formal communication. The interpretation of the Constitution has been challenged in courtrooms for over a hundred years. It is common that communications that need to be precise are often the subject of considerable debate. Engineering drawings are legal documents. Therefore, they should be treated as formal, precise documents. An engineering drawing should fully define the part. Each specification should be measurable.

Engineering drawings are a communications tool. Engineering drawings affect many parts of an organization. They have a major impact on costs.

Consequences of Poor Drawings

Engineering drawings not only need to communicate precisely, they also need to be correct. A drawing error can be very costly to an organization. The following analysis is an example based on a medium-sized manufacturing firm.

Figure 1-2 shows typical costs resulting from a drawing error. If a drawing error is found within the design department, it can be corrected for a few dollars. The cost is simply the time required to fix the error; let's say $1-10 to correct the drawing error.

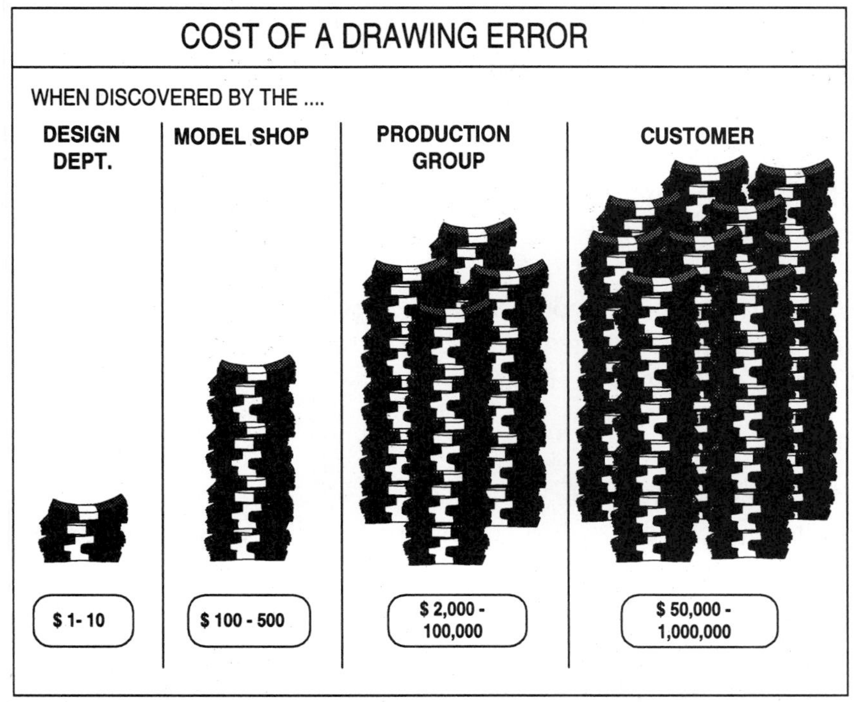

FIGURE 1-2 How Costs of a Drawing Error can Increase as the Drawing Moves Through the Organization

If the drawing error is missed in the design department and is discovered in the model shop, it may cost several hundred dollars to fix the error. This is because, now—in addition to the time to fix the drawing—additional costs may be involved in loss of material, machine time, and machinist's time.

Worse yet, let's say a part described on a drawing that contains an error gets into production. Now the costs escalate quickly. The cost to process the paperwork for fixing the drawing error may be several thousand dollars. In addition, gaging costs, tooling costs, and scrap costs can easily bring the total to over a hundred thousand dollars.

If a drawing error gets into the final product and it's shipped to the customer, the costs that result from the error can be much higher. If a product recall is involved, it can easily cost the organization over a million dollars. If a product liability lawsuit is involved, the costs that result from the drawing error can run into millions of dollars.

Drawing errors cost the organization in four ways:

1. Money
2. Time
3. Material
4. Unhappy customers

INTRODUCTION TO DIMENSIONING

Author's Comment
A tolerance is not associated with dimensions that are identified as reference, maximum or minimum.

What are Dimensions and Tolerances?

A ***dimension*** is a numerical value expressed in appropriate units of measure and used to define the size, location, orientation, form, or other geometric characteristics of a part.

A ***tolerance*** is the total amount that features of the part are permitted to vary from the specified dimension. The tolerance is the difference between the maximum and minimum limits.

Types of Tolerances

Two common methods used to specify tolerances are limit tolerances and plus-minus tolerances.

A ***limit tolerance*** is when a dimension has its high and low limits stated. In a limit tolerance, the high value is placed on top, and the low value is placed on the bottom. Figure 1-3*A* shows an example of a limit tolerance. The high limit for this dimension is 12.5. The low limit for this dimension is 12.0. The tolerance for this dimension is the total amount of variation permitted or 12.5 minus 12.0 = 0.5. When limit tolerances are expressed in a single line, the low limit is stated first, then a dash, followed by the high limit (for example, 12.0-12.5).

A ***plus-minus tolerance*** is the nominal or target value of the dimension is given first, followed by a plus-minus expression of a tolerance. An example of a plus-minus dimension is shown in Figure 1-3*B*. For this dimension, the nominal value is 12.25. The plus-minus tolerance is 0.25. The total tolerance for this dimension is 0.5.

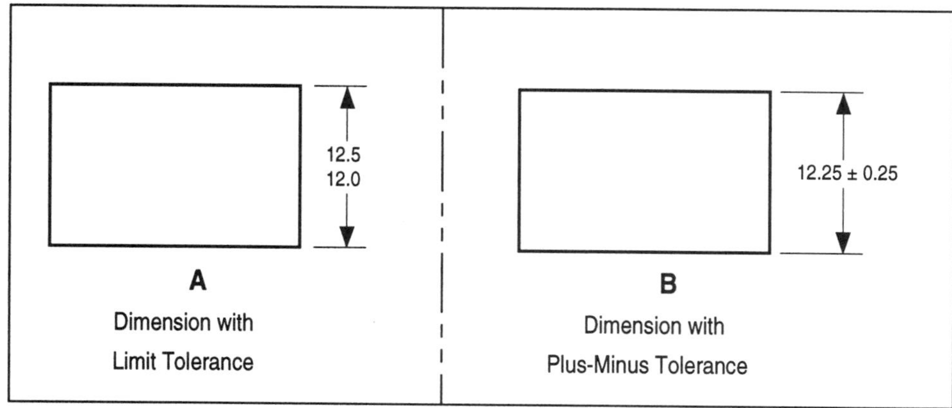

FIGURE 1-3 Examples of Limit Tolerances and Plus-Minus Tolerances

A tolerance for a plus-minus dimension can be expressed in several ways. A ***bilateral tolerance*** is one that allows the dimension to vary in both the plus and minus directions. An ***equal bilateral tolerance*** is where the allowable variation from the nominal value is the same in both directions. Figure 1-4*A* shows an example.

A ***unilateral tolerance*** is where the allowable variation from the target value is all in one direction and zero in the other direction. Figure 1-4*B* shows an example.

An ***unequal bilateral tolerance*** is where the allowable variation is from the target value, and the variation is not the same in both directions. Figure 1-4*C* shows an example.

Author's Comment
Most of industry considers target value the value around which manufacturing centers the process distribution.

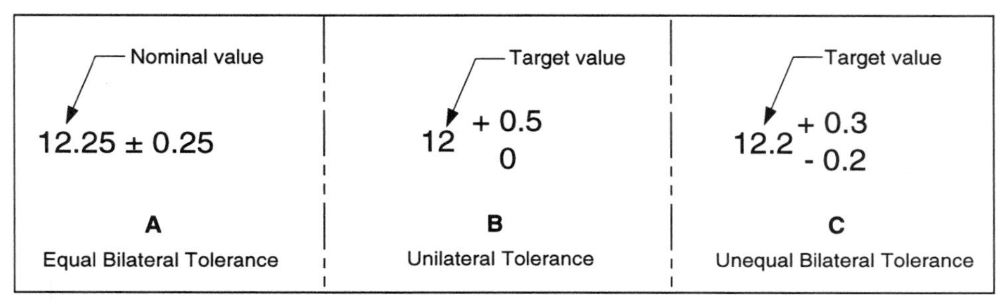

FIGURE 1-4 Examples of Equal, Unilateral, and Unequal Bilateral Tolerances

Metric Dimension Specifications

The dimensions in this text are shown in metric units. The Metric International System of Units (SI) is used. The millimeter is the common unit of measurement used on engineering drawings made to the metric system.

In industry, a general note would be shown on the drawing to invoke the metric system. A typical general note is: "UNLESS OTHERWISE SPECIFIED, ALL DIMENSIONS ARE IN MILLIMETERS."

Three conventions are used when specifying dimensions in metric units. Examples of these conventions are shown in Figure 1-5.

1. When a metric dimension is a whole number, the decimal point and zero are omitted.
2. When a metric dimension is less than one millimeter, a zero precedes the decimal point. For example, the dimension "0.2" has a zero to the left of the decimal point.
3. When a metric dimension is not a whole number, a decimal point with the portion of a millimeter (10ths or 100ths) is specified, as shown in Figure 1-5.

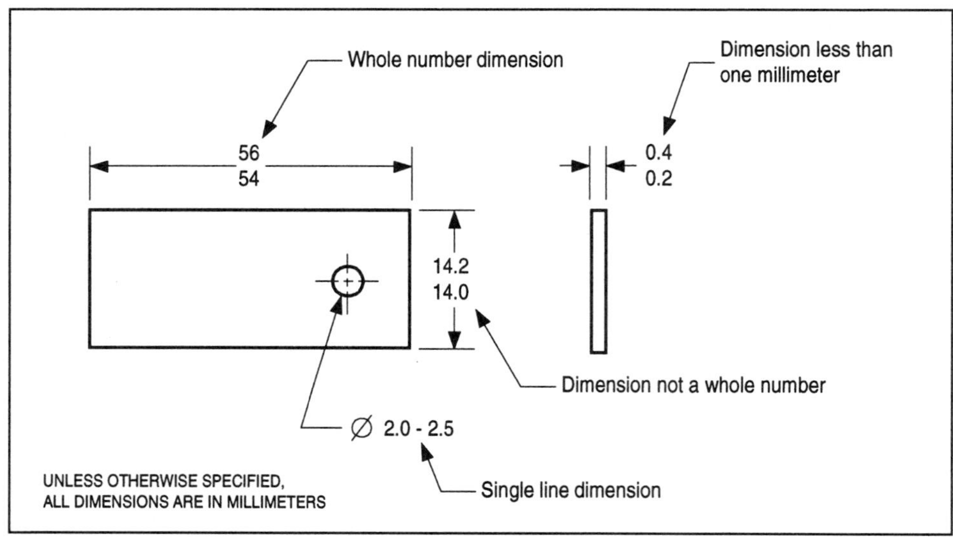

FIGURE 1-5 Metric Unit Specifications

Interpreting Dimensional Limits

All dimensional limits are absolute. In other words, a dimension is considered to be followed with zeros after the last specified digit (significant figure). See Figure 1-6 for examples. To determine part acceptance, the measured value is compared directly to the specified print dimension, and any deviation outside the specified dimension signifies an unacceptable part.

26.02	means	26.020....0
26	means	26.0....0
16.54	means	16.540....0
16.5	means	16.50....0

For this dimension 26.2 / 26.0

A part measuring
26.201 would be rejected
and
25.999 would be rejected

FIGURE 1-6 Interpreting Dimensional Limits

TECHNOTE 1-1 Dimensional Limits

All dimensional limits are absolute. A dimension is considered to be followed by zeros after the last specified digit.

DIMENSIONING STANDARDS

The information in this book is based on ASME Y14.5M-1994. ***ASME Y14.5M-1994*** is a dimensioning and tolerancing standard. ASME stands for American Society of Mechanical Engineers. The Y14.5 is the standard number. "M" is to indicate the standard is metric, and 1994 is the year the standard was officially approved. The Y14.5M-1994 standard is considered the national standard for dimensioning and tolerancing in the United States.

Author's Comment
ASME Y14.5M-1994 is a revision of ANSI Y14.5M-1982.

There is another predominant standard used in parts of the world. The ***International Standards Organization (ISO)*** is an organization that has published an associated series of standards on dimensioning and tolerancing. A list of the ISO dimensioning standards is shown in Appendix C. The ISO dimensioning standards and the Y14.5M-1994 Dimensioning Standard are about 90% common. A comparison chart of these standards is shown in Appendix D.

History of GD&T

As long as people have made things, they have used measurements, drawing methods, and drawings. Drawings existed as far back as six thousand B.C., when a unit of measure in the Nile and Chaldean civilizations was a "royal cubit." For thousands of years it fluctuated anywhere from 18 to 19 inches in length. Then, around four thousand B.C., the royal cubit was standardized at 18.24 inches. This set a pattern that has held true for nearly six thousand years. As long as there are measurements, drawing methods and drawings, there will be controversies, committees, and standards.

Manufacturing as we know it began with the Industrial Revolution in the 1800s. There were, of course, drawings, but these drawings were very different from the ones we use today. A typical drawing from the 1800s was a neatly inked, multi-viewed artistic masterpiece that portrayed the part with almost pictorial precision. Occasionally, the designer would write in a dimension, but generally such things were considered unnecessary.

Why? They were unnecessary because the manufacturing process was different then. There were no assembly lines, no widely dispersed departments or corporate units scattered across the nation or even worldwide as there are today. In those days, manufacturing was a cottage industry employing artisans who did it all, from parts fabrication to final assembly. These craftsmen passed their hard-won skills down from generation to generation. To them, there was no such thing as variation. Nothing less than perfection was good enough.

Of course there was variation, but back then the measuring instruments were not precise enough to identify it. When misfits and assembly problems occurred (which they routinely did), the craftsmen would simply cut-and-try, file-and-fit until the assembly worked perfectly. The total process was conducted under one roof, and communication among craftsmen was immediate and constant: "Keep that on the high side." "That edge has plenty of clearance." "That fit is OK now."

You can see that manufacturing back then was a quality process, but also slow, laborious and consequently quite an expensive one. The advent of the assembly line and other improved technologies revolutionized manufacturing. The assembly line created specialists to take the place of artisans, and these people did not have the time or skills for "file-and-fit."

Improved methods of measurement also helped to do away with the myth of "perfection." Now, engineers understand that variation is unavoidable. Moreover, in every dimension of every part in every assembly, some variation is acceptable without impairing the function of the assembly, as long as the limit of that variation—the "tolerance"—is identified, understood and controlled. This variation led to the development of the plus-minus (or coordinate) system of tolerancing, and to the determination that the logical place to record these tolerances and other information was on the engineering or design drawing.

With this development, drawings became more than just pretty pictures of parts; they became the main means of communication among manufacturing departments that were increasingly less centralized, more specialized, and subject to stricter demands.

Engineering Drawing Standards

To improve the quality of drawings, an effort was made to standardize them. In 1935, after years of discussion, the American Standards Association (ASA) published the first recognized standard for drawings, "American Drawing and Drafting Room Practices." Of its eighteen short pages, just five discussed dimensioning; tolerancing was covered in just two paragraphs.

It was a beginning, but its deficiencies became obvious with the start of World War II. In Britain, wartime production was seriously hampered by high scrap rates due to parts that would not assemble properly. The British determined that this was caused by weakness in the plus-minus system of coordinate tolerancing, and more critically, by the absence of full and complete information on engineering drawings.

Driven by the demands of war, the British innovated and standardized. Stanley Parker of the Royal Torpedo Factory in Alexandria, Scotland, created a positional tolerancing system that called for cylindrical (rather than square) tolerance zones. The British went on to publish a set of pioneering drawings' standards in 1944, and in 1948 they published "Dimensional Analysis of Engineering Design." This was the first comprehensive standard that used fundamental concepts of true position tolerancing.

GD&T in the United States

In the United States, Chevrolet published the *Draftsman's Handbook* in 1940, the first publication with any significant discussion of position tolerancing. In 1945, the U.S. Army published an ordinance manual on dimensioning and tolerancing that introduced the use of symbols (rather than notes) for specifying form and positioning tolerances.

Even so, the second edition of the American Standard Association's "American Standard Drawing and Drafting Room Practice," published in 1946, made minimal mention of tolerancing. That same year, however, the Society of Automotive Engineers (SAE) expanded coverage of dimensioning practices as applied in the aircraft industry in its "SAE Aeronautical Drafting Manual." An automotive version of this standard was published in 1952.

In 1949, the U.S. military followed the lead of the British by publishing the first standard for dimensioning and tolerancing, known as MIL-STD-8. Its successor, MIL-STD-8A, published in 1953, authorized seven basic drawing symbols and introduced a methodology of functional dimensioning.

As a result, there were three different groups in the United States publishing standards for drawings: the ASA, the SAE, and the military. This led to years of turmoil about the inconsistencies among the standards and resulted in slow, but measured progress in uniting those standards.

In 1957, the ASA (in coordination with the British and Canadians) approved the first American standard devoted to dimensioning and tolerancing. The 1959 MIL-STD-8B brought the military standards closer to ASA and SAE standards, and in 1966—after years of debate—the first united standard was published by the American National Standards Institute (ANSI), successor to the ASA. It was known as ANSI Y14.5. This first standard was updated in 1973 to replace notes with symbols in all tolerancing, and an updated standard was also published in 1982. The current Y14.5 standard was published in 1994.

FUNDAMENTAL DIMENSIONING RULES

The ***Fundamental Dimensioning Rules*** are a set of general rules for dimensioning and interpreting drawings. ASME Y14.5M-1994 has defined a set of fundamental rules for this purpose. The ten rules that apply to this text are paraphrased in the list below:

Design Tip
Dimensioning Rule #10 is new in Y14.5. When using Y14.5, assembly drawings must show dimensions from the detail drawings if they are required to be maintained in the assembly.

1. Each dimension shall have a tolerance, except those dimensions specifically identified as reference, maximum, minimum, or stock (commercial stock) size.
2. Dimensioning and tolerancing shall be complete so there is full definition of each part feature.
3. Dimensions shall be selected and arranged to suit the function and mating relationship of a part and shall not be subject to more than one interpretation.
4. The drawing should define a part without specifying manufacturing methods.
5. A 90° angle applies where centerlines and lines depicting features are shown on a drawing at right angles, and no dimension is shown.
6. A 90° basic angle applies where centerlines of features in a pattern—or surfaces shown at right angles on a drawing—are located and defined by basic dimensions, and no angle is specified.
7. Unless otherwise specified, all dimensions are applicable at 20° C (68° F).
8. All dimensions and tolerances apply in the free-state condition. This principle does not apply to non-rigid parts.
9. Unless otherwise specified, all geometric tolerances apply to the full depth, length, and width of the feature.
10. Dimensions and tolerances apply only at the drawing level where they are specified. A dimension specified on a detail drawing is not mandatory for that feature on the assembly drawing.

For more info. . .
See Paragraph 1.4 of Y14.5.

The first three rules establish dimensioning conventions; rule four states that manufacturing methods should not be specified. Rules five and six establish conventions for implied 90° angles. Rules seven, eight, and nine establish default conditions for dimensions and tolerance zones. Rule ten establishes a convention for which drawing level dimensions and tolerances apply.

THE COORDINATE TOLERANCING SYSTEM

Definition

For about one hundred fifty years, a tolerancing approach called "coordinate tolerancing" was the predominant tolerancing system used on engineering drawings. ***Coordinate tolerancing*** is a dimensioning system where a part feature is located (or defined) by means of rectangular dimensions with given tolerances. An example of coordinate tolerancing is shown in Figure 1-7.

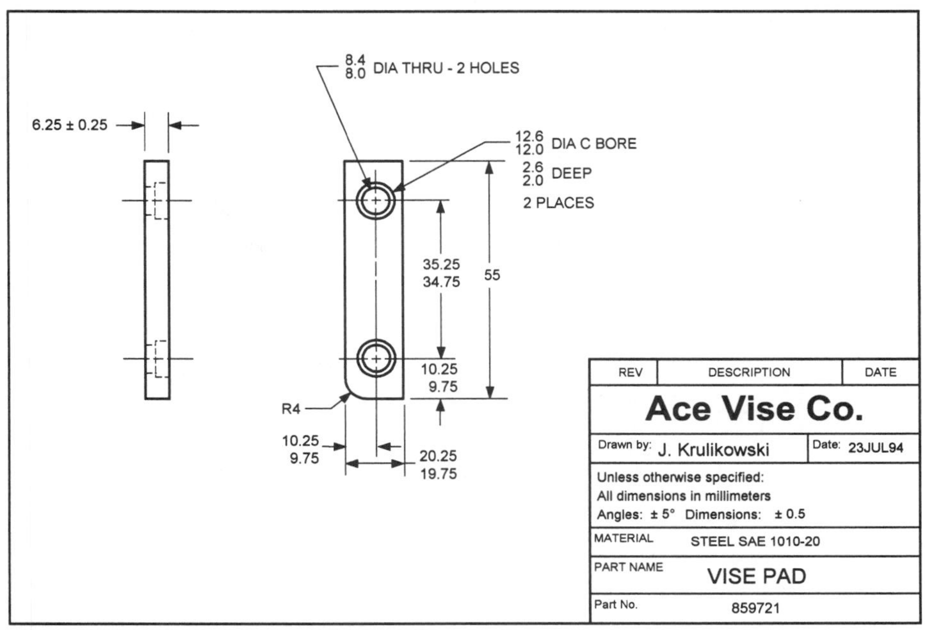

FIGURE 1-7 Coordinate Tolerancing Drawing

SHORTCOMINGS OF COORDINATE TOLERANCING

Coordinate tolerancing was very successful when companies were small, because it was easy to talk to the machinist to explain what the drawing intent was. Over the years, as companies grew in size, parts were obtained from many sources. The ability for the designer and machinist to communicate directly had diminished, and the shortcomings of the coordinate tolerancing system became evident. Coordinate tolerancing simply does not have the completeness to precisely communicate the part requirements.

Coordinate tolerancing contains three major shortcomings. They are:

1. Square or rectangular tolerance zones
2. Fixed-size tolerance zones
3. Ambiguous instructions for inspection

Coordinate Tolerancing and Square (or Illogical) Tolerance Zones

Let's look at the coordinate tolerancing shortcomings in more depth. First, let's examine the tolerance zone for the 8.0-8.4 dia. hole locations from the part in Figure 1-7. The hole location tolerance zone is formed by the max. and min. of the vertical and horizontal location dimensions.

Figure 1-8 shows that a 0.5 square tolerance zone would be formed. The illogical aspect of a square tolerance zone is that the hole can be off its nominal location in the diagonal directions a greater distance than in the vertical and horizontal directions. A more logical and functional approach is to allow the same tolerance for a hole location in all directions, creating a cylindrical tolerance zone.

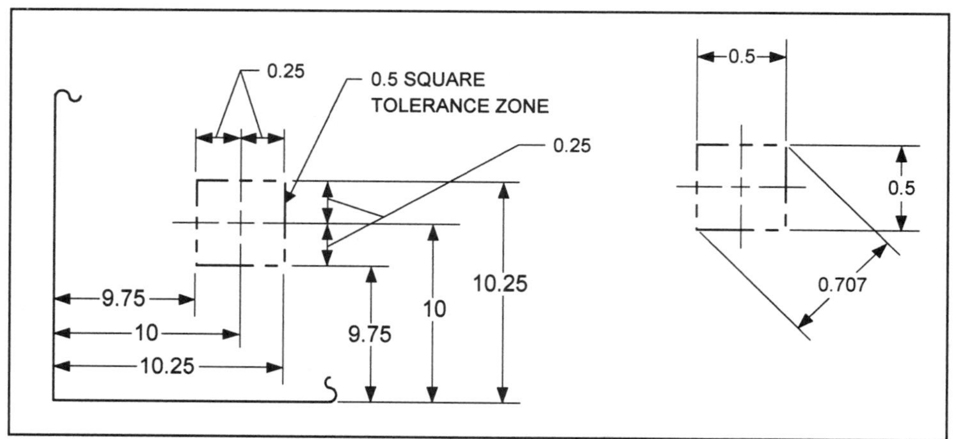

FIGURE 1-8 Square Tolerance Zone that Results from Coordinate Dimensions

Coordinate Tolerancing and Fixed-Size Tolerance Zones

Next, let's discuss how coordinate tolerancing uses fixed-size tolerance zones. The print specification requires the center of the hole to be within a 0.5 square tolerance zone, whether the hole is at its smallest size limit or its largest size limit. When the important function of the holes is assembly, the hole location is most critical when the hole is at its minimum limit of size. If the actual hole size is larger than its minimum size limit, its location tolerance can be correspondingly larger without affecting the part function.

Square and fixed-size tolerance zones can cause functional parts to be scrapped. Since coordinate tolerancing does not allow for cylindrical tolerance zones or tolerance zones that increase with the hole size, lengthy notes would have to be added to a drawing to allow for these conditions.

Coordinate Tolerancing and Ambiguous Instructions for Inspection

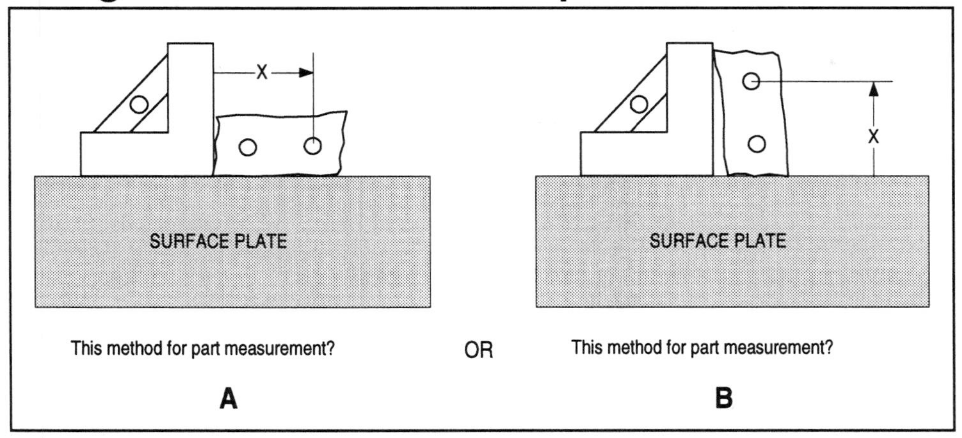

FIGURE 1-9 Methods of Inspection

A third major shortcoming of coordinate tolerancing is that it has ambiguous instructions for inspection. Figure 1-9 shows two logical methods an inspector could use to set up the part from Figure 1-7 for inspecting the holes. The inspector could rest the part on the face first, long side second and the short side third, or the inspector could rest the part on the face first, the short side second and the long side third.

Because there are different ways to hold the part for inspection, two inspectors could get different measurements from the same part. This can result in two problems: good parts may be rejected or, worse yet, bad parts could be accepted as good parts.

The problem is that the drawing does not communicate to the inspector which surfaces should touch the gaging equipment first, second, and third. When using coordinate tolerancing, additional notes would be required to communicate this important information to the inspector.

As you can see, coordinate tolerancing has some very significant shortcomings. That's why its use is rapidly diminishing in industry. However, coordinate tolerancing is not totally obsolete; it does have some useful applications on engineering drawings. The chart in Figure 1-10 shows appropriate uses for coordinate tolerances on engineering drawings.

Coordinate Dimension Usage		
Type of Dimension	**Appropriate Use**	**Poor Use**
Size	X	
Chamfer	X	
Radius	X	
Locating Part Features		X
Controlling Angular Relationships		X
Defining the Form of Part Features		X

FIGURE 1-10 Appropriate Uses for Coordinate Tolerancing

TECHNOTE 1-2 Coordinate Tolerancing

Coordinate tolerancing is a dimensioning system where a part feature is located (or defined) by means of rectangular dimensions with given tolerances. Coordinate tolerancing has three shortcomings:

1. Square or rectangular tolerance zones
2. Fixed-size tolerance zones
3. Ambiguous instructions for inspection

THE GEOMETRIC DIMENSIONING AND TOLERANCING SYSTEM

Definition

Geometric Dimensioning and Tolerancing (GD&T) is an international language that is used on engineering drawings to accurately describe a part. The GD&T language consists of a well-defined set of symbols, rules, definitions, and conventions. GD&T is a precise mathematical language that can be used to describe the size, form, orientation, and location of part features. GD&T is also a design philosophy on how to design and dimension parts. Figure 1-11 shows an example of an engineering drawing that is toleranced with GD&T.

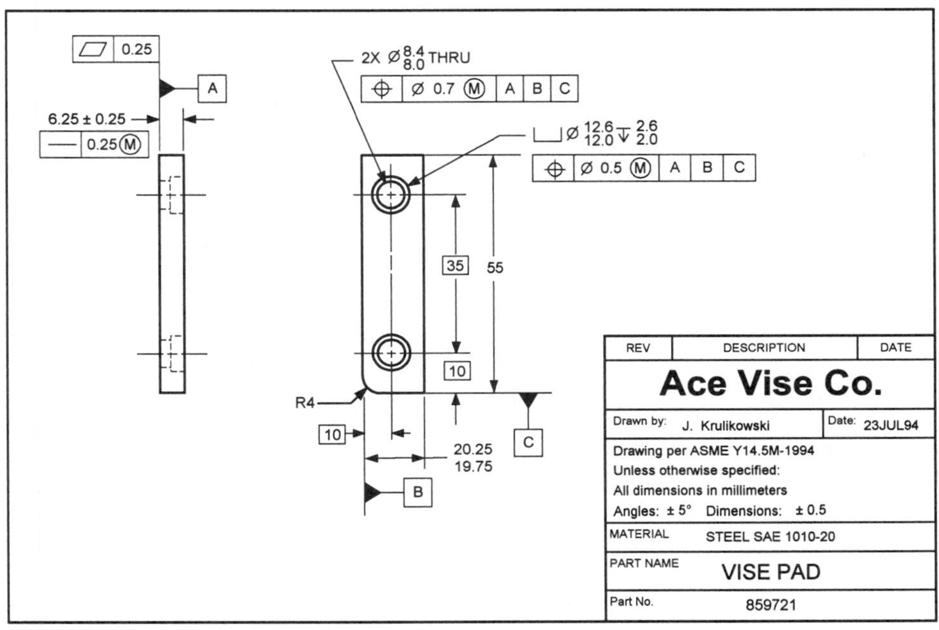

FIGURE 1-11 Engineering Drawing Example

Design Philosophy of Geometric Tolerancing

Geometric tolerancing encourages a dimensioning philosophy called "functional dimensioning." ***Functional dimensioning*** is a dimensioning philosophy that defines a part based on how it functions in the final product. The functional dimensioning philosophy is encouraged in many places throughout the Y14.5 standard. Although functional dimensioning is the philosophy, it does not mean the designer should design the component without taking other factors into consideration. Many companies find it a great advantage to use a process called "simultaneous engineering." ***Simultaneous engineering*** is a process where design is a result of input from marketing, engineering, manufacturing, inspection, assembly, and service. Simultaneous engineering often results in better products at lower cost.

GD&T BENEFITS

- **Improves Communication**
  GD&T can provide uniformity in drawing specifications and interpretation, thereby reducing controversy, guesswork and assumptions. Design, production, and inspection all work in the same language.
- **Provides Better Product Design**
  The use of GD&T can improve your product designs by providing designers with the tools to "say what they mean," and by following the functional dimensioning philosophy.
- **Increases Production Tolerances**
  There are two ways tolerances are increased through the use of GD&T. First, under certain conditions, GD&T provides "bonus"—or extra—tolerance for manufacturing. This additional tolerance can make a significant savings in production costs. Second, by the use of functional dimensioning, the tolerances are assigned to the part based upon its functional requirements. This often results in a larger tolerance for manufacturing. It eliminates the problems that result when designers copy existing tolerances, or assign tight tolerances, because they don't know how to determine a reasonable (functional) tolerance.

COMPARISON BETWEEN GD&T AND COORDINATE TOLERANCING

Sometimes designers think that it is faster to dimension a part with coordinate tolerancing than by using geometric tolerancing. This is not true. Let's take the drawing from Figure 1-7 and add geometric tolerances to eliminate the major shortcomings of the coordinate dimensions.

The first major shortcoming of coordinate tolerancing is "square tolerance zones." Let's look at how geometric tolerancing eliminates this shortcoming. In Figure 1-12, the arrow labeled "A" points to a GD&T symbol. This symbol specifies a cylindrical tolerance zone. The square tolerance zone from the coordinate toleranced version (Figure 1-7) is converted into a cylindrical tolerance zone. Notice that the tolerance value is larger than the 0.5 tolerance allowed in Figure 1-7. Figure 1-13 shows how the cylindrical zone provides additional tolerance in comparison with the square tolerance zone. The additional tolerance gained from using cylindrical tolerance zones can reduce manufacturing costs.

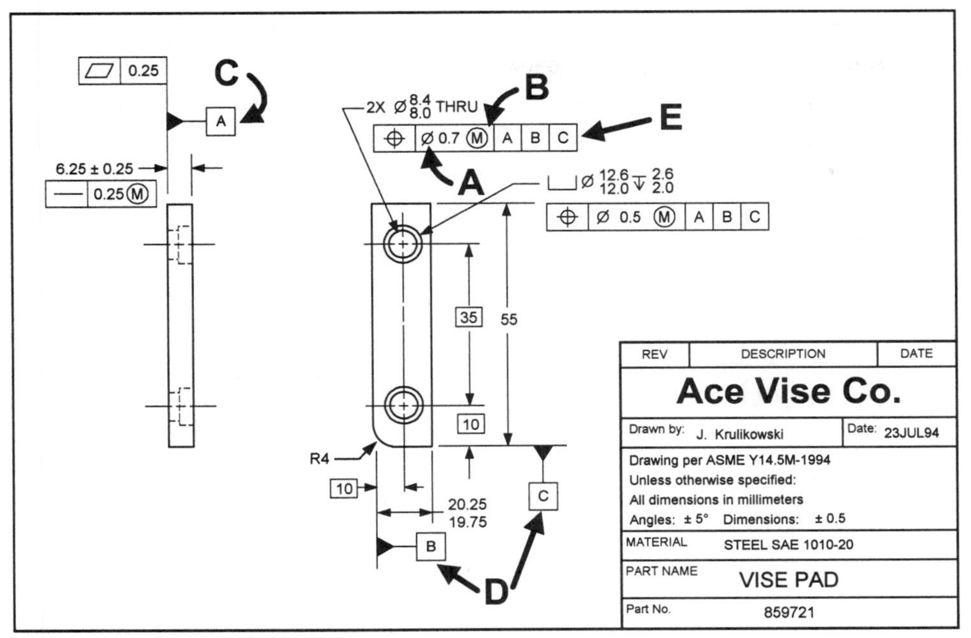

FIGURE 1-12 Vise Pad Drawing Using GD&T

The second major shortcoming of coordinate tolerancing is "fixed-size tolerance zones." Let's look at how geometric tolerancing eliminates this shortcoming. In Figure 1-12 the arrow labeled "B" points to a GD&T symbol. This symbol specifies a tolerance zone that applies when the holes are their smallest diameter. When the holes are larger, this GD&T symbol allows the hole location to have additional tolerance. This additional tolerance allowed by the GD&T symbol can reduce manufacturing costs.

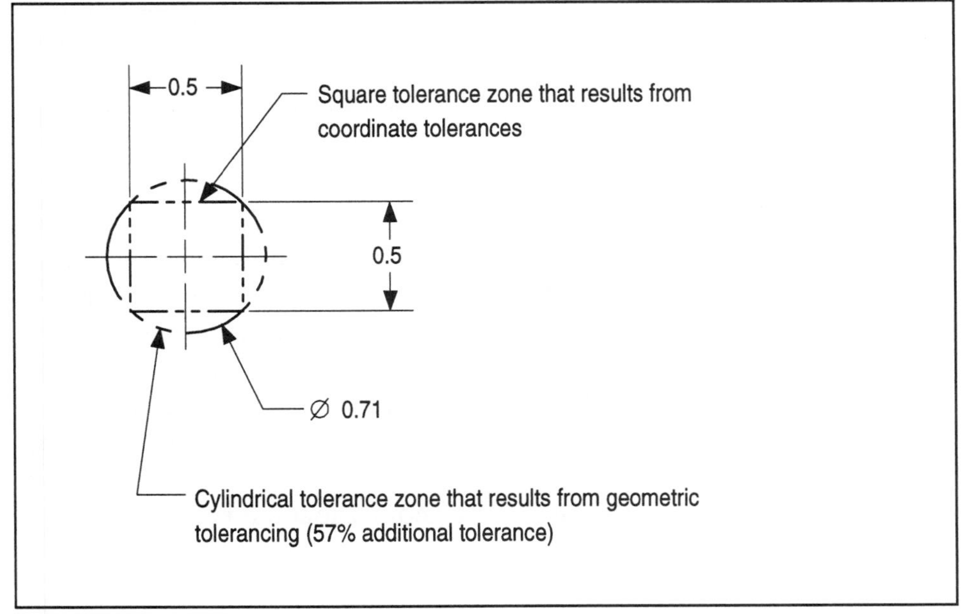

FIGURE 1-13 Cylindrical vs. Square Tolerance Zone

The third major shortcoming of coordinate tolerancing is that it has "ambiguous instructions for inspection." Let's look at how geometric tolerancing eliminates this shortcoming. Geometric tolerancing contains a concept called the "datum system." The datum system allows the designer to communicate the appropriate method of part setup to the inspector. First, symbols are added to the drawing to denote which surfaces touch the gage. See Figure 1-12, arrows labeled "C" and "D." Then, inside the feature control frame (see arrow labeled "E"), the sequence is given for the inspector to address the part to gage surfaces. Using the geometric tolerancing specifications from Figure 1-12, the inspection method would be the one shown in Figure 1-9*A*.

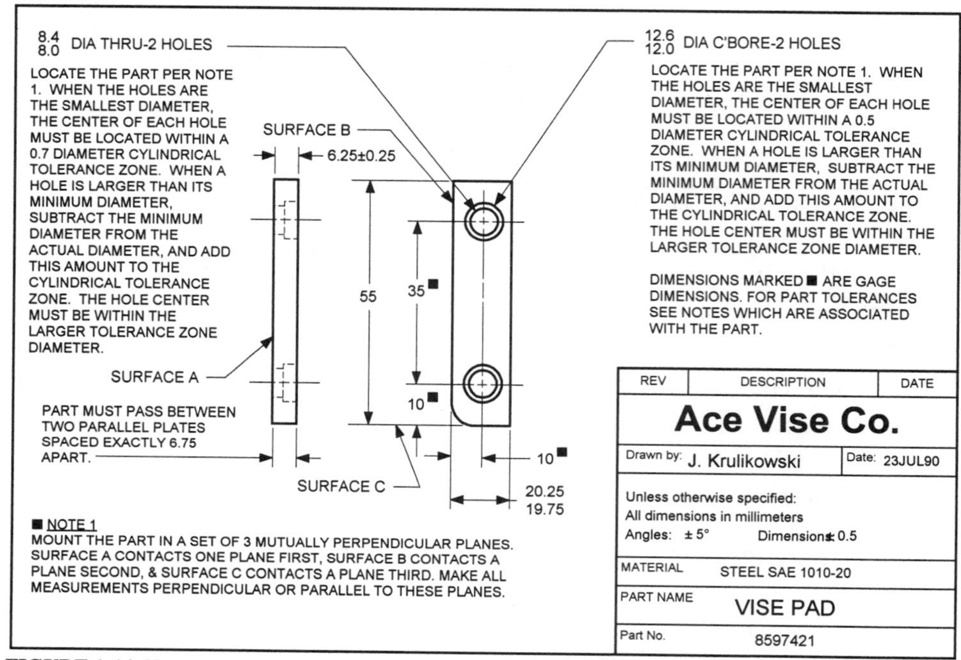

FIGURE 1-14 Notes Required to Make Coordinate Dimensioned Drawing Equivalent to GD&T Drawing

Now, through the use of geometric tolerancing, the dimensioning shortcomings are eliminated. Let's take a look at what the drawing would look like if we tried to accomplish the same level of drawing completeness with coordinate tolerancing. Figure 1-14 shows the vise pad drawing from Figure 1-12. This time the part is dimensioned with coordinate dimensions to the same level of completeness as the GD&T version, but using words instead of symbols. Now, which drawing do you think would be easier to create? When the goal is to dimension both drawings to the same degree of completeness, it is faster to use geometric tolerances.

The differences between coordinate tolerancing and geometric tolerancing are summarized in Figure 1-15. When comparing these tolerancing methods, it is easy to understand why geometric tolerancing is replacing coordinate tolerancing.

DRAWING CONCEPT	COORDINATE TOLERANCING	GEOMETRIC TOLERANCING
TOLERANCE ZONE SHAPE	**CONDITION** » Square or rectangular tolerance zones for hole locations **RESULTS** » Less tolerance available for hole » Higher manufacturing costs	**CONDITION** » Can use diameter symbol to allow round tolerance zones **RESULTS** » 57% more tolerance for hole location » Lower manufacturing costs
TOLERANCE ZONE FLEXIBILITY	**CONDITION** » Tolerance zone is fixed in size **RESULTS** » Functional parts scrapped » Higher operating costs	**CONDITION** » Use of MMC modifier allows tolerance zones to increase under certain conditions **RESULTS** » Functional parts used » Lower operating costs
EASE OF INSPECTION	**CONDITION** » Implied datum allows choices for set up when inspecting the part **RESULTS** » Multiple inspectors may get different results » Good parts scrapped » Bad parts accepted	**CONDITION** » The datum system communicates one set up for inspection **RESULTS** » Clear instructions for inspection » Eliminates disputes over part acceptance

FIGURE 1-15 Comparison Between Coordinate Tolerancing and Geometric Tolerancing

THE GREAT MYTH OF GD&T

Author's Comment
This myth is often spread by people who do not understand geometric tolerancing.

Even though geometric tolerancing has been accepted by many companies and individuals, it is still associated with a great myth. The ***Great Myth of GD&T*** is the misconception that geometric tolerancing raises product costs.

The myth stems from two factors. The first is the fear of the unknown; it is common to be skeptical of things that are not well understood. When a part dimensioned with GD&T is sent out for a cost estimate, people tend to inflate their assessment of how much the part will cost simply because they are fearful that the drawing contains requirements they may not be able to easily meet. Geometric tolerancing gets the blame for the higher cost, but in reality, geometric tolerancing probably allowed the part more tolerance, and the drawing user did not understand how to read the drawing.

The second factor that helps to create the myth is poor design practices. Many drawings contain tolerances that are very difficult to achieve in production, regardless of what dimensioning system is used. This stems from designers who simply do not use due care in assigning the tolerances. Somehow the language of GD&T gets the blame. It's not the fault of the language; it is the fault of the designer.

The fact is, that when properly used, **GD&T SAVES MONEY**. The great myth about geometric tolerancing can be eliminated with a better understanding of geometric tolerancing by both drawing makers and drawing users. Simply put, knowledge is the key to eliminating the myth.

Let's review a few FACTS about geometric tolerancing:

- GD&T increases tolerances with cylindrical tolerance zones.
- GD&T allows additional (bonus) tolerances.
- GD&T allows the designer to communicate more clearly.
- GD&T eliminates confusion at inspection.

VOCABULARY LIST

New Terms Introduced in this Chapter

ASME Y14.5M-1994
Bilateral tolerance
Coordinate tolerancing system
Dimension
Engineering drawing
Equal bilateral tolerance
Functional dimensioning
Fundamental Dimensioning Rules
Geometric Dimensioning and Tolerancing (GD&T)
Great Myth of GD&T
International Standards Organization (ISO)
Limit tolerance
Plus-minus tolerance
Simultaneous engineering
Tolerance
Unequal bilateral tolerance
Unilateral tolerance

Study Tip
Read each term. If you don't recall the meaning of a term, look it up in the chapter.

ADDITIONAL RELATED TOPICS

Topic	ASME Y14.5M-1994 Reference
• ASME Y14.5M & ISO standards comparison	Appendix D (pg. 372)
• Metric limits and fits	Paragraph 2.2.1
• Plated or coated parts	Paragraph 2.4.1

Author's Comment
These topics, plus advanced coverage of many of the topics introduced in this text, will be covered in my new book on advanced GD&T concepts.

QUESTIONS AND PROBLEMS

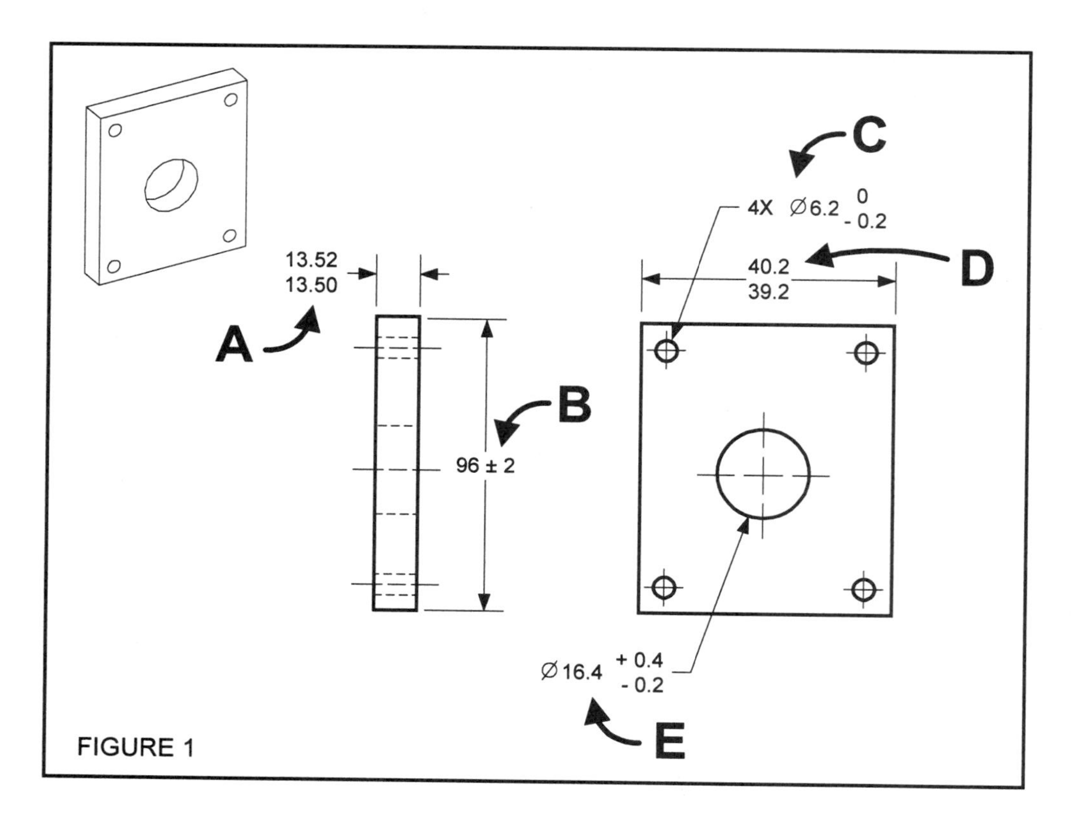

FIGURE 1

The questions refer to Figure 1.

1. Dimension *A* contains a __________________ tolerance. (limit / plus-minus)
2. Dimension *B* contains a __________________ tolerance. (limit / plus-minus)
3. Dimension *C* contains a __________________ tolerance. (limit / plus-minus)
4. Dimension *D* contains a __________________ tolerance. (limit / plus-minus)
5. Dimension *B* contains an __________________ bilateral tolerance. (equal / unequal)
6. Dimension *C* contains a __________________ tolerance. (unilateral / bilateral)
7. Dimension *E* contains an __________________ bilateral tolerance. (equal / unequal)
8. What is the convention for a whole number metric dimension? __________________

____

9. What is the convention for a metric dimension less than one millimeter? __________________

____

10. Fill in the chart using the drawing from Figure 1.

Dimension	Max/Min limits	If the measured value was. . .	This dimension would be Accepted	Rejected	Why
A		13.52001			
B		93.9999			
C		6.27001			
D		40.1999			
E		16.80			

11. What does ASME Y14.5M-1994 stand for?
    ASME ______
    Y14.5 ______
    M ______
    1994 ______

12. What is coordinate tolerancing?
    ______
    ______
    ______

13. The three major shortcomings of coordinate tolerancing are:
    ______
    ______
    ______

14. The "Great Myth of GD&T" is . . . .
    ______

Chapter **2**

Introduction to Geometric Tolerancing Symbols and Terms

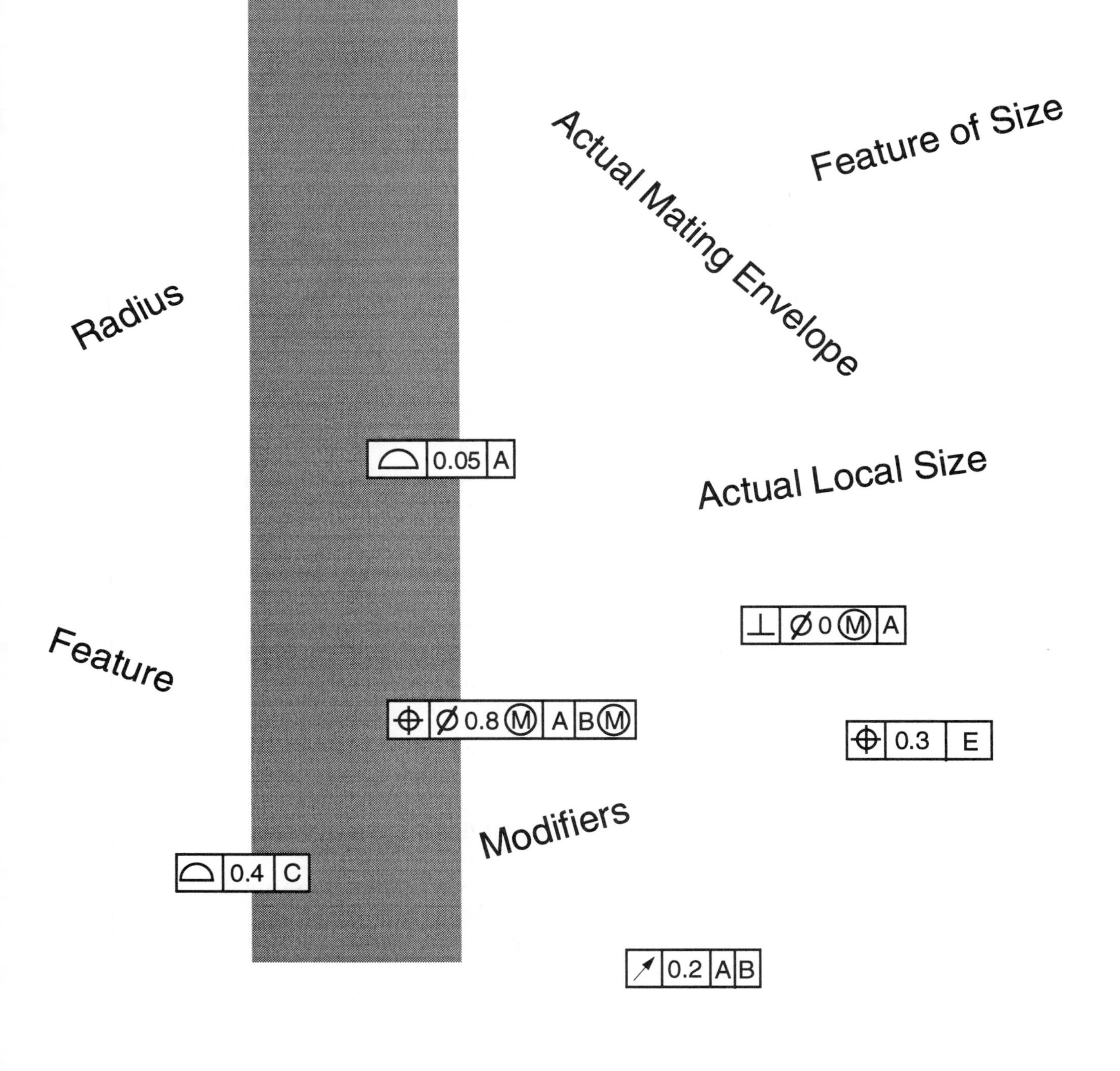

INTRODUCTION

Studying geometric tolerancing is like constructing a building; if you want the building to be strong and last a long time, you must begin by laying a solid foundation. Likewise, if you want to build an understanding of geometric tolerancing that will be strong and long lasting, you should begin by establishing a solid foundation. The terms and symbols in this chapter lay the foundation for understanding the concepts in the following chapters.

CHAPTER GOALS AND OBJECTIVES

There are Two Goals in this Chapter:

2-1. Understand eight key terms and how they affect the interpretation of a drawing.

2-2. Understand the modifiers and symbols used in geometric tolerancing.

Performance Objectives that Demonstrate Mastery of These Goals

Upon completion of this chapter, each student should be able to:

Study Tip

Take a few minutes to fully understand these objectives. When reading this chapter, look for information to help you master these objectives.

Goal 2-1 (pp. 29-35)

- Define a feature.
- Define a FOS.
- Describe cylindrical and planar features of size.
- Distinguish between feature of size and non-feature of size dimensions.
- Define actual local size.
- Define the actual mating envelope of an external feature of size.
- Define the actual mating envelope of an internal feature of size.
- Describe the maximum material condition of a feature of size.
- Describe the least material condition of a feature of size.
- Describe the term, "regardless of feature size."
- Identify the maximum and least material condition of a feature of size.

Goal 2-2 (pp. 36-40)

- Identify the eight common modifiers used in geometric tolerancing.
- Describe the tolerance conditions for a radius.
- Describe the tolerance conditions for a controlled radius.
- Name the fourteen geometric characteristic symbols.
- List the five categories of geometric controls.
- Identify the parts of a feature control frame.

DEFINITIONS

Features and Features of Size

This section contains definitions of six important GD&T terms. These terms are used throughout the text. A ***feature*** is a general term applied to a physical portion of a part, such as a surface, hole, or slot. An easy way to remember this term is to think of a feature as a part surface. The part in Figure 2-1 contains seven features: the top and bottom, the left and right sides, the front and back, and the hole surface.

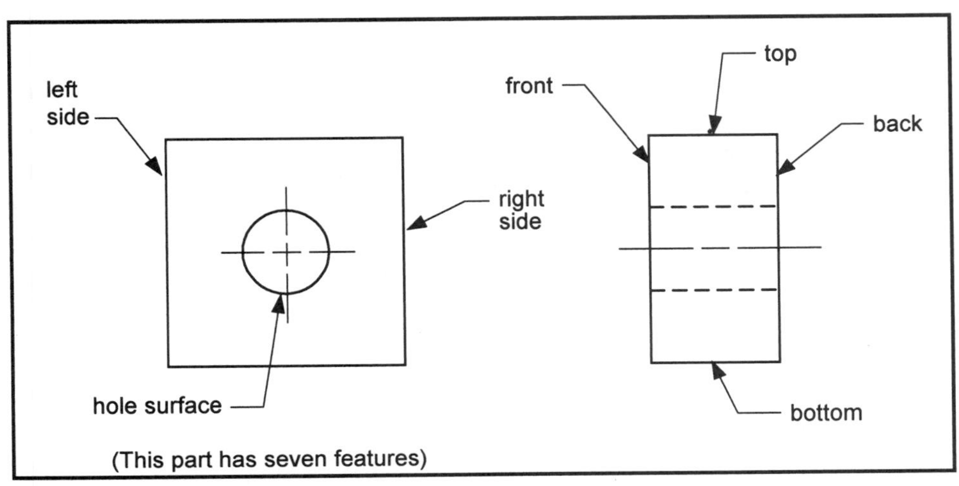

FIGURE 2-1 Examples of Features

> **TECHNOTE 2-1 Feature**
>
> A feature is any surface on a part.

A ***feature of size (FOS)*** is one cylindrical or spherical surface, or a set of two opposed elements or opposed parallel surfaces, associated with a size dimension. A key part of the FOS definition is that the surfaces or elements *must be opposed*. An axis, median plane or centerpoint can be derived from a feature of size.

> **TECHNOTE 2-2 Feature of Size**
>
> A feature of size. . .
>
> - Contains opposing elements or surfaces.
> - Can be used to establish an axis, median plane, or centerpoint.
> - Is associated with a size dimension.

Figure 2-2 shows several examples of features of size. Notice that in each case, the feature of size contains opposed surfaces or elements and could be used to derive an axis, median plane, or centerpoint.

Every feature of size contains one or more features—surface(s)—within it. A ***cylindrical FOS*** contains one feature: the cylindrical surface. A ***planar FOS*** is a FOS that contains two features: the two parallel plane surfaces. ("Feature" is a general term; it is often used when referring to a FOS.)

Internal and External Features of Size

There are two types of features of size—external and internal. External features of size are comprised of part surfaces (or elements) that are external surfaces, like a shaft diameter or the overall width or height of a planar part. In Figure 2-2, the 34-36 dimension and the 24.0-24.2 dimension are size dimensions for external features of size. An internal FOS is comprised of part surfaces (or elements) that are internal part surfaces, such as a hole diameter or the width of a slot. In Figure 2-2, the 4.2-4.8 diameter hole and the 10.2-10.8 diameter hole are size dimensions for internal features of size.

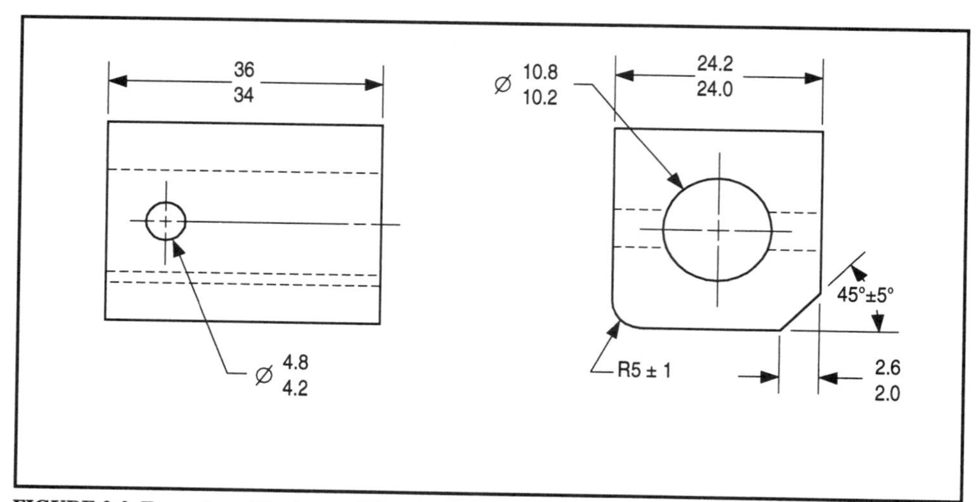

FIGURE 2-2 Examples of Features of Size and Non-Features of Size

Author's Comment

Usually (99% of the time), the opposing elements will both be the same type—for example, two opposing surfaces or two opposing line elements, instead of a surface on one side opposing a line element on the other side.

Feature of Size Dimensions

Let's look at how the feature of size concept relates to dimensions. A ***feature of size dimension*** is a dimension that is associated with a feature of size. A ***non-feature of size dimension*** is a dimension that is not associated with a feature of size. In Figure 2-2 there are four feature of size dimensions and three non-feature of size dimensions. Whether a dimension is or is not a feature of size dimension is an important concept in geometric tolerancing. Later in the course, you will learn that certain rules automatically apply when a feature of size dimension is specified on a drawing.

Actual Local Size and Actual Mating Envelope

The next term is actual local size. ***Actual local size*** is the value of any individual distance at any cross section of a FOS. The actual local size is a two-point measurement, taken with an instrument like a caliper or micrometer, that is checked at a point along the cross section of the part. A FOS may have several different values of actual local size.

The term, "actual mating envelope" is defined according to the type of feature of size being considered. The ***actual mating envelope (AME) of an external feature of size*** is a similar perfect feature counterpart of the smallest size that can be circumscribed about the feature so it just contacts the surfaces at the highest points.

For example, a similar perfect counterpart could be:

- a smallest cylinder of perfect form
- two parallel planes of perfect form at a minimum separation

. . . that just contact(s) the highest points of the surfaces. AME is a variable value; it is derived from an actual part. See Figure 2-3 for examples.

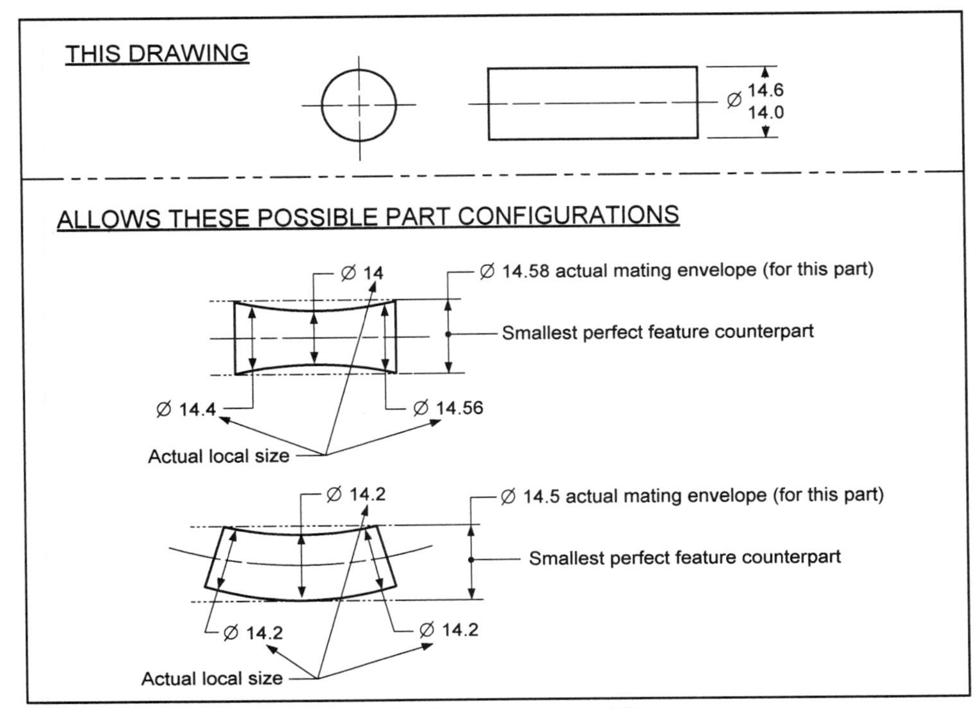

FIGURE 2-3 Actual Mating Envelope of an External Feature of Size

If a feature of size is controlled by an orientation or positional tolerance, with the MMC or LMC modifier used, the AME is relative to the appropriate datums.

The ***actual mating envelope (AME) of an internal feature of size*** is a similar perfect feature counterpart of the largest size that can be inscribed within the feature so that it just contacts the surfaces at their highest points. A similar perfect feature counterpart could be a largest cylinder of perfect form. It could also be two parallel planes of perfect form at maximum separation that just contact the highest points of the surfaces. AME is a variable value; it is derived from an actual part. See Figure 2-4 for examples.

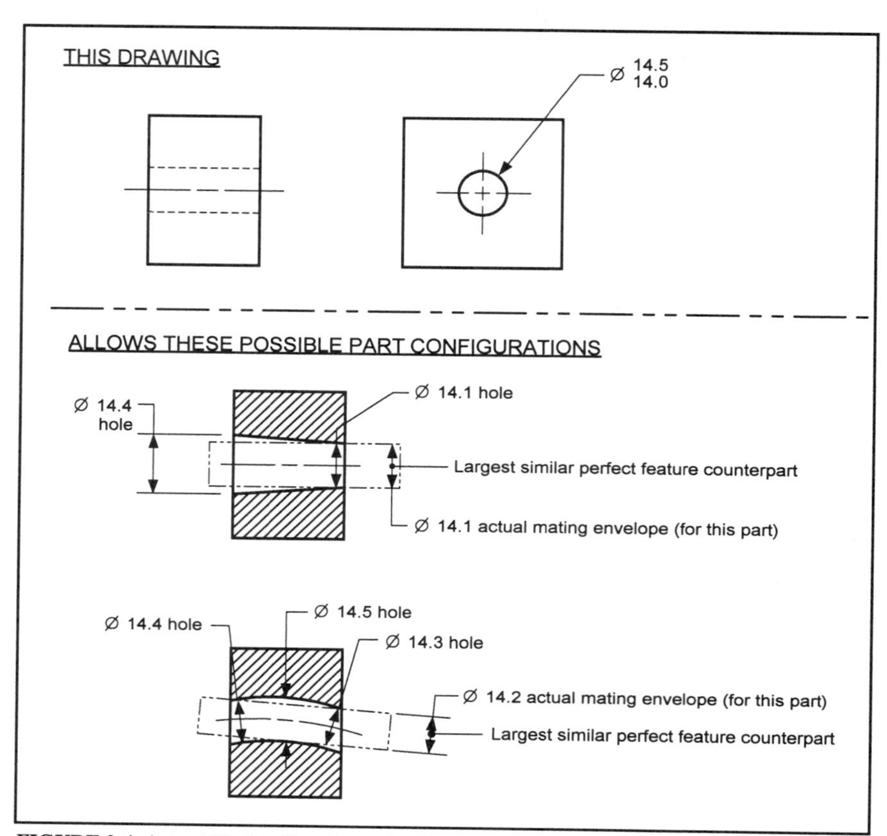

FIGURE 2-4 Actual Mating Envelope of an Internal Feature of Size

If a feature of size is controlled by an orientation or positional tolerance, with the MMC or LMC modifier, the actual mating envelope is relative to the appropriate datums.

TECHNOTE 2-3 Actual Mating Envelope

- Actual mating envelope is a variable value, derived from an actual part.
- For an external feature, the actual mating envelope is the smallest perfect feature counterpart that can be circumscribed about the feature.
- For an internal feature, the actual mating envelope is the largest perfect feature counterpart that can be inscribed within the feature.

MATERIAL CONDITIONS

A key concept in geometric tolerancing is the ability to specify tolerances at various part feature material conditions. A geometric tolerance can be specified to apply at the largest size, smallest size, or actual size of a feature of size. This section provides definitions of the three common material conditions used in GD&T.

Author's Comment
These material condition concepts can only be used when referring to a feature of size such as a hole, shaft diameter, tab, etc.

Maximum Material Condition (MMC)

Maximum material condition is the condition in which a feature of size contains the maximum amount of material everywhere within the stated limits of size—for example, the largest shaft diameter or smallest hole diameter. Figure 2-5 shows examples of maximum material condition.

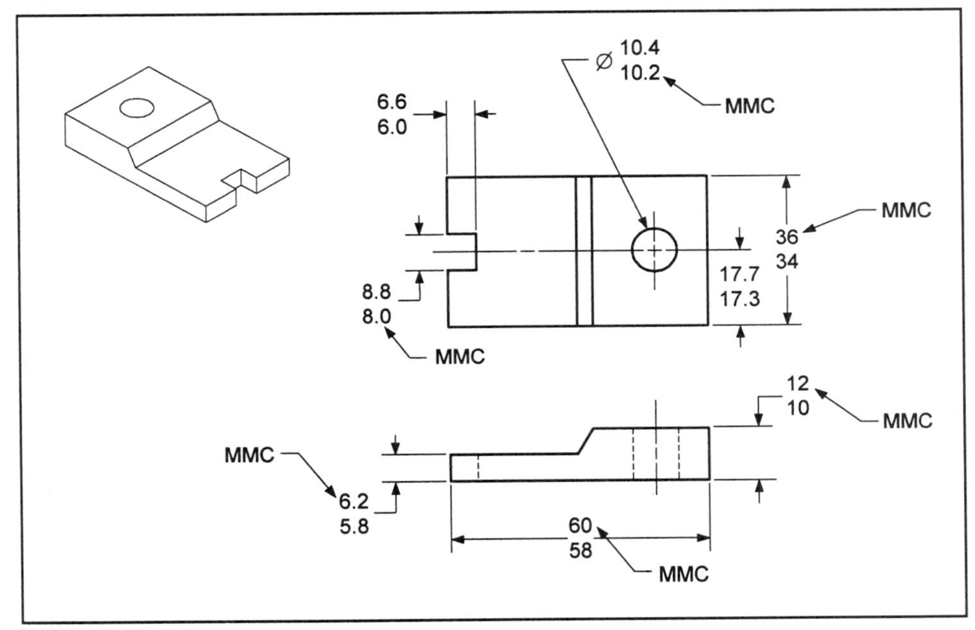

FIGURE 2-5 Maximum Material Condition

TECHNOTE 2-4 Maximum Material Condition

- The maximum material condition of an external feature of size (i.e., shaft) is its largest size limit.
- The maximum material condition of an internal feature of size (i.e., hole) is its smallest size limit.

Least Material Condition (LMC)

Least material condition is the condition in which a feature of size contains the least amount of material everywhere within the stated limits of size—for example, the smallest shaft diameter or the largest hole diameter. Figure 2-6 shows examples of the least material condition.

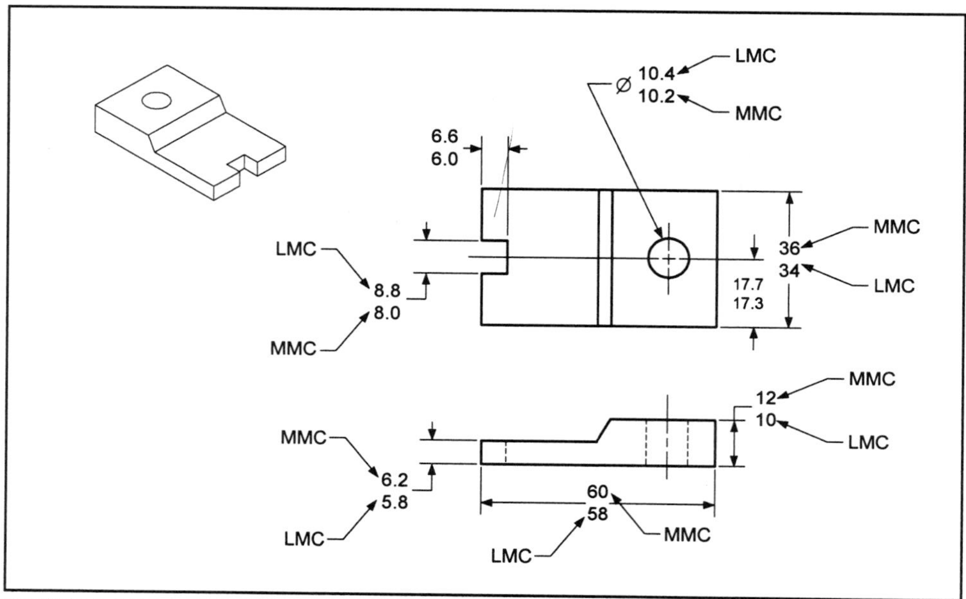

FIGURE 2-6 Examples of Maximum and Least Material Condition

> **TECHNOTE 2-5 Least Material Condition**
>
> - The least material condition for an external feature of size (i.e., shaft) is its smallest size limit.
> - The least material condition for an internal feature of size (i.e., hole) is its largest size limit.

For more info. . .
See Rule #2 and Rule #2A on page 54.

Regardless of Feature Size (RFS)

Regardless of feature size is the term that indicates a geometric tolerance applies at any increment of size of the feature within its size tolerance. Another way to visualize RFS is that the geometric tolerance applies at whatever size the part is produced. There is no symbol for RFS because it is the default condition for all geometric tolerances.

Material Conditions and Part Dimensions

Every feature of size has a maximum and least material condition. Limit dimensions directly specify the maximum and least material condition of a feature of size. When a drawing contains plus-minus dimensions, the material conditions may have to be derived from the dimensions. Figure 2-7 shows examples of material conditions.

Design Tip

Geometric tolerances specified at maximum or least material condition have a significant cost advantage over geometric tolerances specified regardless of feature size.

Material Condition Usage

Each material condition is used for different functional reasons. Geometric tolerances are often specified to apply at MMC when the function of a FOS is assembly. Geometric tolerances are often specified to apply at LMC to insure a minimum distance on a part. Geometric tolerances are often specified to apply at RFS to insure symmetrical relationships.

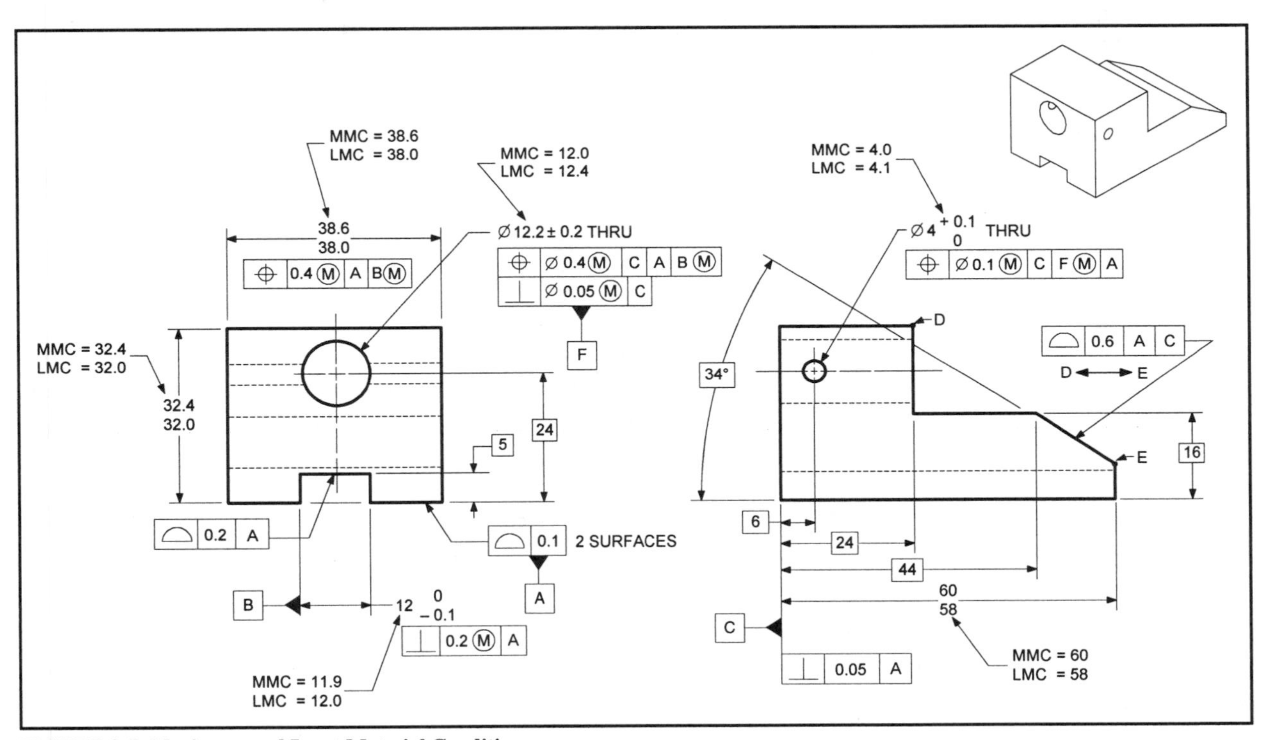

FIGURE 2-7 Maximum and Least Material Conditions

MODIFIERS

In the language of geometric tolerancing there are a set of symbols called "modifiers." ***Modifiers*** communicate additional information about the drawing or tolerancing of a part. There are eight common modifiers used in geometric tolerancing. They are shown in Figure 2-8.

The first two modifiers, MMC and LMC were explained earlier in this chapter. The projected tolerance zone modifier and the tangent plane modifier are placed inside the feature control frame, as are the MMC and LMC modifier symbols.

For more info. . .
The projected tolerance zone modifier is explained in Chapter 9. The tangent plane modifier is explained in Chapter 7.

The next two modifiers, the "P" and "T" in circles, stand for projected tolerance zone and tangent plane. The projected tolerance zone modifier changes the location of the tolerance zone on the part. It projects the tolerance zone so that it exists above the part. The tangent plane modifier denotes that only the tangent plane of the toleranced surface needs to be within this tolerance zone.

The diameter symbol is used two ways: inside a feature control frame as a modifier to denote the shape of the tolerance zone, or outside the feature control frame to simply replace the word, "diameter." The radius and controlled radius modifiers are always used outside the feature control frame. They are explained in the next section.

The modifier for reference is simply the method of denoting that information is for reference only. The information is not to be used for manufacturing or inspection. To designate a dimension or other information as reference, the reference information is enclosed in parentheses.

Author's Comment
Additional modifiers and the concept of RFS are explained in Section 3 of Y14.5.

TERM	ABBREVIATION	SYMBOL
MAXIMUM MATERIAL CONDITION	MMC	Ⓜ
LEAST MATERIAL CONDITION	LMC	Ⓛ
PROJECTED TOLERANCE ZONE	—	Ⓟ
TANGENT PLANE	—	Ⓣ
DIAMETER	DIA	⌀
RADIUS	—	R
CONTROLLED RADIUS	—	CR
REFERENCE	—	( )

FIGURE 2-8 Modifiers

RADIUS AND CONTROLLED RADIUS

Arcs are dimensioned with a radius symbol on drawings. The Y14.5M-1994 standard contains two radius symbols: radius and controlled radius.

Author's Comment
The controlled radius is a new symbol introduced in Y14.5.

Radius

A ***radius*** is a straight line extending from the center of an arc or a circle to its surface. The symbol for a radius is "R." When the "R" symbol is used, it creates a zone defined by two arcs (the minimum and maximum radii). The part surface must lie within this zone. Figure 2-9 shows a radius tolerance zone. The part surface may have flats or reversals within the tolerance zone.

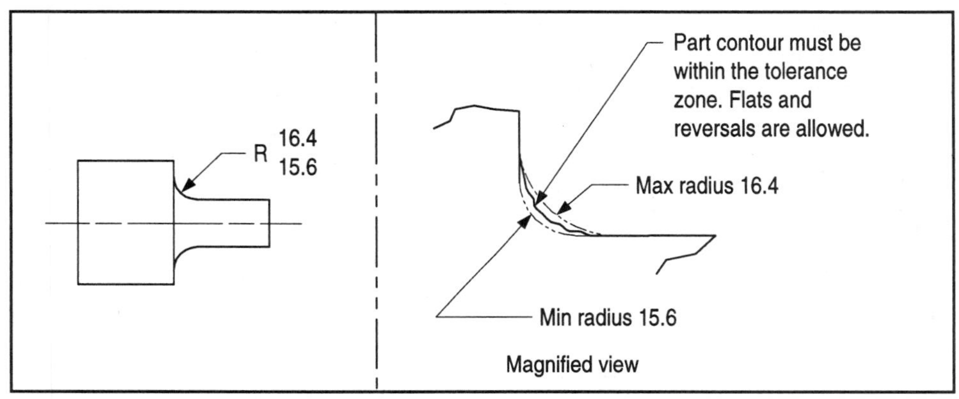

FIGURE 2-9 Radius Example

Controlled Radius

A ***controlled radius*** is a radius with no flats or reversals allowed. The symbol for a controlled radius is "CR." When the "CR" symbol is used, it creates a tolerance zone defined by two arcs (the minimum and maximum radii). The part surface must be within the crescent-shaped tolerance zone and be an arc without flats or reversals. Figure 2-10 shows a controlled radius tolerance zone.

Design Tip
A controlled radius is a stringent requirement; it should only be used in special cases. One example is when the part stresses are very high and reversals in the radiused surface would produce higher additional stresses.

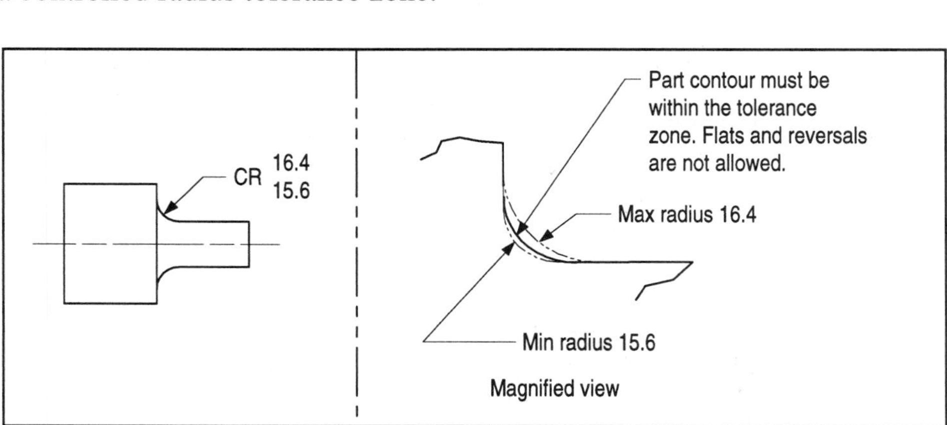

FIGURE 2-10 Controlled Radius Example

TECHNOTE 2-6 Radius & Controlled Radius

- When an "R" symbol is specified, flats or reversals are allowed.
- When a "CR" symbol is specified, flats or reversals are not allowed.

INTRODUCTION TO GEOMETRIC TOLERANCES

Author's Comment
The symmetry symbol did not exist in the 1982 version of Y14.5.

Geometric Characteristic Symbols

Geometric characteristic symbols are a set of fourteen symbols used in the language of geometric tolerancing. They are shown in Figure 2-11. The symbols are divided into five categories: form, profile, orientation, location, and runout. The chart in Figure 2-11 shows that certain geometric symbols never use a datum reference and other geometric symbols always use a datum reference. Furthermore, some geometric symbols may or may not use a datum reference. The size and proportions for the geometric symbols are given in Appendix B.

CATEGORY	CHARACTERISTIC	SYMBOL	USES A DATUM REFERENCE
FORM	STRAIGHTNESS	⏤	NEVER
	FLATNESS	⏥	
	CIRCULARITY (ROUNDNESS)	○	
	CYLINDRICITY	⌭	
PROFILE	PROFILE OF A LINE	⌒	SOMETIMES
	PROFILE OF A SURFACE	⌓	
ORIENTATION	ANGULARITY	∠	ALWAYS
	PERPENDICULARITY	⊥	
	PARALLELISM	//	
LOCATION	POSITION	⌖	
	CONCENTRICITY	◎	
	SYMMETRY	⌯	
RUNOUT	CIRCULAR RUNOUT	↗	
	TOTAL RUNOUT	⌰	

FIGURE 2-11 Geometric Characteristic Symbols

Feature Control Frame

Geometric tolerances are specified on a drawing through the use of a feature control frame. A ***feature control frame*** is a rectangular box that is divided into compartments within which the geometric characteristic symbol, tolerance value, modifiers, and datum references are placed. The compartments of a feature control frame are shown in Figure 2-12.

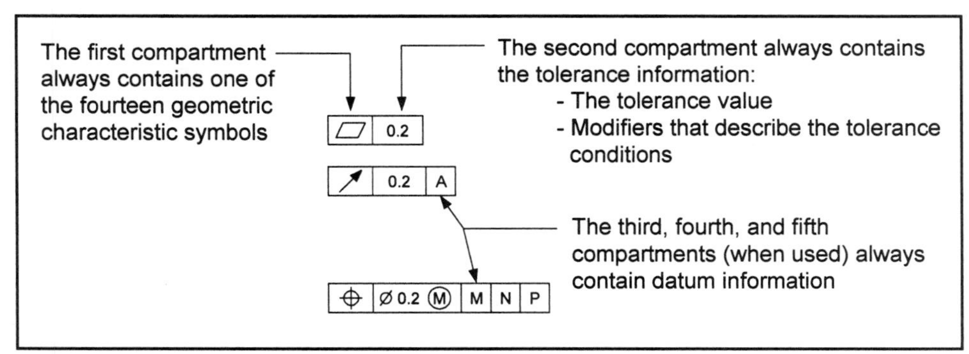

FIGURE 2-12 Parts of a Feature Control Frame

The first compartment of the feature control frame is called the geometric characteristic portion. It contains one of the fourteen geometric characteristic symbols.

The second compartment of the feature control frame is referred to as the tolerance portion. The tolerance portion of a feature control frame may contain several bits of information. For example, if the tolerance value is preceded by a diameter symbol Ø , the shape of the tolerance zone is a cylinder. If a diameter symbol is not shown in the front of the tolerance value, the shape of the tolerance zone is either parallel planes, parallel lines, or a uniform boundary in the case of profile. The tolerance value specified is always a total value.

When specifying a non-datum related control, the feature control frame will have two compartments. When specifying a datum related control, the feature control frame may have up to five compartments: the first for a geometric characteristic symbol, one for tolerance information, and up to three additional compartments for datum references. The third, fourth, and fifth compartments of the feature control frame are referred to as the datum reference portion of the feature control frame.

The typical placement of feature control frames is shown in Figure 2-13.

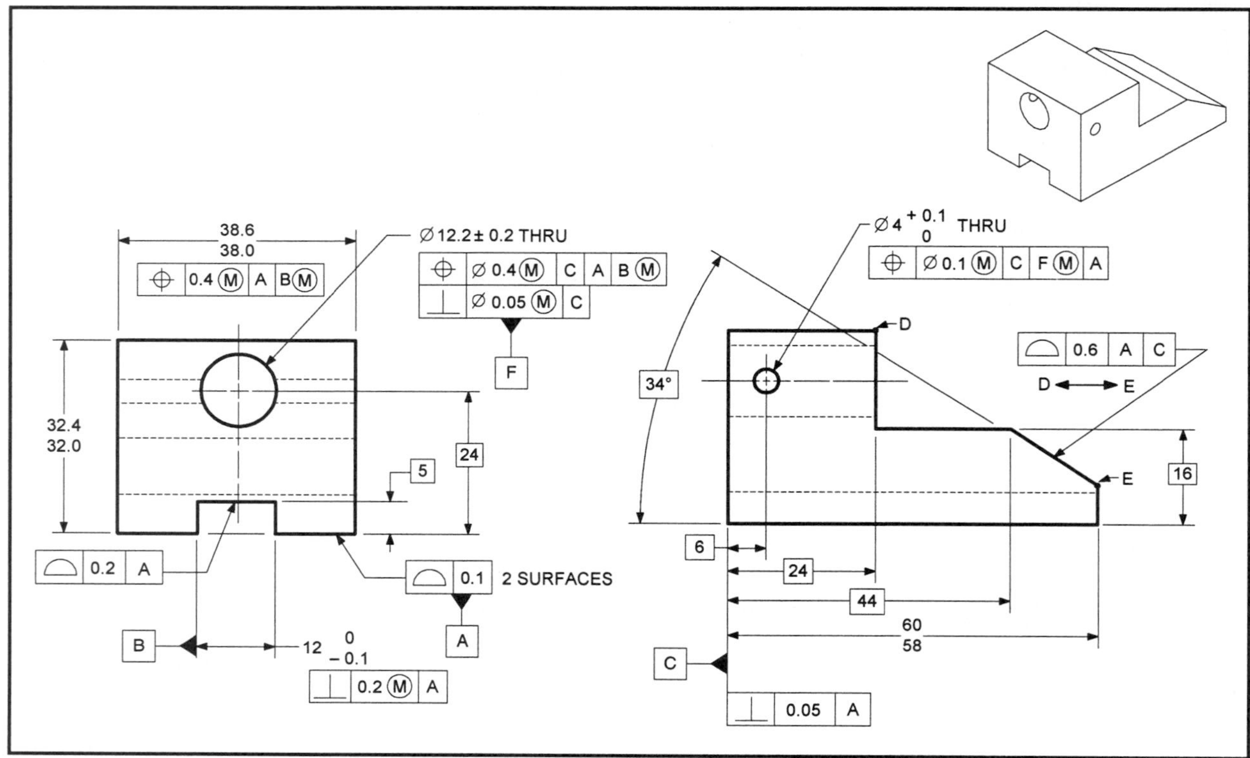

FIGURE 2-13 Placement of Feature Control Frames

VOCABULARY LIST

New Terms Introduced in this Chapter

Actual local size
Actual mating envelope (AME) (external FOS)
Actual mating envelope (AME) (internal FOS)
Controlled radius (CR)
Cylindrical FOS
Feature
Feature control frame
Feature of size (FOS)
Feature of size dimension
Geometric characteristic symbols
Least material condition (LMC)
Maximum material condition (MMC)
Modifiers
Non-feature of size dimension
Planar FOS
Radius (R)
Regardless of feature size (RFS)

Study Tip
Read each term. If you don't know its meaning, look it up in the chapter.

ADDITIONAL RELATED TOPICS

Topic	ASME Y14.5M-1994 Reference
• Statistical tolerancing	Paragraph 2.16
• Spherical diameter symbol	Paragraph 3.3.7
• Arc length symbol	Paragraph 3.3.9
• Free-state symbol	Paragraph 3.3.19
• Surface texture symbols	Paragraph 3.3.21

Author's Comment
These topics, plus advanced coverage of many of the topics introduced in this text, will be covered in my new book on advanced GD&T concepts.

QUESTIONS AND PROBLEMS

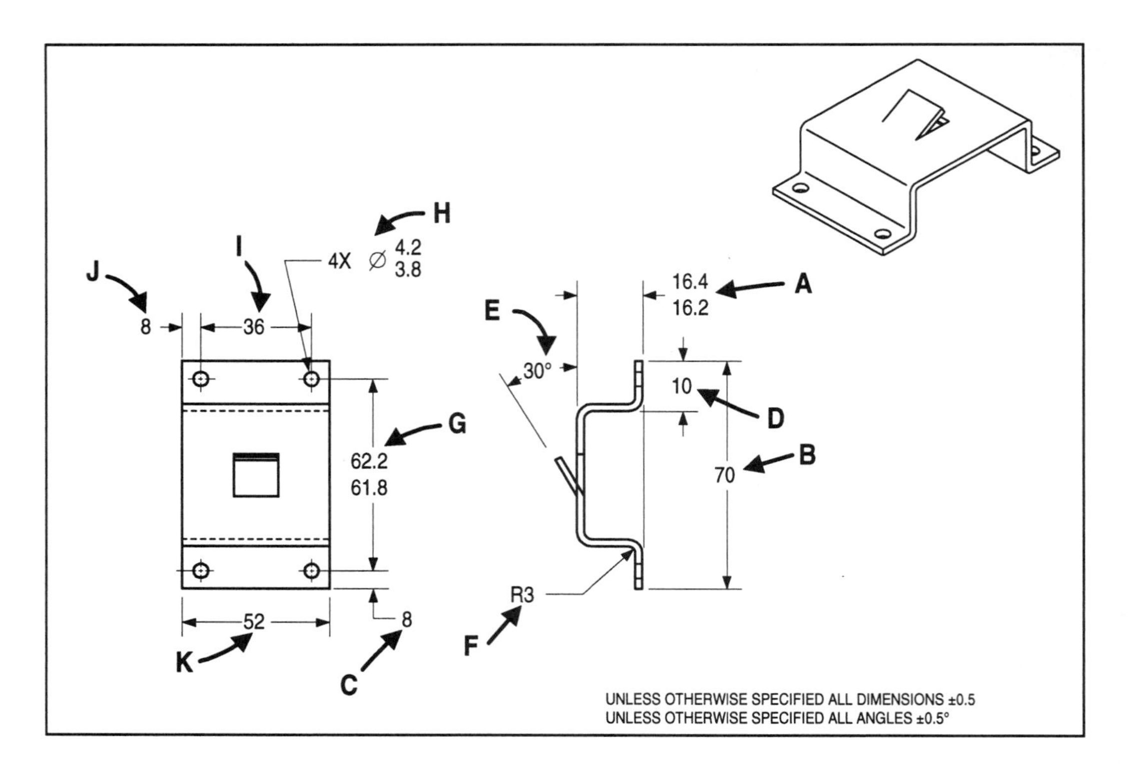

1. Using the drawing above, indicate if each letter is associated with a feature of size dimension or a non-feature of size dimension.

Letter	Feature of size dimension	Non-feature of size dimension
A		
B		
C		
D		
E		
F		
G		
H		
I		
J		
K		

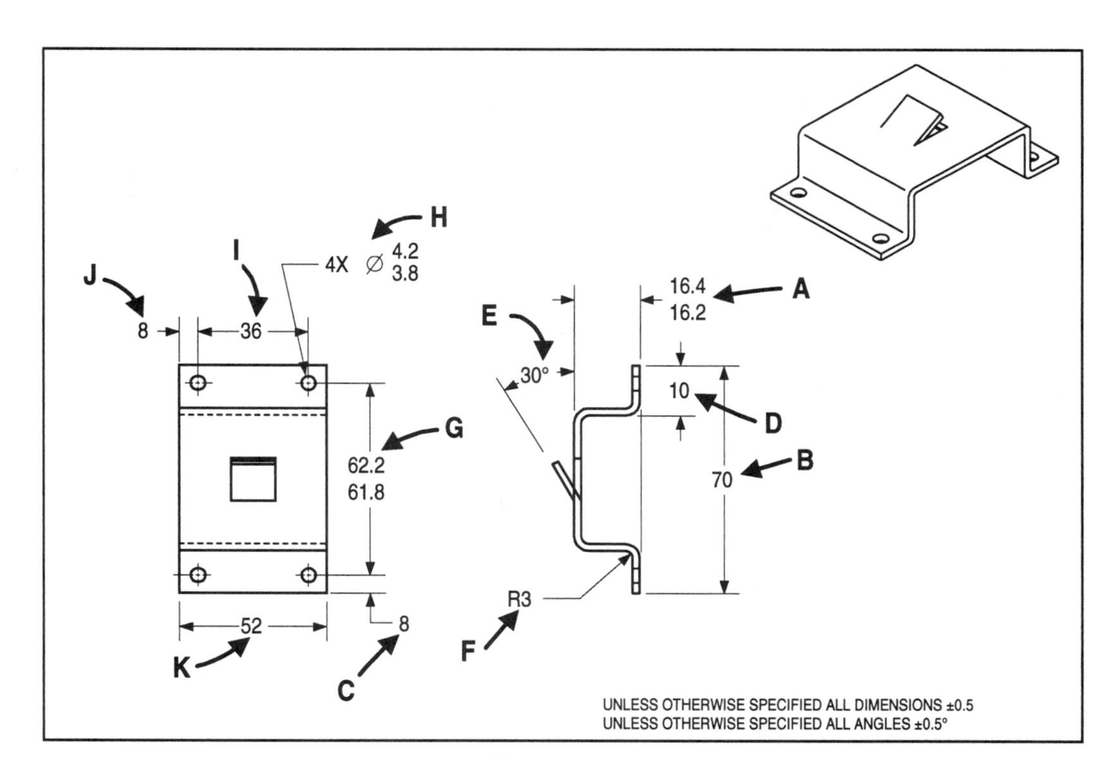

2. Use the drawing above to fill in the value of the MMC and LMC for each dimension (or indicate, "does not apply").

Letter	MMC	LMC	Does not apply
A			
B			
C			
D			
E			
F			
G			
H			
I			
J			
K			

Use the word list on the right to fill in the blanks. (There are more words than blanks.)

3. Actual local size is the value of any ___________ distance at any _____________ of a feature.

4. In a feature of size, the surfaces or elements must be _____________.

5. There are two types of features of size, _____________ and _____________ .

6. Actual mating envelope is a _____________ value.

7. For an internal feature of size, the actual mating envelope is the _____________ perfect feature counterpart that can be inscribed about the feature.

8. The smallest diameter of a hole is its _____________ material condition.

9. When a radius is specified, flats or reversals are _____________.

10. An angle is a _____________ dimension.

11. A_______________FOS is a FOS that contains two parallel plane surfaces.

12. A radius without flats and reversals is referred to as a ____________ ___________.

13. The five categories of geometric characteristic symbols are: Form, _____________, _____________, _____________, and _____________.

Word List
Actual local
Actual mating
Allowed
Angularity
Controlled radius
Cross section
Datum reference portion
External feature of size
Feature of size
Fixed
Geometric characteristic portion
Individual
Internal
Largest
Least
Limit
Location
Maximum
Non-feature of size
Not allowed
Opposed
Orientation
Permissible
Planar
Position
Profile
Runout
Size
Smallest
Smooth radius
Tolerance portion
Variable

14. Label the intended contents of a feature control frame.

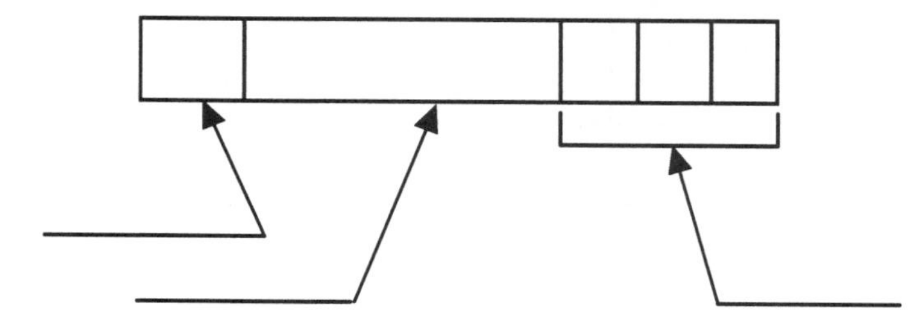

Chapter **3**

Rules and Concepts of GD&T

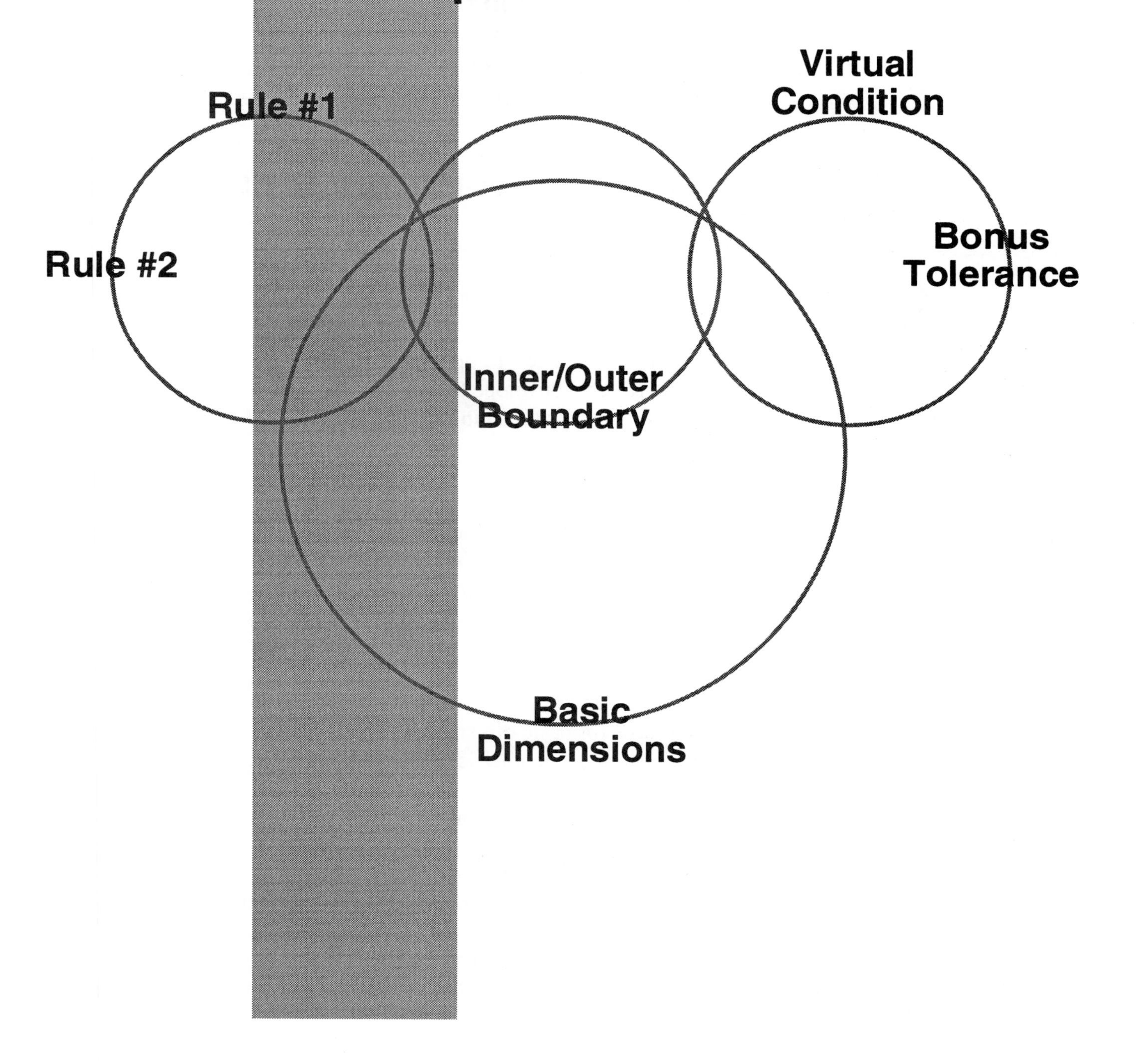

INTRODUCTION

This chapter continues to lay the foundation for further meaningful study. It introduces two important rules and three key concepts that are used throughout the course.

CHAPTER GOALS AND OBJECTIVES

There are Two Goals in this Chapter:

3-1. Understand Rule #1 and Rule #2.

3-2. Understand the concepts of basic dimensions, virtual condition, inner and outer boundary, worst-case boundary, and bonus tolerance.

Performance Objectives that Demonstrate Mastery of These Goals:

Upon completion of this chapter, each student should be able to:

Study Tip
Take a few minutes to fully understand these objectives. When reading this chapter, look for information to help you master these objectives.

Goal 3-1 (pp. 47-54)

- Explain Rule #1.
- Determine when Rule #1 applies to a dimension.
- Draw the Rule #1 envelope boundary.
- Explain the interrelationship between Rule #1 and its size dimension.
- Explain how Rule #1 affects the interrelationship between features of size.
- List two ways Rule #1 can be overridden.
- List two exceptions to Rule #1.
- Explain how to inspect a FOS that is controlled by Rule #1.
- Explain Rule #2.

Goal 3-2 (pp. 54-65)

- Define basic dimensions.
- List two uses for basic dimensions.
- Determine when a geometric tolerance applies to a feature or a feature of size.
- Explain the concept of virtual condition.
- Explain the concept of inner and outer boundary.
- Explain the term, "worst-case boundary."
- Calculate the virtual condition of a feature of size (with GD&T applied).
- Explain the concept of bonus tolerance.
- Calculate the amount of bonus tolerance available for a geometric tolerance.

RULES

There are two general rules in ASME Y14.5M-1994. The first rule establishes default conditions for features of size. The second rule establishes a default material condition for feature control frames.

Rule #1

Rule #1 is referred to as the "Individual Feature of Size Rule." It is a key concept in geometric tolerancing. Rule #1 is a dimensioning rule used to ensure that features of size will assemble with one another. When Rule #1 applies, the maximum boundary (or envelope) for an external FOS is its MMC. The minimum envelope for an internal FOS is its MMC. To determine if two features of size will assemble, the designer can then compare the MMCs of the features of size. The Y14.5 definition for Rule #1 is shown below.

> ***Rule #1***: Where only a tolerance of size is specified, the limits of size of an individual feature prescribe the extent to which variations in its form—as well as in its size—are allowed.

In industry, Rule #1 is often paraphrased as "perfect form at MMC" or the "envelope rule."

There are two components to Rule #1: the envelope principle and the effects on the form of a FOS as it departs from MMC. When Rule #1 applies, the limits of size define the size as well as the form of an individual FOS.

For example, let's look at how Rule #1 affects the diameter of a pin. When the pin diameter is at MMC, the pin must have perfect form. For a pin diameter, perfect form means perfect straightness and perfect roundness. This would allow the pin to fit through a boundary equal to its MMC. If the size of the pin was less than its MMC, the pin could contain form error (straightness and roundness error) equal to the amount the pin departed from MMC.

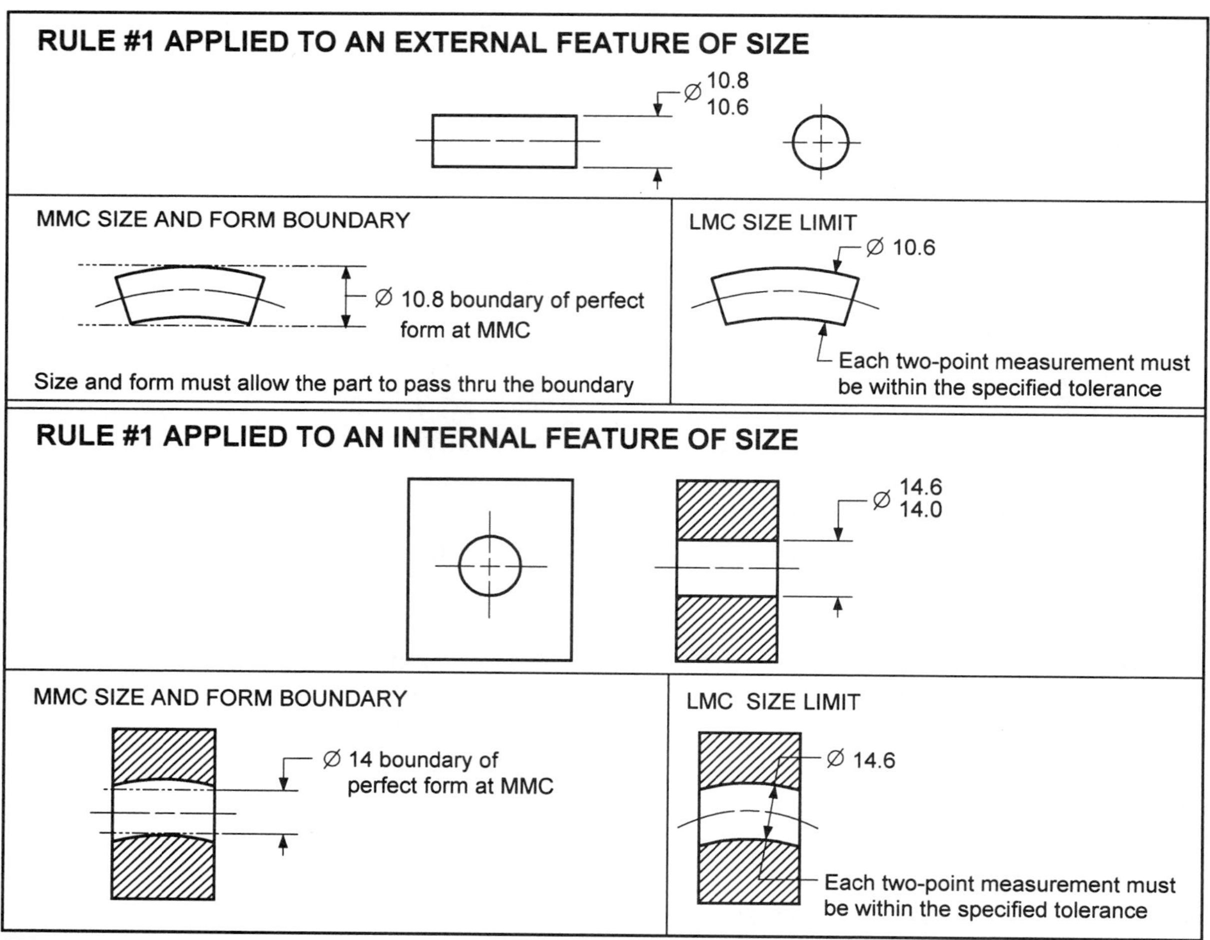

FIGURE 3-1 Rule #1 Examples

An example of the effects of Rule #1 on an external and an internal FOS is shown in Figure 3-1.

The form of a FOS is controlled by its limits of size, as described below:

- The surfaces of a feature of size shall not extend beyond a boundary (envelope) of perfect form at MMC.
- When the actual local size of a FOS has departed from MMC toward LMC, the form is allowed to vary by the same amount.
- The actual local size of an individual feature of size must be within the specified tolerance of size.
- There is no requirement for a boundary of perfect form at LMC. If a feature of size is produced at LMC, it can vary from true form by the amount allowed by the MMC boundary.

An example of the effects of Rule #1 on a planar FOS is shown in Figure 3-2.

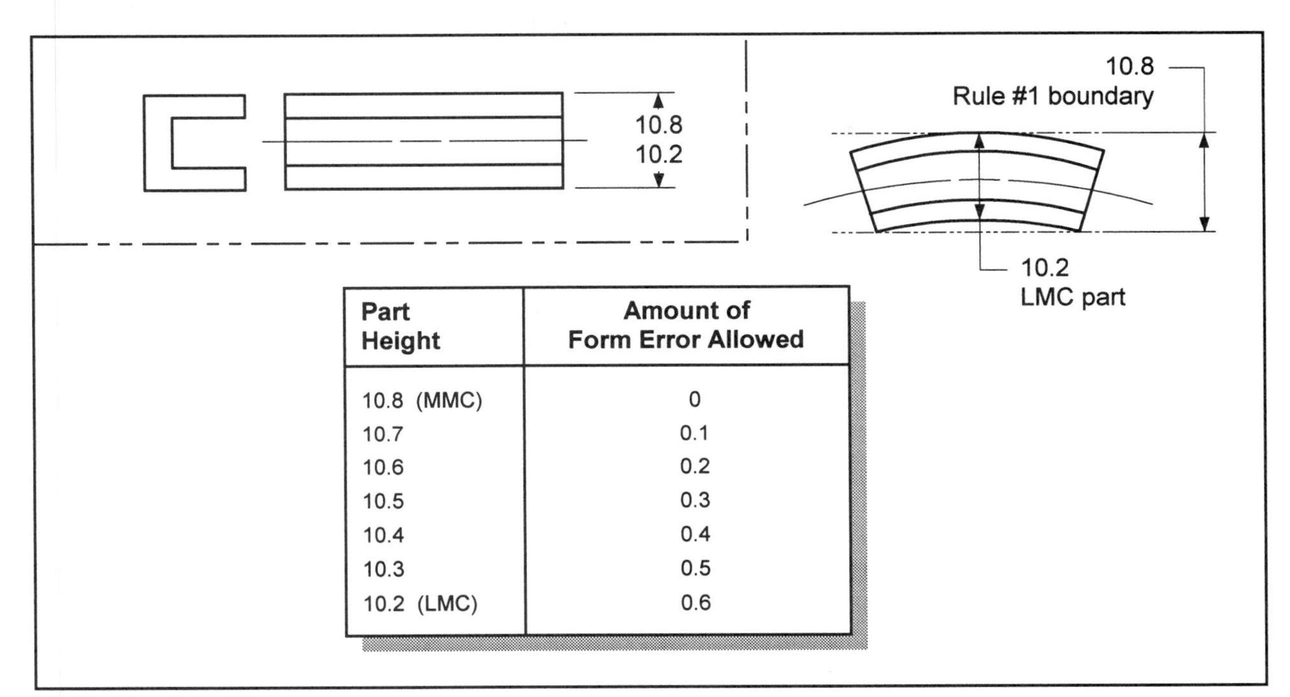

Part Height	Amount of Form Error Allowed
10.8 (MMC)	0
10.7	0.1
10.6	0.2
10.5	0.3
10.4	0.4
10.3	0.5
10.2 (LMC)	0.6

FIGURE 3-2 Rule #1 Boundary

In Rule #1, the words "perfect form" mean perfect flatness, straightness, circularity, and cylindricity. In other words, if a feature of size is produced at MMC, it is required to have perfect form. If a feature of size is not at MMC, then a form error is allowed. On a planar FOS (as shown in Figure 3-2) perfect form refers to perfect flatness and perfect straightness. If the height of the part in Figure 3-2 was at 10.7, then a form error equal to the amount of departure from MMC (10.8 - 10.7 = 0.1) would be allowed. If the part was produced at LMC, a form error equal to the amount of departure from MMC (0.6) would be permitted. For example, if the part height was 10.2 (everywhere), the flatness of the bottom of the block would be limited to 0.6.

TECHNOTE 3-1 Rule #1

Rule #1: For features of size, where only a tolerance of size is specified, the surfaces shall not extend beyond a boundary (envelope) of perfect form at MMC.

How to Override Rule #1

Rule #1 applies whenever a feature of size is specified on a drawing. There are two ways Rule #1 can be overridden:

If a straightness control is applied to a feature of size, Rule #1 is overridden.
If a note such as "PERFECT FORM AT MMC NOT REQUIRED" is specified next to a FOS dimension, it exempts the FOS dimension from Rule #1.

TECHNOTE 3-2 How to Override Rule #1

There are two ways to override Rule #1:

1. A straightness control applied to a FOS
2. A special note applied to a FOS

Rule #1 Limitation

A part often contains multiple features of size. Rule #1 does not affect the location, orientation, or relationship between features of size. Features of size shown perpendicular, symmetrical, or coaxial must be controlled for location or orientation to avoid incomplete drawing specifications. Often, implied 90° angles are covered by a general angular tolerance note or by an angular tolerance in the drawing titleblock.

In Figure 3-3 there are four features of size, marked *A, B, C,* & *D*. Rule #1 applies independently to each FOS. The angles between these features of size (angles *E, F,* & *G*) are not controlled by Rule #1.

For more info. . .
Paragraph 2.7.3 of Y14.5 shows several methods for specifying the relationship between individual features.

TECHNOTE 3-3 Rule #1 Limitation

Rule #1 does not control the location, orientation, or relationship between features of size.

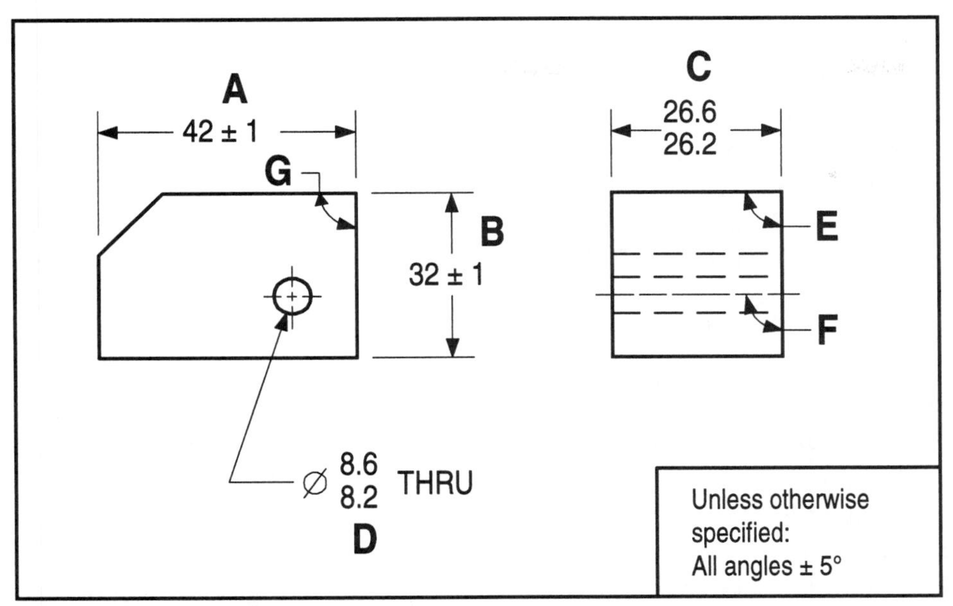

FIGURE 3-3 Interrelationship Between Features of Size

There are two exceptions to Rule #1. First, Rule #1 does not apply to a FOS on a part subject to free-state variation in the unrestrained condition. In simple terms, Rule #1 does not apply to flexible parts that are not restrained. The second exception is that Rule #1 does not apply to stock sizes, such as bar stock, tubing, sheet metal, or structural shapes. Paragraph 2.7.1.3 in ASME Y14.5M-1994 explains why. It states, ". . . standards for these items govern the surfaces that remain in the as-furnished condition on the finished part."

TECHNOTE 3-4 Rule #1 Exceptions

There are two exceptions to Rule #1:

1. A FOS on a non-rigid part
2. Stock sizes

Inspecting a Feature of Size

When inspecting a FOS that is controlled by Rule #1, both its size and form need to be verified. The MMC size and the Rule #1 envelope can be verified with a Go gage. A ***Go gage*** is a gage that is intended to fit into (for an internal FOS) or fit over (for an external FOS) the FOS. A Go gage is made to the MMC limit of the FOS and has perfect form. A Go gage can verify the MMC limit and Rule #1 form envelope of a FOS. To fully verify the Rule #1 effects, a Go gage must be at least as long as the FOS it is verifying. Figure 3-4 shows examples of a Go gages for a pin and a hole.

The minimum size (LMC) of a FOS can be measured with a No-Go gage. A ***No-Go gage*** is a gage that is not intended to fit into or over a FOS. A No-Go gage is made to the LMC limit of the FOS. A No-Go gage makes a two-point check; a caliper or snap gage could be used as a No-Go gage. The two-point check is made at various points along the cross section to insure that the FOS does not violate the LMC limit. Figure 3-4 shows examples of No-Go gages for a pin and a hole.

Author's Comment
Theory vs. reality of inspection: When inspecting a part dimension, the measurements taken (reality) verify that the part is as close to the theoretical definition as practical. In most cases, the method and number of measurements taken involve judgment by the inspector.

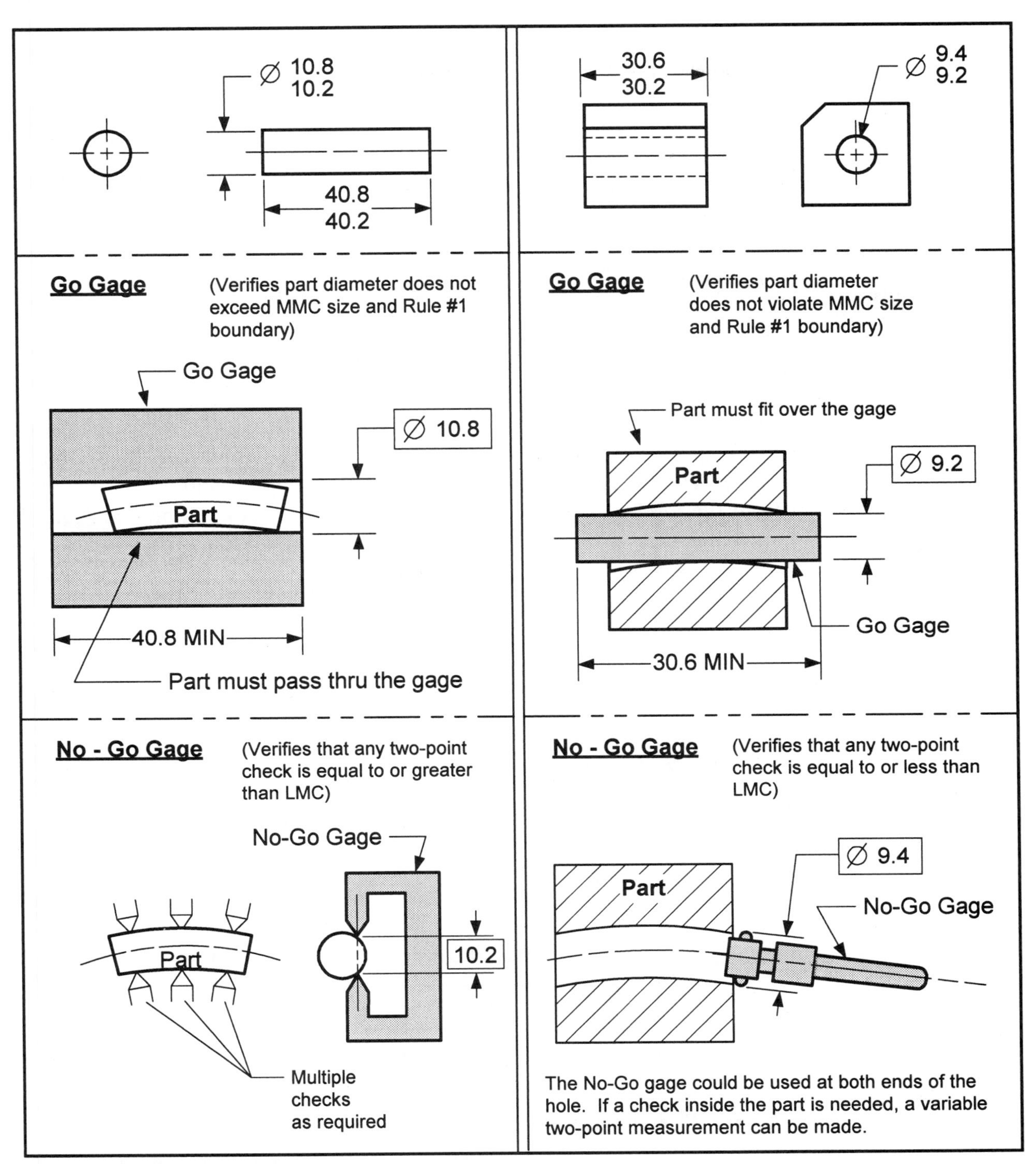

FIGURE 3-4 Go Gage and No-Go Gage Examples

Rule #2

Rule #2 is called "the all applicable geometric tolerances rule."

> ***Rule #2***: RFS applies, with respect to the individual tolerance, datum reference, or both, where no modifying symbol is specified. MMC or LMC must be specified on the drawing where required.

Author's Comment
In past versions of Y14.5, RFS was not implied for all geometric tolerances. For positional tolerances, RFS had to be stated.

Certain geometric tolerances always apply RFS and cannot be modified to MMC or LMC.

Where a geometric tolerance is applied on an RFS basis, the tolerance is limited to the specified value regardless of the actual size of the feature.

Rule #2a is an alternative practice of Rule #2. Rule #2a states that, for a tolerance of position, RFS may be specified in feature control frames if desired and applicable. In this case, the RFS symbol would be the symbol from the 1982 version of Y14.5. Figure 3-5 shows examples of Rule #2 and Rule #2a.

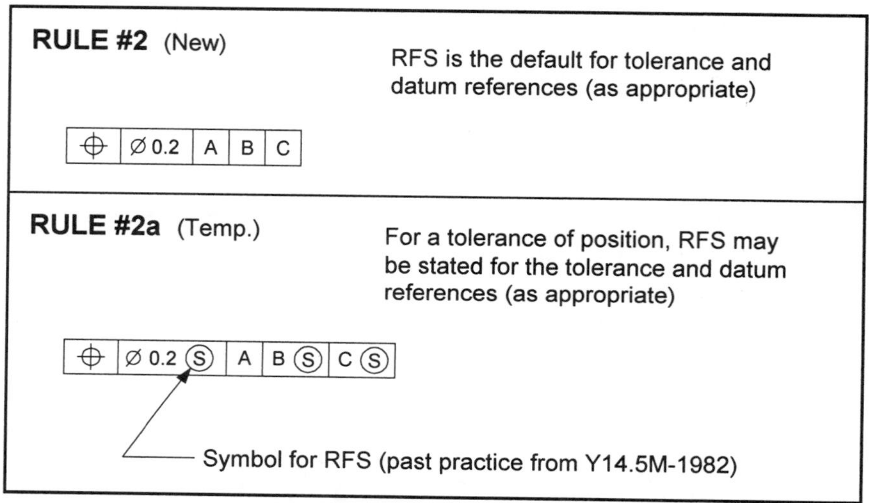

FIGURE 3-5 Rule #2 and Alternate Rule #2a

INTRODUCTION TO BASIC DIMENSIONS

Basic Dimension

A ***basic dimension*** is a numerical value used to describe the theoretically exact size, true profile, orientation, or location of a feature or gage information (i.e., datum targets). On engineering drawings there are two uses for basic dimensions. One is to define the theoretically exact location, size, orientation, or true profile of a part feature; the other use is to define gage information (example: datum targets). When a basic dimension is used to define part features, it provides the nominal location from which permissible variations are established by geometric tolerances.

Basic dimensions are usually specified by enclosing the numerical value in a rectangle (as shown in Figure 3-6) or in a general note, such as, "Untoleranced dimensions are basic."

Author's Comment
Another way to think about basic dimensions: the basic dimension is the goal, and a geometric tolerance specifies the amount of acceptable variation from the goal.

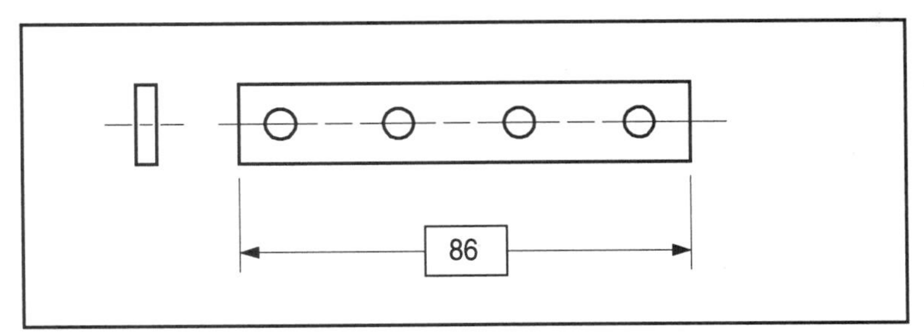

FIGURE 3-6 Basic Dimension Symbol

In simple terms, a basic dimension locates a geometric tolerance zone or defines gage information (example: datum targets). When basic dimensions are used to describe part features, they *must* be accompanied by geometric tolerances to specify how much tolerance the part feature may have. A good way to look at this is that the basic dimension only specifies half the requirement. To complete the specification, a geometric tolerance must be added to the feature involved with the basic dimension. Figure 3-7 shows basic dimensions along with their associated geometric tolerances.

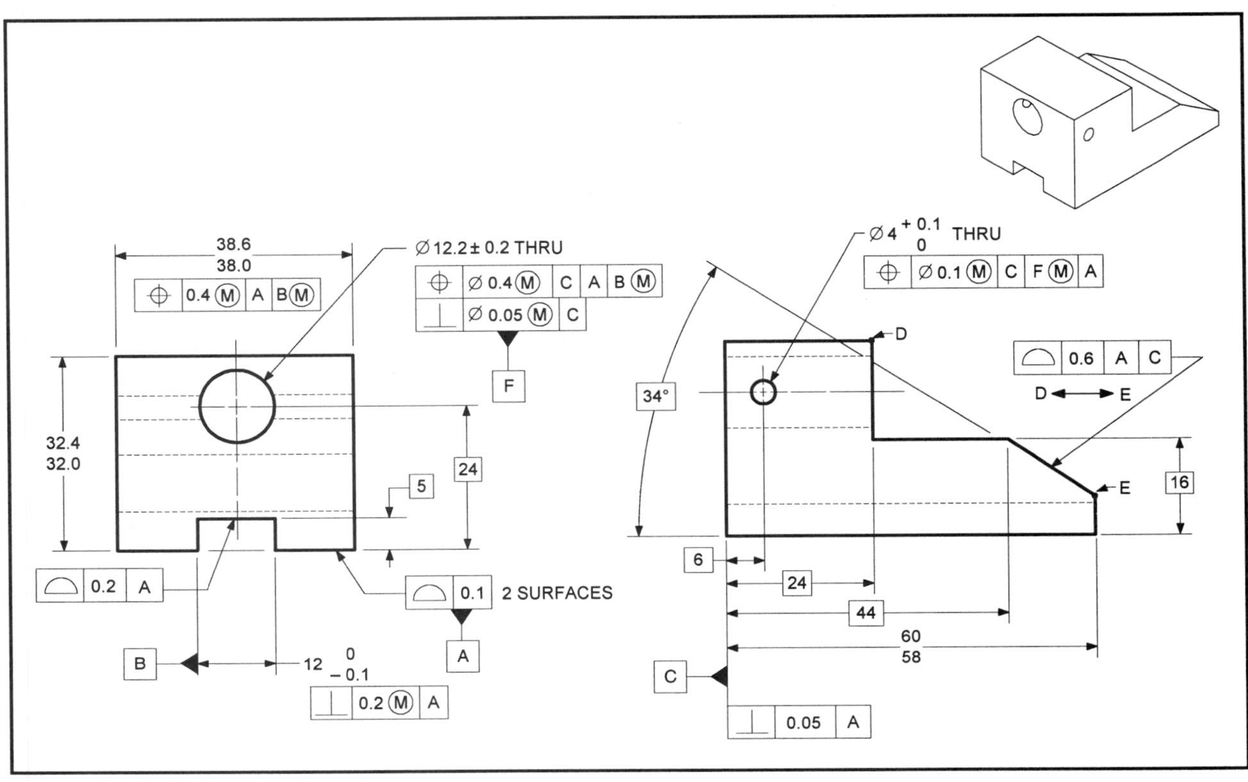

FIGURE 3-7 Basic Dimension Examples

No geometric control is used on basic dimensions that specify datum targets. When basic dimensions are used to specify datum targets, they are considered gage dimensions. Gage-makers' tolerances (a very small tolerance compared to product tolerances) apply to gage dimensions. Figure 3-8 shows basic dimensions that locate datum targets.

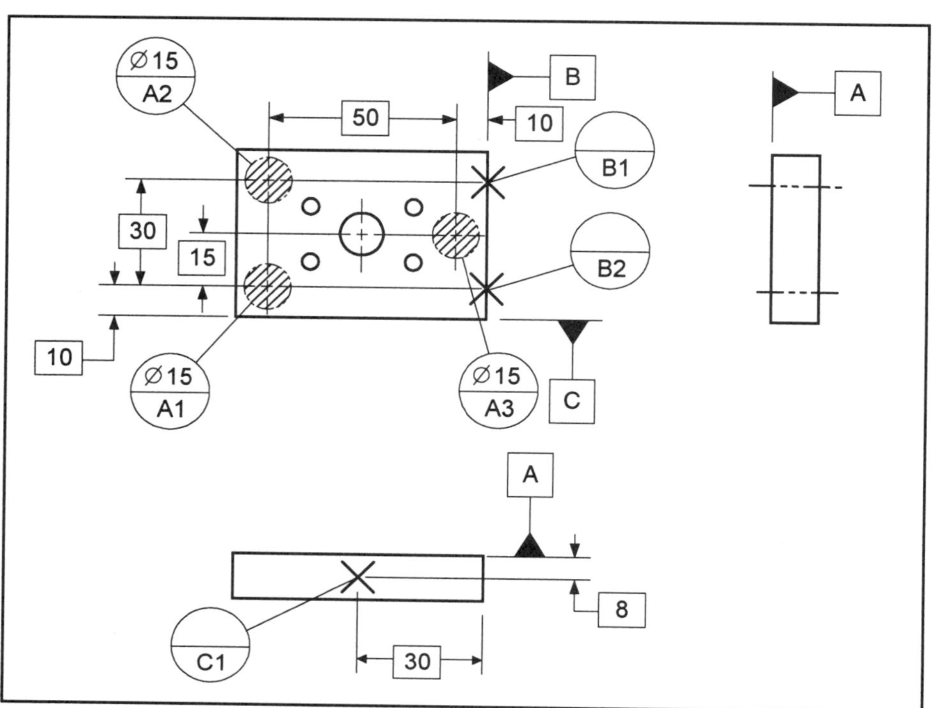

FIGURE 3-8 Basic Dimensions Used to Locate Datum Targets

TECHNOTE 3-5 Basic Dimensions

Basic dimensions. . .

- can be used to define the theoretically exact location, orientation, or true profile of part features or gage information.
- that define part features must be accompanied by a geometric tolerance.
- that define gage information do not have a tolerance shown on the print.
- are theoretically exact (but gage-makers' tolerances *do* apply).

For more info. . .
To learn about implied basic dimensions, see page 213.

Titleblock tolerances do not apply to basic dimensions. Sometimes basic dimensions are used to define a part feature, and the geometric tolerance for the basic dimensions is unintentionally left off the drawing. The drawing user is tempted to apply the tolerance from the general tolerances specified in the titleblock. This is not correct. Basic dimensions must get their tolerances from a geometric tolerance or from a special note.

INTRODUCTION TO: VIRTUAL CONDITION AND BOUNDARY CONDITIONS

Depending upon its function, a FOS is controlled by a size tolerance and one or more geometric controls. Various material conditions (MMC, LMC, or RFS) may also be applied. In each case, consideration must be given to the collective effects of the size, specified material condition, and geometric tolerance of the FOS. The terms that apply to these conditions are virtual condition, inner boundary, and outer boundary.

Definitions

Virtual condition (VC) is a worst-case boundary generated by the collective effects of a feature of size at MMC or at LMC and the geometric tolerance for that material condition. The VC of a FOS includes effects of the size, orientation, and location for the FOS. The virtual condition boundary is related to the datums that are referenced in the geometric tolerance used to determine the VC.

Inner boundary (IB) is a worst-case boundary generated by the smallest feature of size minus the stated geometric tolerance (and any additional tolerance, if applicable).

Outer boundary (OB) is a worst-case boundary generated by the largest feature of size plus the stated geometric tolerance (and any additional tolerance, if applicable).

Worst-case boundary (WCB) is a general term to refer to the extreme boundary of a FOS that is the worst-case for assembly. Depending upon the part dimensioning, a worst-case boundary can be a virtual condition, inner boundary, or outer boundary.

Author's Comment
In this text, the term "inner boundary" is only used on internal features of size, and the term "outer boundary" is only used on external features of size.

Feature of Size Boundary Conditions

If there are no geometric controls applied to a FOS, the WCB is the outer or inner boundary. The outer or inner boundary is equal to the MMC boundary as defined by Rule #1. See Figure 3-9.

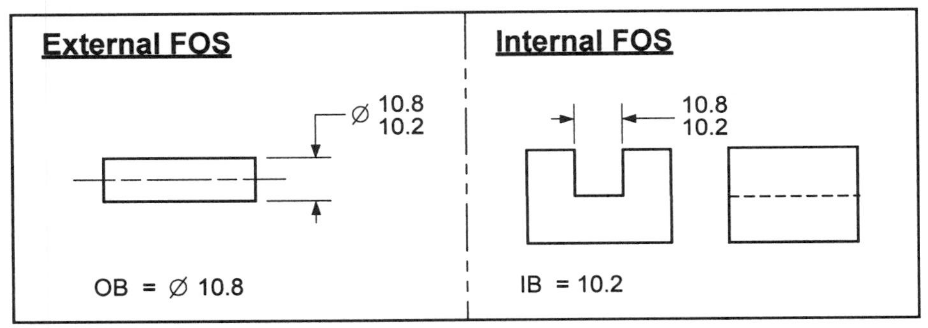

FIGURE 3-9 Worst-Case Boundary When no Geometric Tolerances are Specified

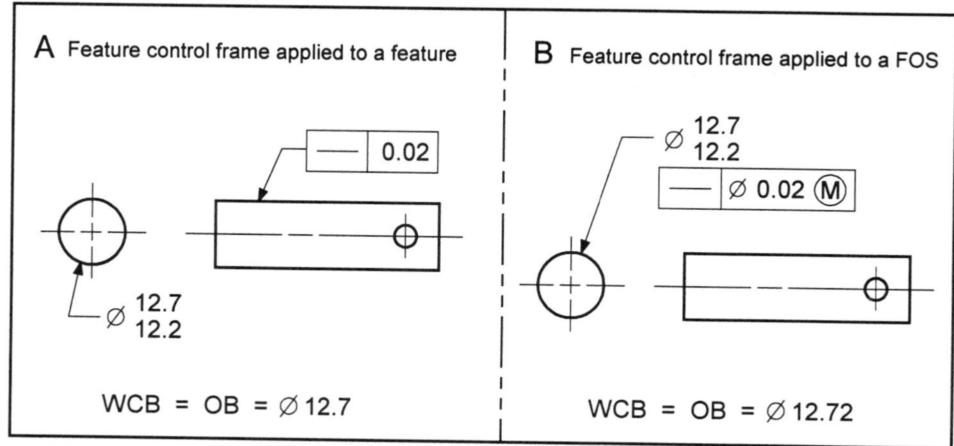

FIGURE 3-10 Feature Control Frame Placement

Whether a geometric control is applied to a feature, a surface, or a FOS (an axis or centerplane) can be determined by the location of the feature control frame on the drawing. When a feature control frame is associated with a surface, it applies to the feature. See Figure 3-10*A*. When a feature control frame is associated with a FOS dimension, or placed beneath or behind the FOS dimension, it applies to the FOS. See Figure 3-10*B*. If a feature control frame is applied to a FOS, then the WCB is affected.

TECHNOTE 3-6 Worst-Case Boundary (WCB)

If a feature control frame is applied to a feature (a surface), it does not affect its WCB. If a feature control frame is applied to a FOS (an axis or centerplane), it does affect its WCB.

MMC Virtual Condition

When a geometric tolerance that contains an MMC modifier in the tolerance portion of the feature control frame is applied to a FOS, the virtual condition (worst-case boundary) of the FOS is affected. The virtual condition (or WCB) is the extreme boundary that represents the worst-case for functional requirements, such as clearance or assembly with a mating part.

In the case of an external FOS such as a pin or a shaft, the VC (or WCB) is determined by the following formula:

VC = MMC + Geometric Tol

In the case of an internal FOS, such as a hole, the VC (or WCB) is determined by the following formula:

VC = MMC - Geometric Tol

The virtual condition of an external FOS (at MMC) is a constant value, and can also be referred to as the "outer boundary" or "worst-case boundary" in assembly calculations. The virtual condition of an internal FOS (at MMC) is a constant value and can also be referred to as the "inner boundary" or "worst-case boundary" in assembly calculations. Figure 3-11 shows examples of virtual condition calculations (at MMC).

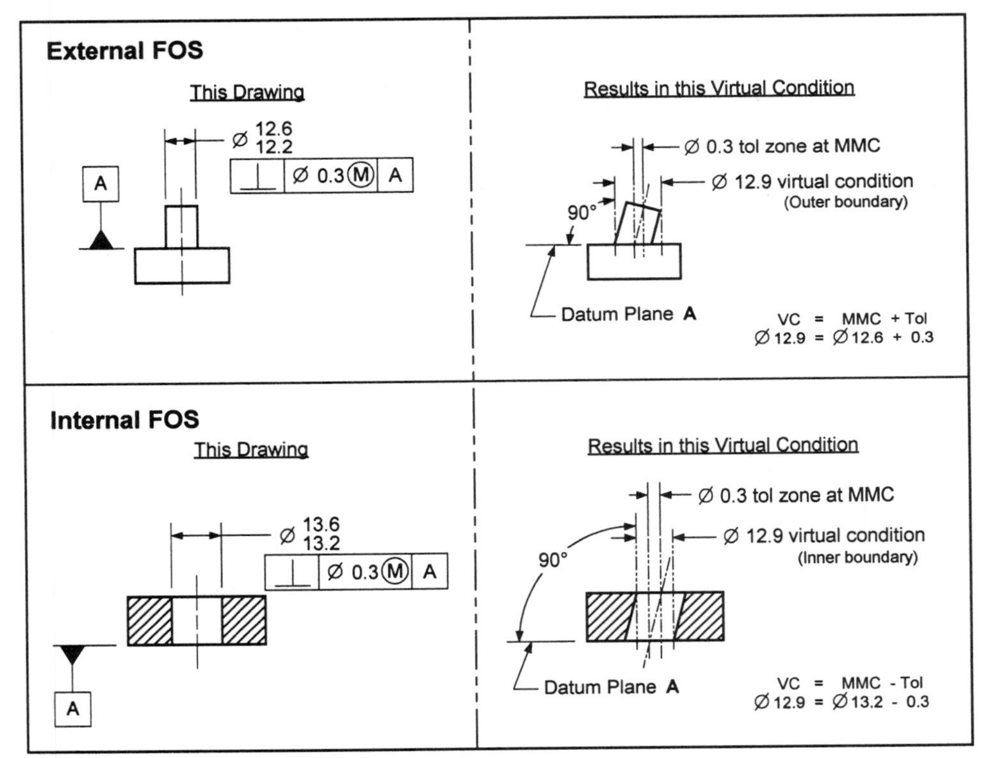

FIGURE 3-11 MMC Virtual Condition Examples

LMC Virtual Condition

When a geometric tolerance that contains an LMC modifier in the tolerance portion of the feature control frame is applied to a FOS, the virtual condition of the FOS is affected. The virtual condition is the extreme boundary that represents the worst case for functional requirements, such as wall thickness, alignment, or minimum machine stock on a part.

In the case of an external FOS, such as a pin or a shaft, the VC is determined by the following formula:

VC = LMC - Geometric Tol

In the case of an internal FOS, such as a hole, the VC is determined by the following formula:

VC = LMC + Geometric Tol

The virtual condition of an external FOS (at LMC) is always a constant value and can also be referred to as the "inner boundary" in calculations. The virtual condition of an internal FOS (at LMC) is always a constant boundary and can also be referred to as the "outer boundary" in calculations. Figure 3-12 shows examples of virtual condition calculations (at LMC).

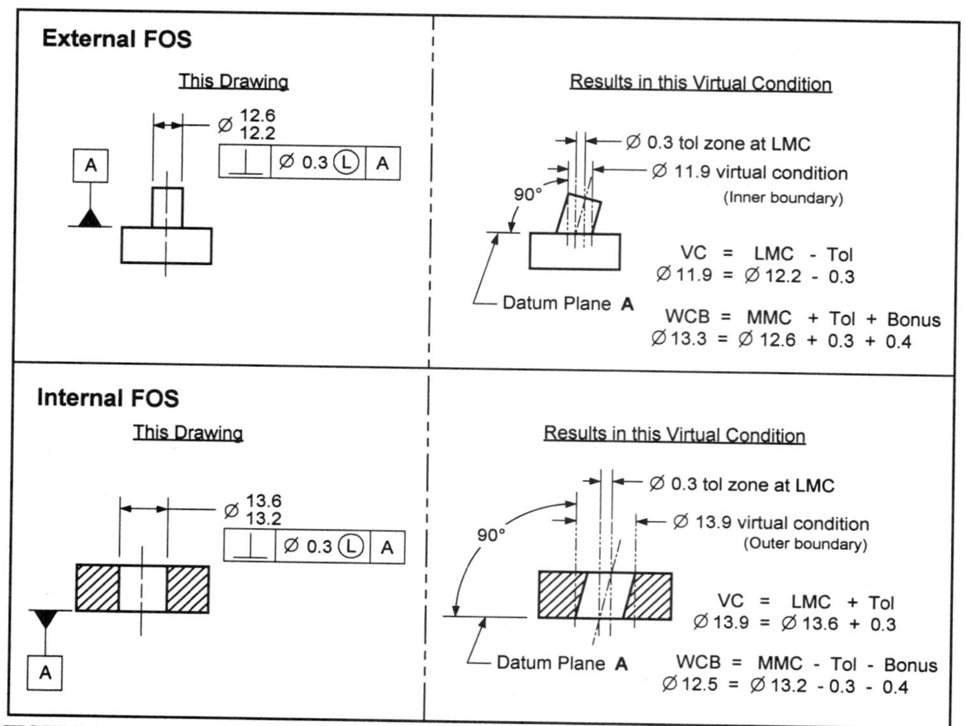

FIGURE 3-12 LMC Virtual Condition Examples

RFS Inner and Outer Boundary

When a geometric tolerance that contains no modifiers (RFS default per Rule #2) in the tolerance portion of the feature control frame is applied to a FOS, the inner or outer boundary (or worst-case boundary) of the FOS is affected. On an external FOS, the term "outer boundary" or "worst-case boundary" is used. Outer boundary (OB) for an external FOS is the worst-case boundary generated by the largest feature (MMC) plus the geometric tolerance. On an internal FOS, the term "inner boundary" is used. Inner boundary (IB) for an internal FOS is the smallest feature (MMC) minus the geometric tolerance.

In the case of an external FOS, such as a pin or a shaft, the OB (or WCB) is determined by the following formula:

OB = MMC + Geometric Tol

In the case of an internal FOS, such as a hole, the IB (or WCB) is determined by the following formula:

IB = MMC - Geometric Tol

Figure 3-13 shows examples of outer and inner boundary calculations (RFS).

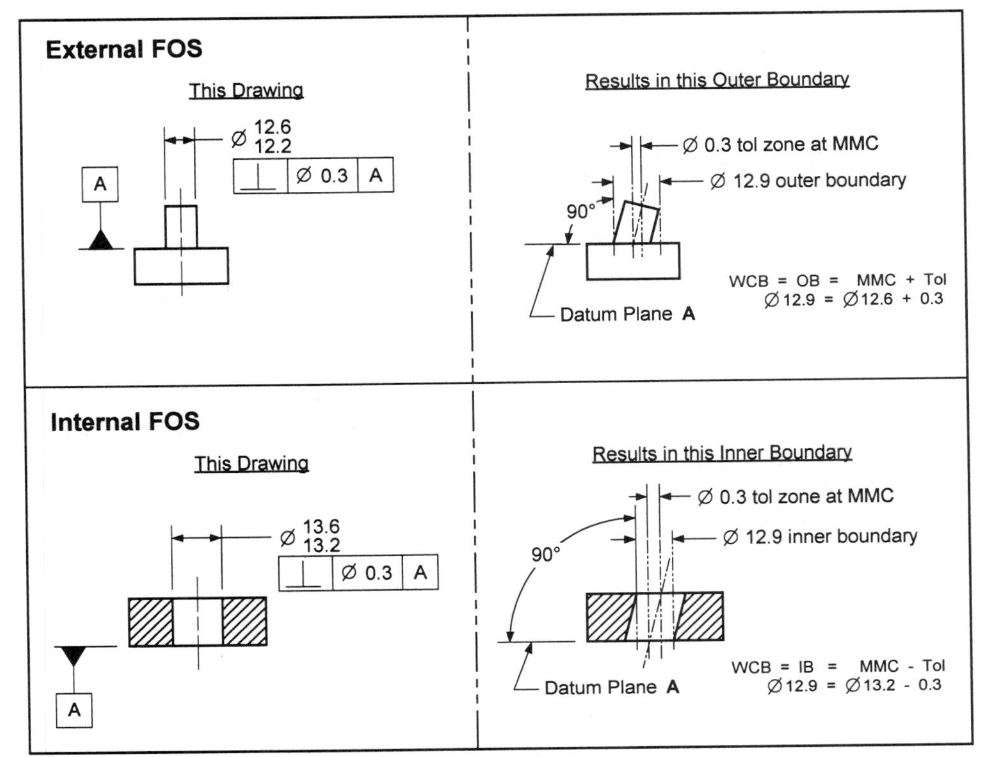

FIGURE 3-13 RFS Inner and Outer Boundary Examples

Worst-Case Boundary Formulas

The worst-case boundary formulas can become difficult to remember with all the different part specifications possible. Figure 3-14 is a chart that shows the proper formula for a given part specification.

Part Specification	FOS Type	The worst case boundary (WCB) is equal to...
FOS with no GD&T specified	Internal	IB = MMC
	External	OB = MMC
FOS with GD&T specified at RFS	Internal	IB = MMC - Tol*
	External	OB = MMC + Tol*
FOS with GD&T specified at MMC	Internal	VC = IB = MMC - Tol*
	External	VC = OB = MMC + Tol*
FOS with GD&T specified at LMC	Internal	IB = MMC - Tol* - Bonus
	External	OB = MMC + Tol* + Bonus
* Tol represents the geometric control tolerance value		

FIGURE 3-14 WCB Formula Chart

Multiple Virtual Conditions

On complex industrial drawings, it is common to have multiple geometric controls applied to a FOS. When this happens, the feature of size may have several virtual conditions. Figure 3-15 shows an example of a FOS with two virtual conditions. Panel *A* shows the size tolerance requirements of Rule #1. Panel *B* shows the virtual condition that results from the perpendicularity control. This control produces a 10.3 dia. boundary relative to datum plane *A*. Panel *C* shows the virtual condition that results from the positional control. This control produces a 10.4 dia. boundary relative to datums *A, B, & C*. When a geometric tolerance is applied to a FOS—other than a straightness control—the requirements of Rule #1 still apply.

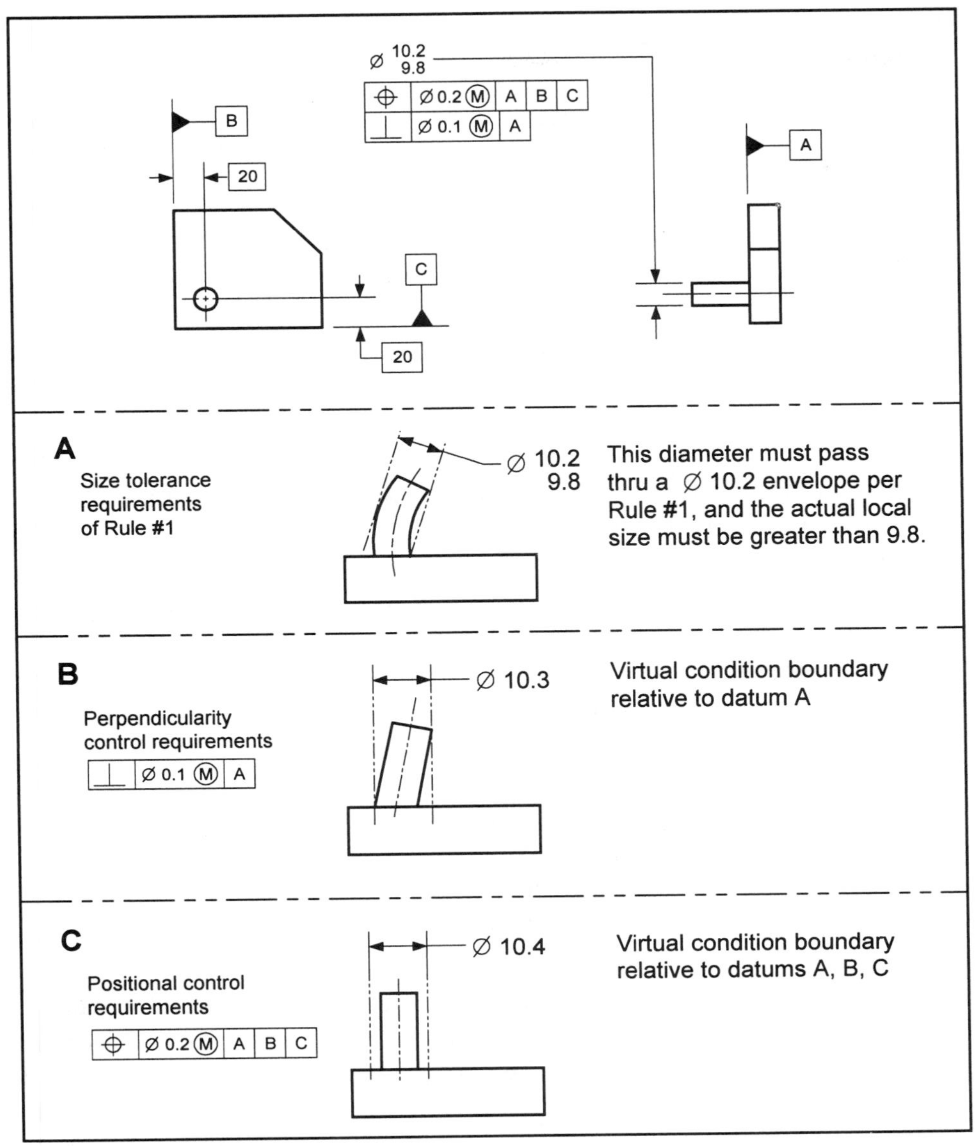

FIGURE 3-15 Multiple Virtual Conditions

TECHNOTE 3-7 Virtual Condition Facts

Three important points about virtual condition are:

1. A virtual condition boundary (or WCB) is a constant value.
2. When a geometric tolerance is applied to a FOS, and the virtual condition is calculated, the size tolerance requirements still apply.
3. A FOS may have several virtual conditions.

INTRODUCTION TO BONUS TOLERANCE

Bonus tolerance is an important concept in dimensioning parts. Bonus tolerances can reduce manufacturing costs significantly. This section is an introduction to the bonus tolerance concept.

Author's Comment

The MMC modifier is frequently used when the function of the FOS is assembly. When the function is assembly, the worst-case is when the FOS is at MMC with the full tolerance applied, or—in other words—at its virtual condition. Since the fixed gage is made to the virtual condition, it represents the worst-case mating part. Therefore, if the part fits the gage, it will fit the mating part. The bonus concept works because the gage is fixed, and when the part departs from MMC, it can have more tolerance and still fit in the gage.

Definition

Bonus tolerance is an additional tolerance for a geometric control. Whenever a geometric tolerance is applied to a FOS, and it contains an MMC (or LMC) modifier in the tolerance portion of the feature control frame, a bonus tolerance is permissible. This section introduces bonus tolerance based on the MMC modifier. When the MMC modifier is used in the tolerance portion of the feature control frame, it means that the stated tolerance applies when the FOS is at its maximum material condition. When the actual mating size of the FOS departs from MMC (towards LMC), an increase in the stated tolerance—equal to the amount of the departure—is permitted. This increase, or extra tolerance, is called the bonus tolerance. Figure 3-16 shows how bonus tolerance is calculated in a straightness application. When the MMC modifier is used, it means that the geometric characteristic can be verified with a fixed gage.

A functional gage is a gage that is built to a fixed dimension (the virtual condition) of a part feature. A part must fit into (or onto) the gage. A functional gage does not provide a dimensional measurement; it only indicates if the part is or is not to the print specification.

In Figure 3-16, the functional gage is designed to the virtual condition (2.7) of the FOS. Since the gage opening is constant, the thinner the washer becomes, the more straightness tolerance it could have and still pass through the gage.

For more info. . .

To learn about functional (fixed) gages, see page 235.

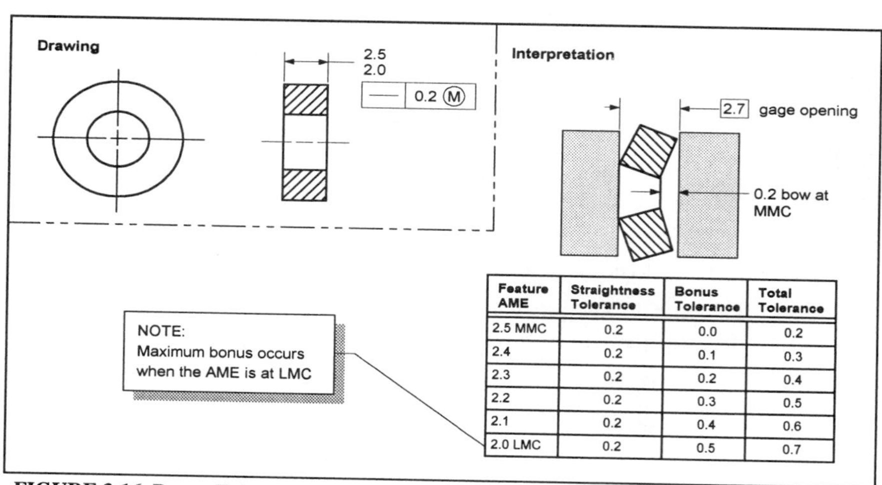

Feature AME	Straightness Tolerance	Bonus Tolerance	Total Tolerance
2.5 MMC	0.2	0.0	0.2
2.4	0.2	0.1	0.3
2.3	0.2	0.2	0.4
2.2	0.2	0.3	0.5
2.1	0.2	0.4	0.6
2.0 LMC	0.2	0.5	0.7

FIGURE 3-16 Bonus Tolerance Examples

Although a straightness control is used for the example in Figure 3-16, the bonus tolerance concept applies to any geometric control that uses the MMC (or LMC) modifier in the tolerance portion of the feature control frame.

TECHNOTE 3-8 Bonus Tolerance

- Bonus tolerance is an additional tolerance for a geometric control.
- Bonus tolerance is only permissible when an MMC (or LMC) modifier is shown in the tolerance portion of a feature control frame.
- Bonus tolerance comes from the FOS tolerance.
- Bonus tolerance is the amount the actual mating size departs from MMC (or LMC).

Author's Comment

On an external FOS, if the AME is larger than the MMC, no bonus tolerance is available. On an internal FOS, if the AME is smaller than the MMC, no bonus tolerance is available.

Figure 3-17 shows how to determine the amount of bonus tolerance permissible in an application. The MMC modifier in the tolerance portion of the feature control frame denotes that a bonus tolerance is permissible. The maximum amount of bonus tolerance permissible is equal to the difference between the MMC and the LMC of the toleranced FOS.

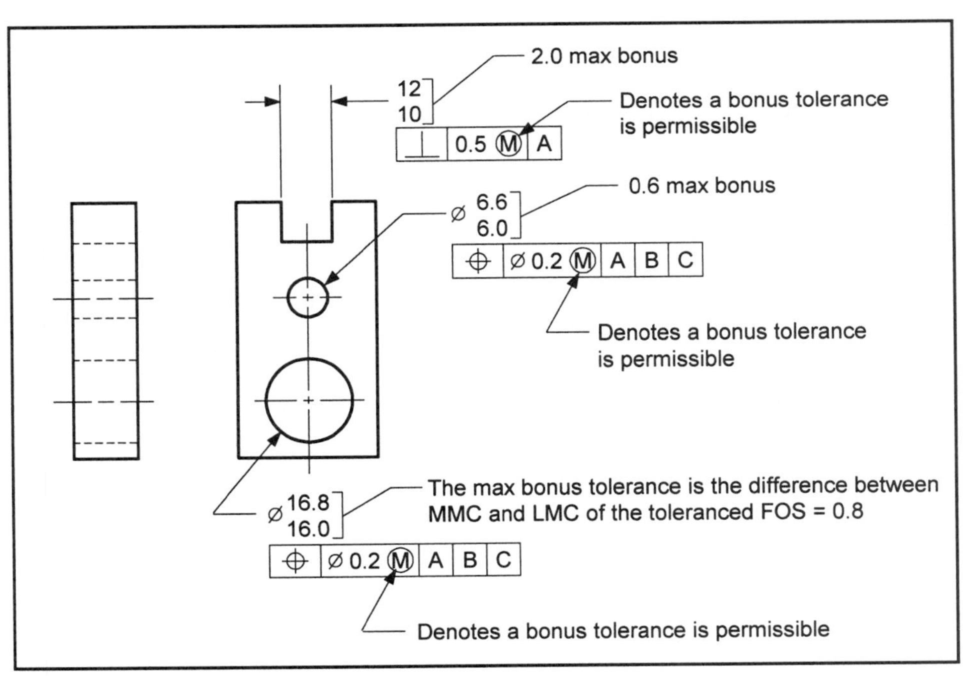

FIGURE 3-17 Bonus Tolerance

Study Tip
Read each term. If you don't know its meaning, look it up in the chapter.

VOCABULARY LIST

New Terms Introduced in this Chapter

Basic dimension
Bonus tolerance
Go gage
Inner boundary (IB)
No-Go gage
Outer boundary (OB)
Rule #1
Rule #2
Virtual condition (VC)
Worst-case boundary (WCB)

Author's Comment
These topics, plus advanced coverage of many of the topics introduced in this text, will be covered in my new book on advanced GD&T concepts.

ADDITIONAL RELATED TOPICS

Topic	ASME Y14.5M-1994 Reference
• Resultant condition	Paragraph 1.3.23
• Screw thread default	Paragraph 2.9
• Gear and spline default	Paragraph 2.10

QUESTIONS AND PROBLEMS

1. Write the definition of Rule #1.

____

____

2. List two ways to override Rule #1.

____

____

3. On the drawing below, circle the dimensions that are controlled by Rule #1.

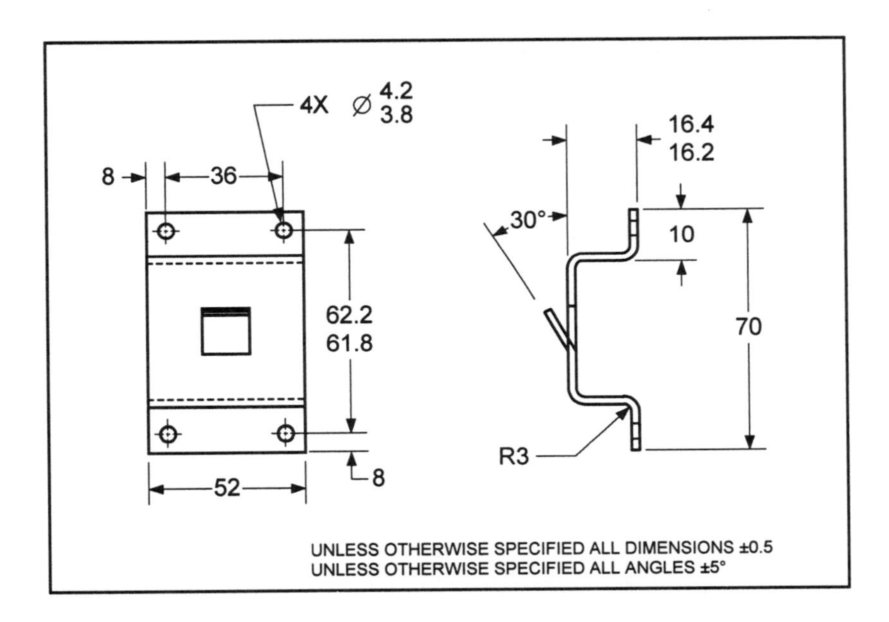

4. Fill in the value of the outer boundary as established by Rule #1.

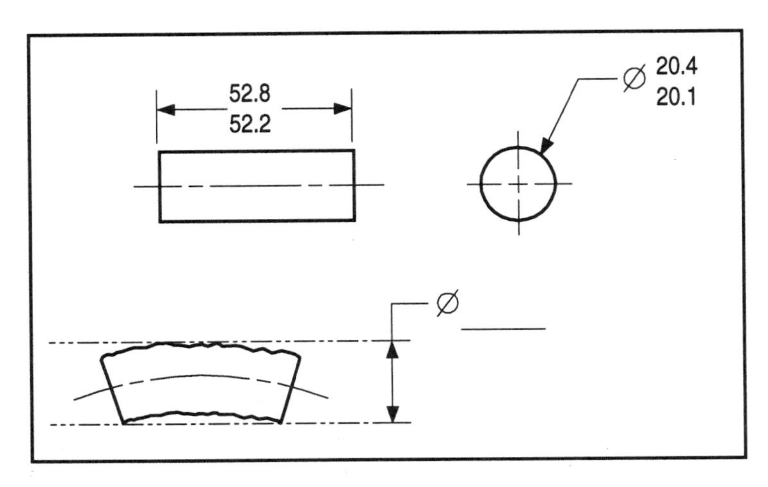

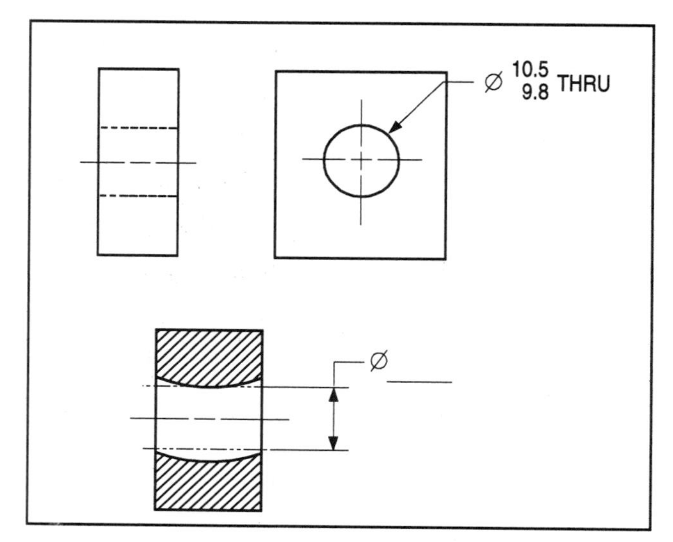

5. On the drawing above, fill in the value of the inner boundary as established by Rule #1.

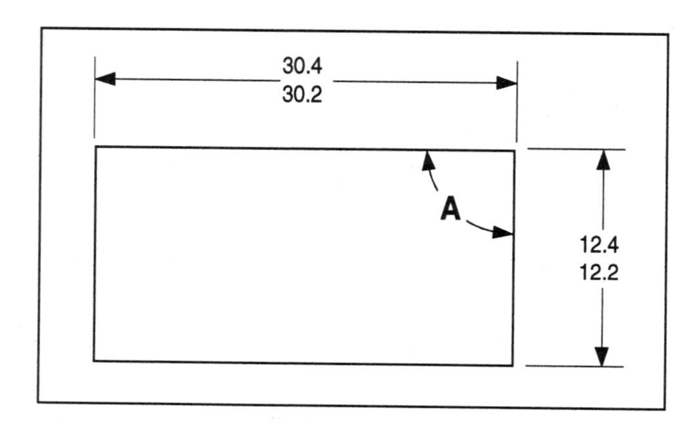

6. On the drawing above, the angle marked "A" is an implied 90° angle (Rule #1 applies to both features of size). If the part was produced at MMC, what would control the tolerance on angle *A*? ______________________________

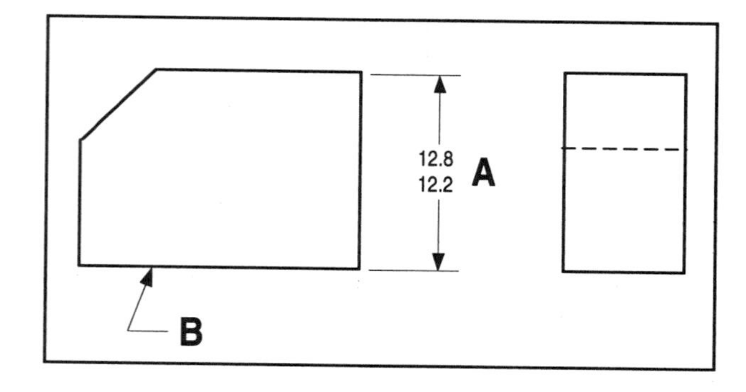

If dimension A was	The allowable form error on surface B is
12.8	
12.7	
12.6	
12.5	
12.4	
12.3	
12.2	

7. Fill in the chart using the drawing above.

8. When inspecting a FOS controlled by Rule #1, two checks are required. Describe how each check could be done.
   a. MMC size + Rule #1 envelope ______________________________

   ____

   b. LMC size ______________________________

   ____

9. Write the definition of Rule #2.

   ____

   ____

   ____

10. A basic dimension is ______________________________

    ____

11. Two uses for basic dimensions are:

    ____

    ____

12. Virtual condition is ______________________________

    ____

13. Bonus tolerance is ______________________________

    ____

14. Inner boundary is ______________________________

    ____

15. Outer boundary is ______________________________

    ____

16. Worst-case boundary is ______________________________

    ____

17. Use the drawing to fill in the chart below.

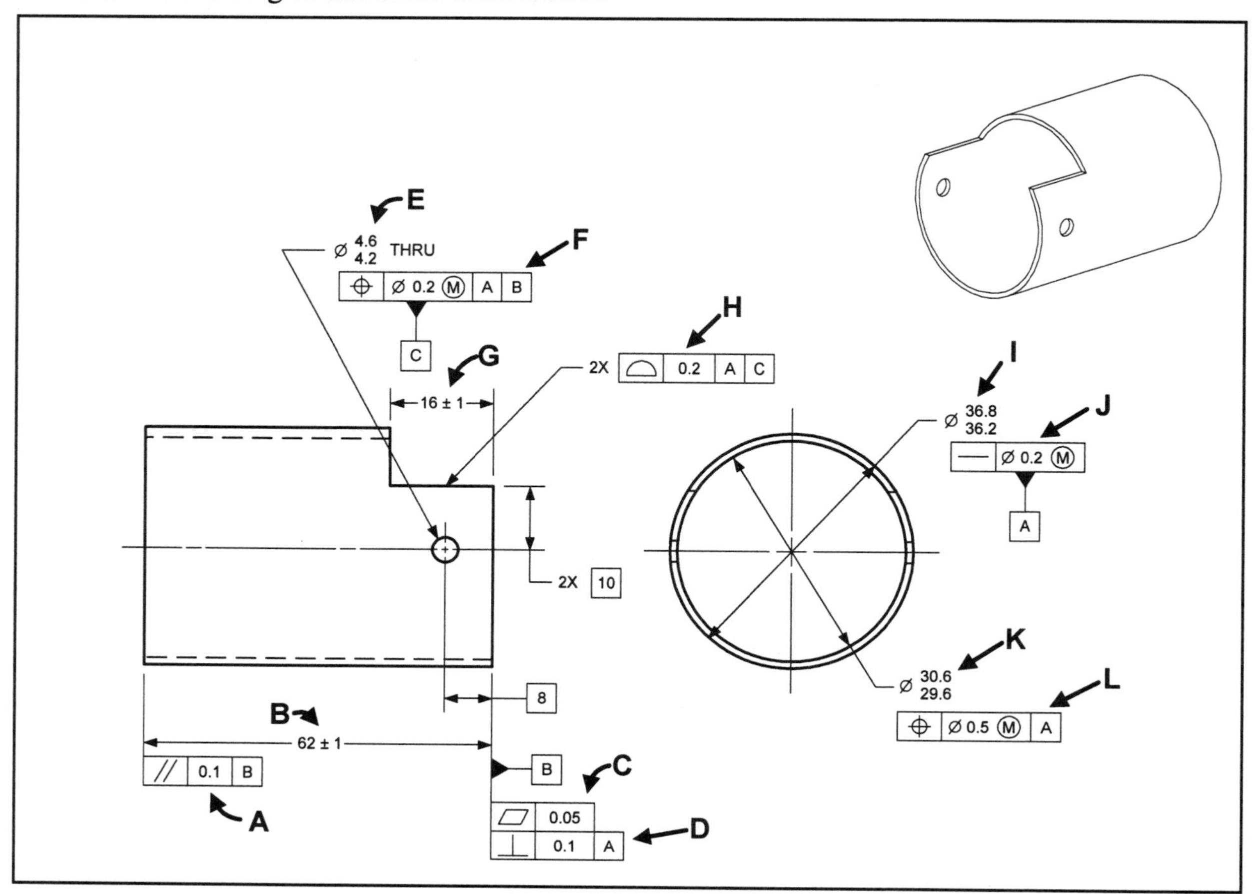

Use N/A for not applicable

Letter	Letter identifies a. . .			If a FOS dimension is identified, Rule #1 applies		If a FOS dimension is identified, VC, OB, or IB is. . .	If a feature control frame is identified, It applies to a. . .		If a feature control frame is identified, The amount of bonus tolerance permissible is. . .
	FOS Dimension	Non-FOS Dimension	Feature Control Frame	YES	NO		Feature	FOS	
A									
B									
C									
D									
E									
F									
G									
H									
I									
J									
K									
L									

Chapter 4

Form Controls

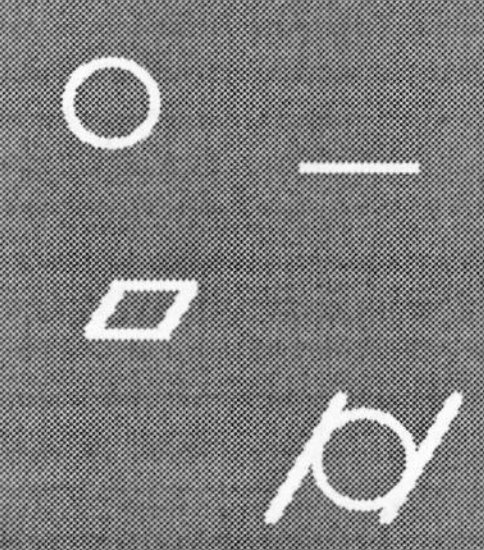

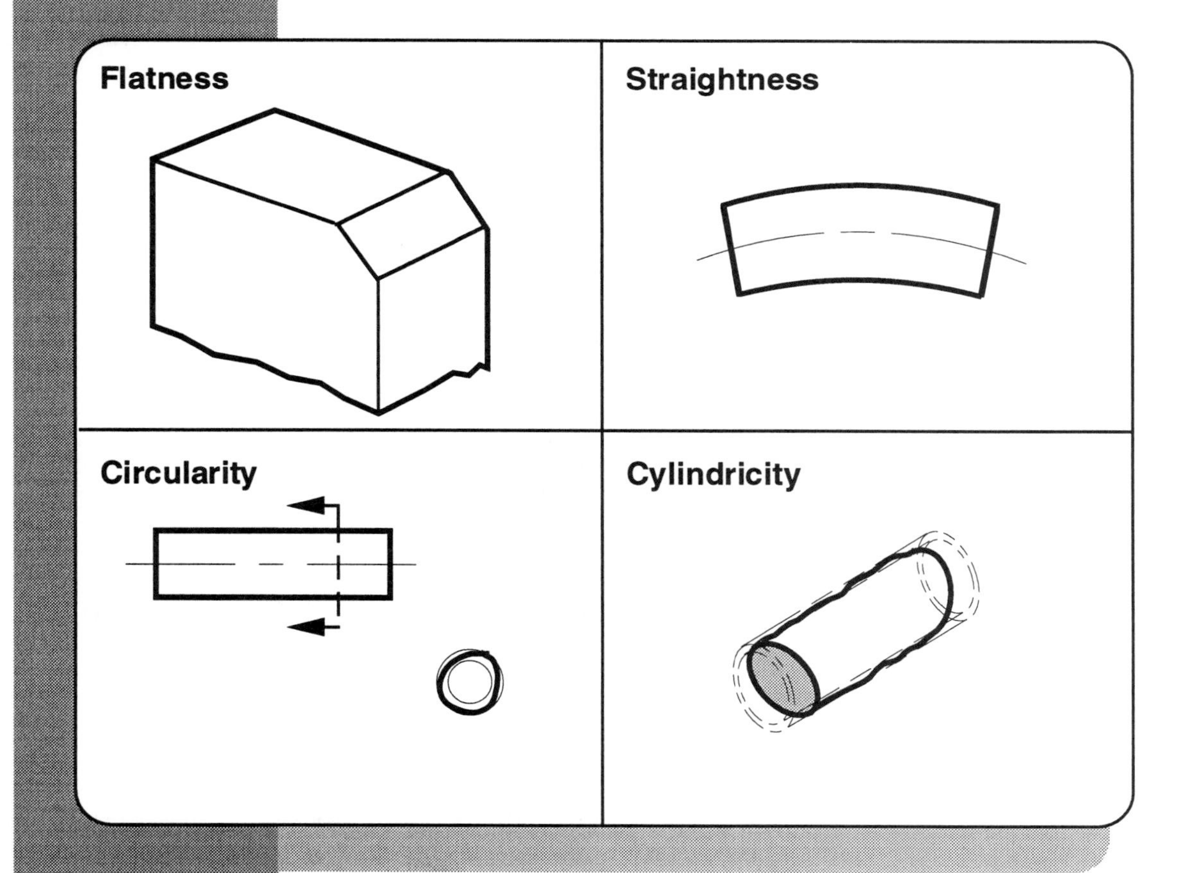

INTRODUCTION

This chapter explains the concepts involved in defining the form of a part surface. Form controls limit the flatness, straightness, circularity, or cylindricity of part surfaces. Form is a characteristic that limits the shape error of a part surface (or in some cases, an axis or centerplane) relative to its perfect counterpart. For example, a form characteristic of a planar surface is flatness. Flatness defines how much a surface can vary from its perfect plane.

Cylindrical surfaces can have three different form characteristics: straightness, circularity, and cylindricity. Straightness defines how much a line element can vary from a straight line. Circularity defines how much circular elements can vary from a perfect circle, and cylindricity defines how much a surface can vary from a perfect cylinder. Form controls are used to define the shape of a feature in relation to itself; therefore, they never use a datum reference. The four form controls, their symbols, and examples are shown in Figure 4-1.

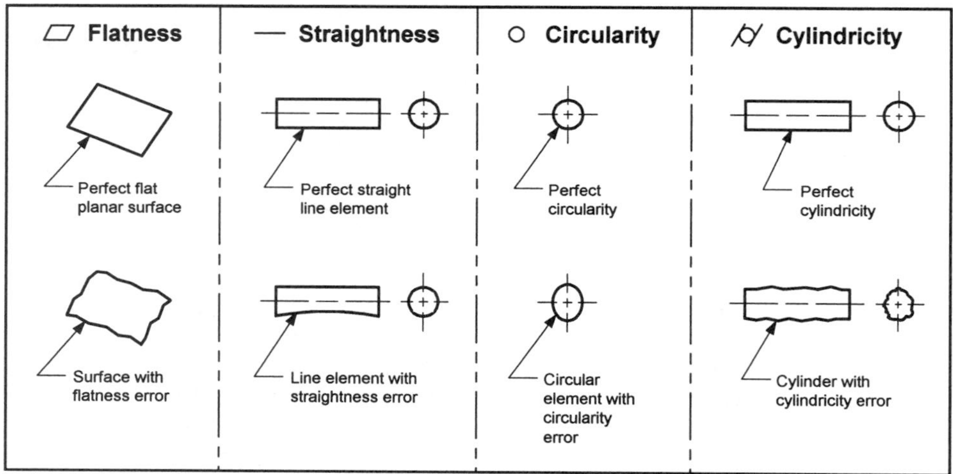

FIGURE 4-1 Form Controls

CHAPTER GOALS AND OBJECTIVES

There are Four Goals in this Chapter:

4-1. Interpret the flatness control.
4-2. Interpret the straightness control.
4-3. Interpret the circularity control.
4-4. Interpret the cylindricity control.

Performance Objectives that Demonstrate Mastery of These Goals

Upon completion of this chapter, each student should be able to:

Goal 4-1 (pp. 74-78)

- Describe what flatness is.
- Describe the tolerance zone for a flatness control.
- Describe how Rule #1 provides an indirect flatness control.
- Determine if a flatness specification is legal.
- Describe how the reference plane is established for flatness.
- Describe how a flatness control can be inspected.

Goal 4-2 (pp. 79-88)

- Describe what straightness is.
- Describe the tolerance zone for a straightness control.
- Describe how Rule #1 provides an indirect straightness control.
- Determine if a straightness specification is legal.
- Describe one method of inspecting straightness applied to a surface.
- Determine if a straightness control is applied to a surface or a FOS.
- Explain how Rule #1 provides an indirect straightness control for a FOS.
- Describe the virtual condition of a FOS with a straightness control applied to its size dimension.
- Calculate the amount of bonus in straightness at MMC applications.
- Draw the gage for a straightness at MMC application.

Study Tip

Take a few minutes to fully understand these objectives. When reading this chapter, look for information to help you master these objectives.

Goal 4-3 (pp. 89-93)

- Describe circularity.
- Describe the tolerance zone for a circularity control.
- Describe how Rule #1 provides an indirect circularity control.
- Determine if a circularity specification is legal.
- Describe how a circularity control can be inspected.

Goal 4-4 (pp. 94-98)

- Describe cylindricity.
- Describe the tolerance zone for cylindricity.
- Describe how Rule #1 provides an indirect cylindricity control.
- Determine if a cylindricity specification is legal.
- Describe how a cylindricity control can be inspected.

FLATNESS CONTROL

Definition

Flatness is the condition of a surface having all of its elements in one plane. A ***flatness control*** is a geometric tolerance that limits the amount of flatness error a surface is allowed. The tolerance zone for a flatness control is three-dimensional. It consists of two parallel planes within which all the surface elements must lie. The distance between the parallel planes is equal to the flatness control tolerance value. Flatness (as well as other form controls) is measured by comparing a surface to its own true counterpart. In the case of flatness, the first plane of the tolerance zone (a theoretical reference plane) is established by contacting the three high points of the controlled surface. The second plane of the tolerance zone is parallel to the first plane and offset by the flatness tolerance value. All the points of the controlled surface must lie within the tolerance zone. An example of a flatness tolerance zone is shown in Figure 4-2.

For more info. . .
See Paragraph 6.4.2 of Y14.5.

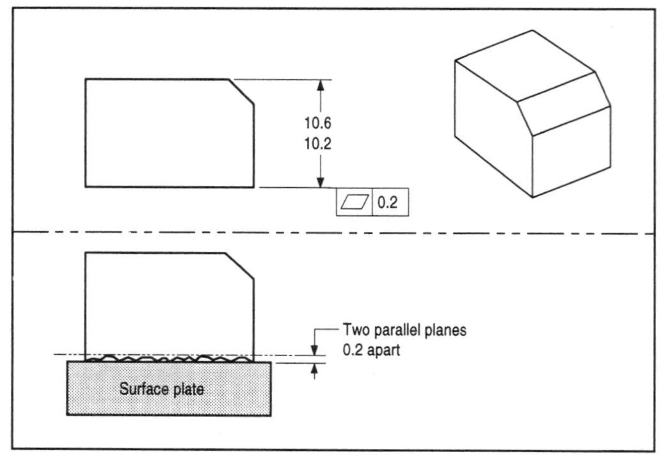

FIGURE 4-2 Flatness Tolerance Zone

A flatness control is always applied to a planar surface. Therefore, a flatness control can never use an MMC or LMC modifier. These modifiers can only be used when a geometric control is applied to a feature of size. Also, flatness cannot override Rule #1. Flatness is a separate requirement and verified separately from the size tolerance and Rule #1 requirements.

TECHNOTE 4-1 Flatness Tolerance Zone

A flatness control tolerance zone is two parallel planes spaced apart by the flatness tolerance value. The first plane of the tolerance zone is established by contacting the three high points of the toleranced surface.

Rule #1's Effects on Flatness

Whenever Rule #1 applies to a feature of size that consists of two parallel planes (ie. tab or slot), an automatic indirect flatness control exists for both surfaces. This indirect control is a result of the interrelationship between Rule #1 (perfect form at MMC) and the size dimension. When the feature of size is at MMC, both surfaces must be perfectly flat. As the feature departs from MMC, a flatness error equal to the amount of the departure is allowed. Since Rule #1 provides an automatic indirect flatness control, a flatness control should not be used unless it is a refinement of the dimensional limits of the surface. Figure 4-3 shows an example of the effects of Rule #1 on flatness.

Design Tip

Rule #1 is an indirect form control. Flatness effects of Rule #1 are not inspected; they are a result of the boundary and size limitations. If it is desired to have the flatness of a surface inspected, a flatness control should be specified.

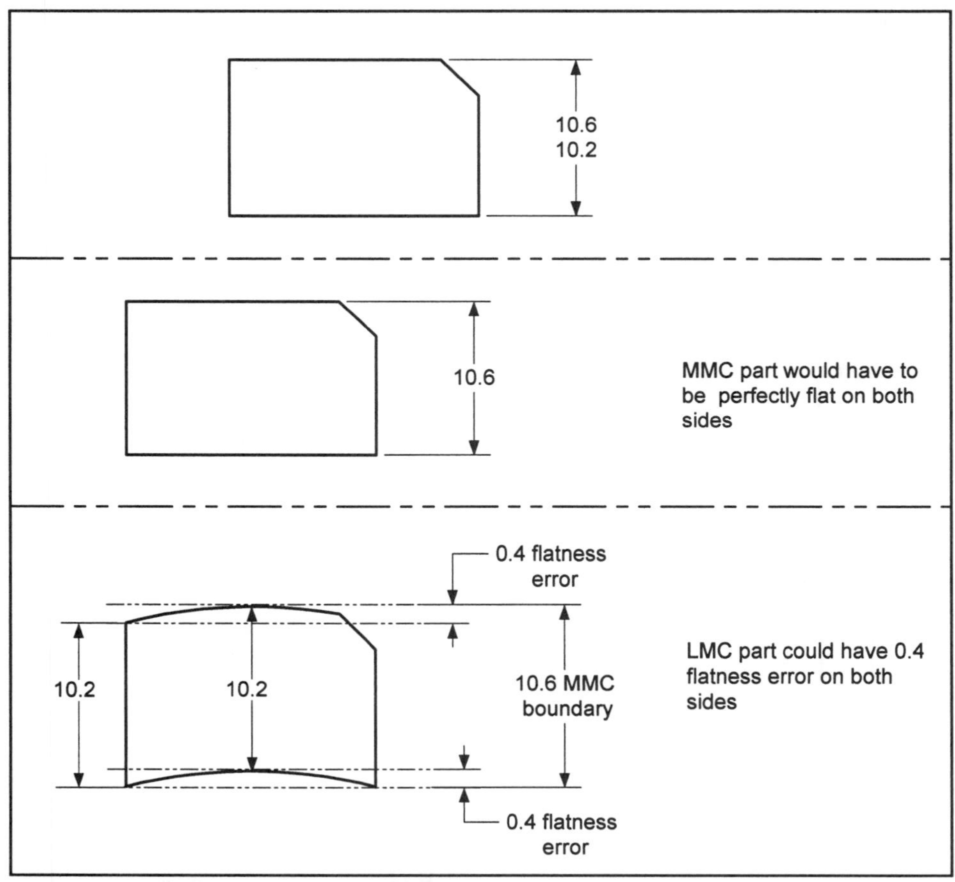

FIGURE 4-3 Rule #1 as an Indirect Flatness Control

TECHNOTE 4-2 Rule #1's Effects on Flatness

Whenever Rule #1 applies to a planar FOS. . .

- It provides an automatic indirect flatness control for both surfaces.

Flatness Control Application

Some examples of when a designer uses a flatness control on a drawing are to provide a flat surface:

- For a gasket or seal
- To attach a mating part
- For better contact with a datum plane

When these types of applications are involved, the indirect flatness control that results from Rule #1 is often not sufficient to satisfy the functional requirements of the part surface. This is when a flatness control is specified on a drawing. Figure 4-4 shows an application of a flatness control.

Design Tip

Whenever a flatness control is specified on a surface of a planar FOS, its tolerance value must be less than the FOS dimension tolerance.

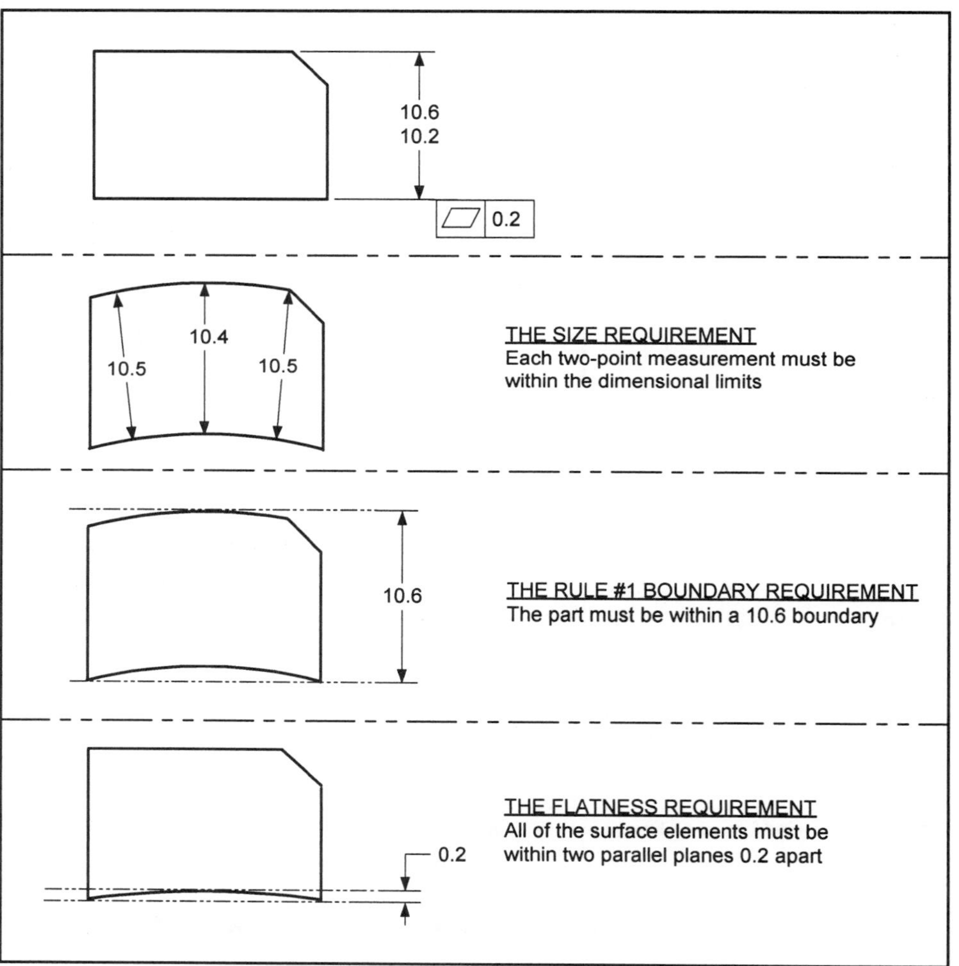

FIGURE 4-4 Flatness Control Application

In this flatness application, three separate part requirements must be verified: the size tolerance, the Rule #1 boundary, and the flatness requirements. The flatness control applies to the bottom of this part. The flatness of the top surface is controlled by the Rule #1 effects.

Indirect Flatness Controls

There are several geometric controls that can indirectly affect the flatness of a surface; they are Rule #1, perpendicularity, parallelism, angularity, total runout, and profile of a surface. When any of these controls are used on a surface, they also limit the flatness of the surface. However, indirect form controls are not inspected. If it is desired to have the flatness of a surface inspected, a flatness control should be specified on the drawing. If a flatness control is specified, its tolerance value must be less than the tolerance value of any indirect flatness controls that affect the surface.

Legal Specification Test for a Flatness Control

For a flatness control to be a legal specification, it must satisfy the following conditions:

- No datum references can be specified in the feature control frame.
- No modifiers can be specified in the feature control frame.
- The control must be applied to a planar surface.
- The flatness control tolerance value must be less than any other geometric control that limits the flatness of the surface.
- The flatness control tolerance value must be less than the size tolerance associated with the surface.

Figure 4-5 shows a legal specification flowchart for a flatness control.

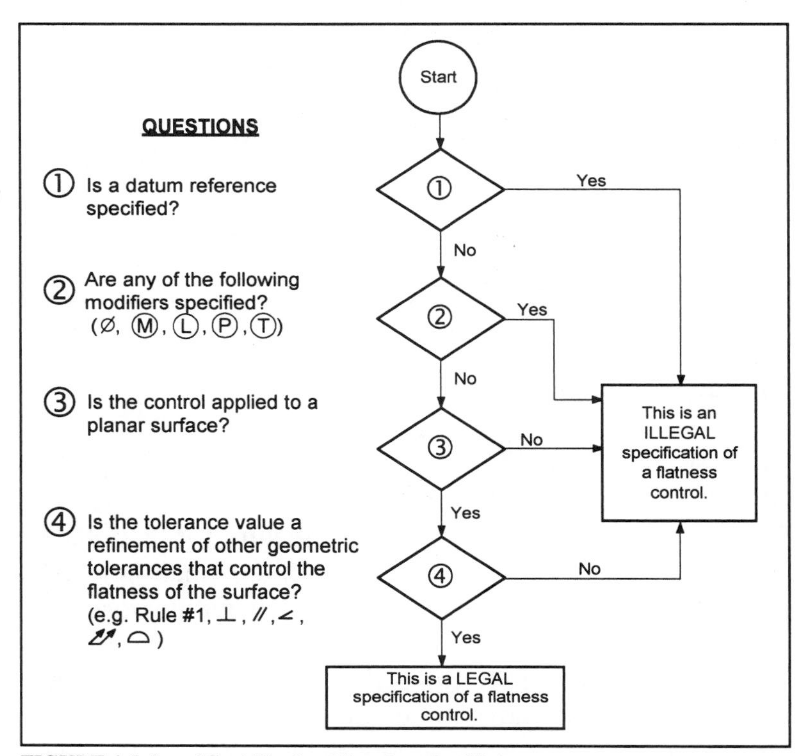

FIGURE 4-5 Legal Specification Flowchart for Flatness

Inspecting Flatness

Figure 4-4 shows a part with a flatness specification. When inspecting this part, three separate parameters must be checked: the size, the Rule #1 boundary, and the flatness requirement.

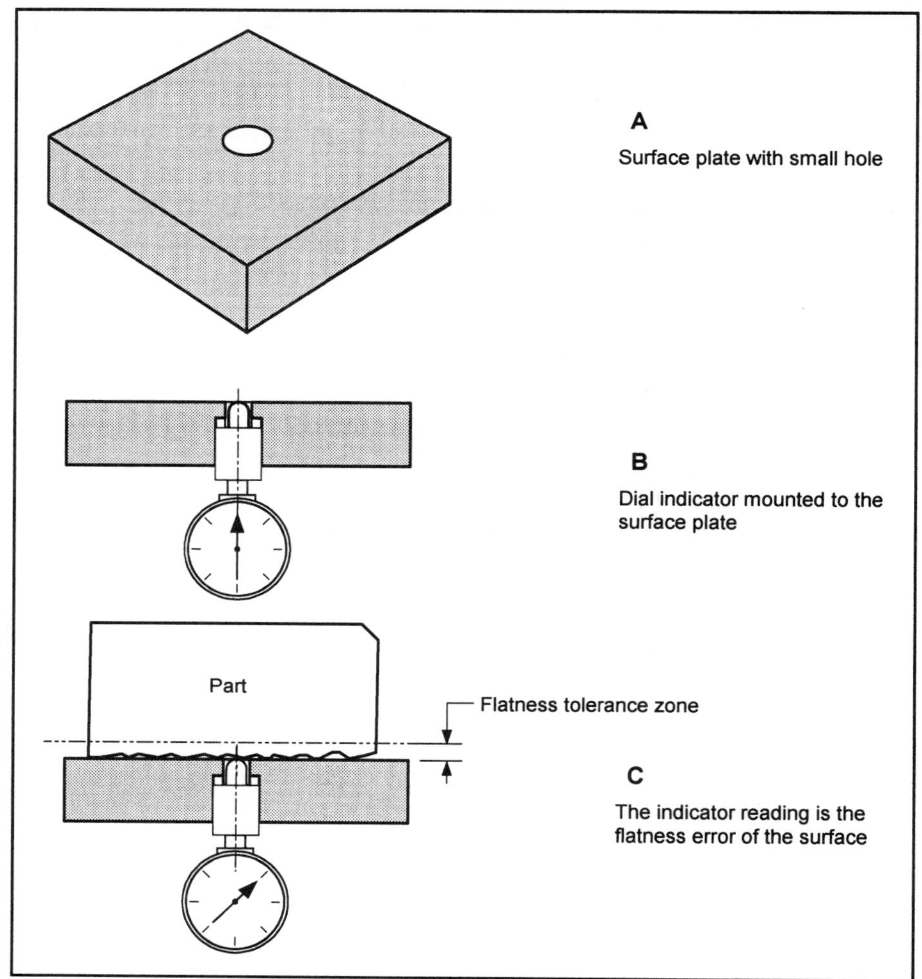

FIGURE 4-6 Inspecting Flatness

The flatness control could be inspected as follows:

Establish the first plane of the tolerance zone by placing the part surface on a surface plate that has a small hole (see Figure 4-6*A*). The surface plate becomes the true counterpart of the controlled feature (or one plane of the flatness tolerance zone). A dial indicator is set in the small hole as shown in Figure 4-6*B*. The tip of the dial indicator traces a path across the entire part surface (see Figure 4-6*C*). The dial indicator measures the distance between the true counterpart of the controlled feature and the low points of the surface, then the part is moved over the hole at random. If the FIM (full indicator movement) is larger than the flatness tolerance value at any point on the path, then the surface flatness is not within its specification.

STRAIGHTNESS AS A SURFACE ELEMENT CONTROL

Definition

Straightness of a line element is the condition where each line element (or axis or centerplane) is a straight line. A ***straightness control directed to a surface*** is a geometric tolerance that limits the amount of straightness error allowed in each surface line element. The tolerance zone for a straightness control (as a surface line element control) is two-dimensional; it consists of two parallel lines for each line element of the surface. The distance between the parallel lines is equal to the straightness tolerance value. The first line element of the tolerance zone is established by the two high points of a line element of a surface. The second line element of the tolerance zone is parallel to the first line element and offset by the straightness tolerance value. A straightness tolerance zone may be located anywhere between the dimensional limits of the surface. All the points of each controlled line element must lie within the tolerance zone.

For more info. . .
See Paragraph 6.4.1 of Y14.5.

When straightness is applied to surface elements, the MMC or LMC modifiers are not used. An example of straightness as a surface line element's control is shown in Figure 4-7.

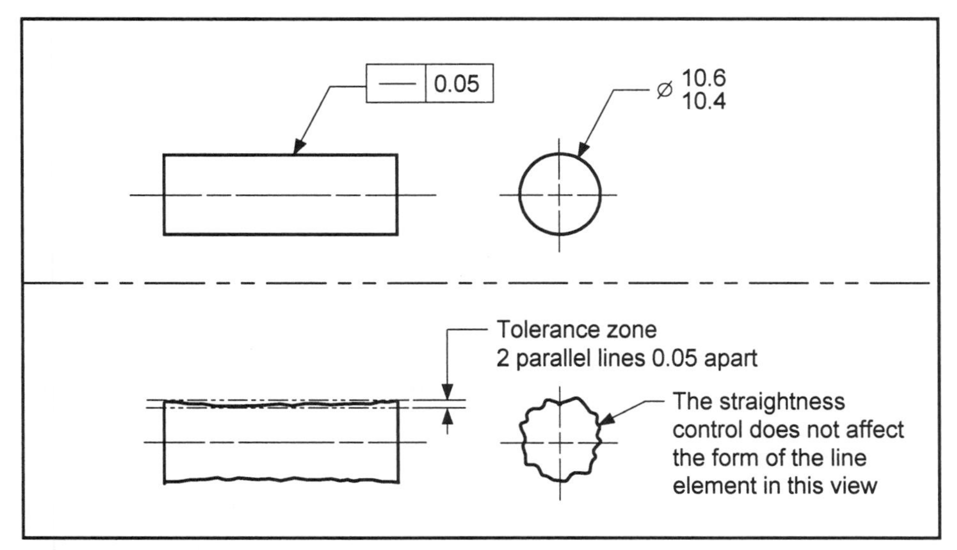

FIGURE 4-7 Straightness of a Surface Line Element

TECHNOTE 4-3 Straightness of a Line Element

The tolerance zone for a straightness control applied to surface elements is two parallel lines spaced apart a distance equal to the straightness tolerance value.

In Figure 4-7, the straightness control is applied to the surface element of the pin. When straightness is applied as a surface element control, the following conditions apply:

- The tolerance zone applies to the surface elements.
- The tolerance zone is two parallel lines.
- Rule #1 applies.
- The outer/inner boundary is not affected.
- No modifiers may be specified.
- The tolerance value specified must be less than the size tolerance.

Design Tip

Rule #1 is an indirect form control. Surface straightness effects of Rule #1 are not inspected; they are a result of the boundary and size limitations. If it is desired to have the straightness of surface elements inspected, a straightness control should be specified.

Rule #1's Effects on Surface Straightness

Whenever Rule #1 is in effect, an automatic indirect straightness control exists for the surface line elements. This indirect control is a result of the interrelationship between Rule #1 and the size dimension. When the feature of size is at MMC, the line elements must be perfectly straight. As the FOS departs from MMC, a straightness error equal to the amount of the departure is allowed (see Figure 4-8). Since Rule #1 provides an automatic indirect straightness control, a straightness control should not be used unless its tolerance value is less than the total size tolerance. Figure 4-8 shows an example of the effects of Rule #1 on straightness.

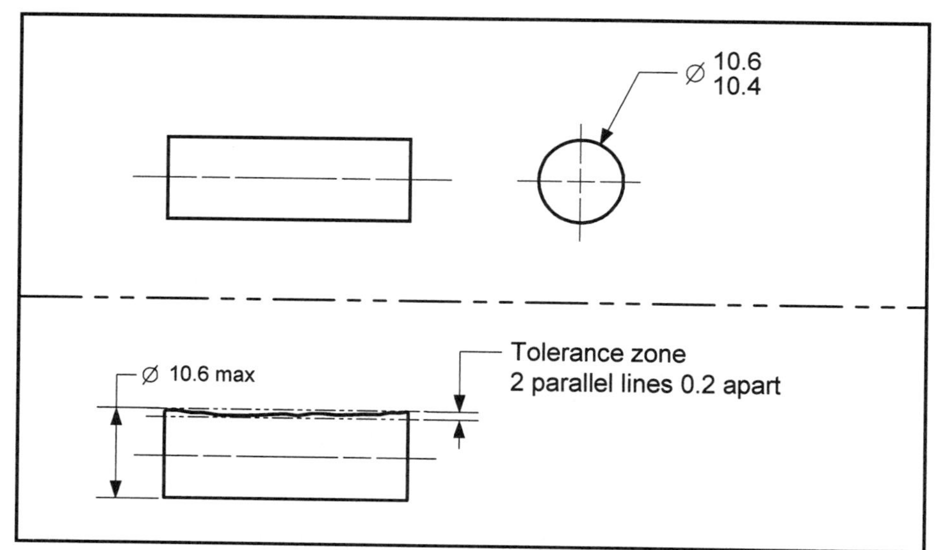

FIGURE 4-8 Indirect Surface Straightness that Results from Rule #1

TECHNOTE 4-4 Rule #1's Effects on Straightness

Whenever Rule #1 applies to a FOS:
- It provides an automatic indirect straightness control for its surface elements.

Legal Specification Test for a Straightness Control Applied to Surface Elements

For a straightness control applied to surface elements to be a legal specification, it must satisfy the following conditions:

- No datum references can be specified in the feature control frame.
- The control must be directed to the surface elements.
- No modifiers can be specified in the feature control frame.
- The straightness control must be applied in the view where the con trolled elements are shown as a line.
- The tolerance value specified must be less than any other geometric controls that limit the form of the surface.
- The tolerance value specified must be less than the size tolerance.

Figure 4-9 shows a legal specification flowchart for a straightness control applied to surface elements.

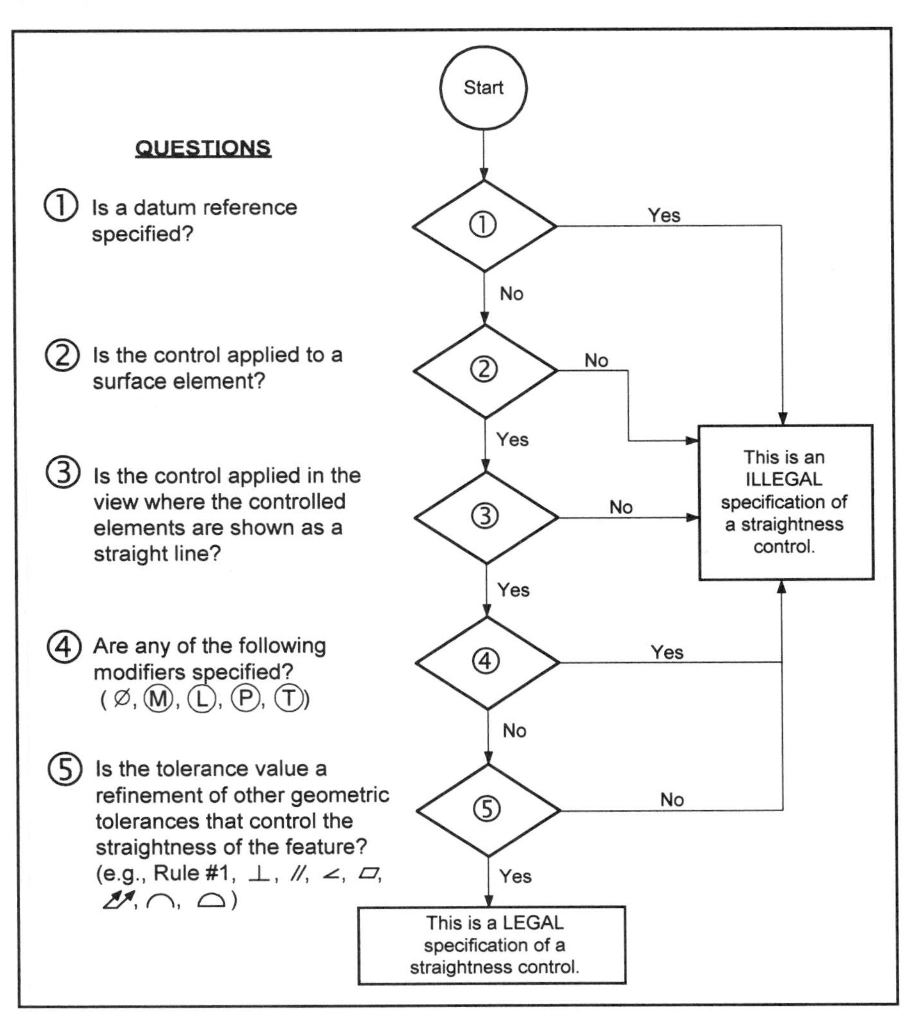

FIGURE 4-9 Legal Specification Flowchart for Straightness Applied to Surface Elements

Verifying Straightness Applied to Surface Elements

Figure 4-10*A* shows a part with a straightness specification. When inspecting this part, several separate parameters must be checked: the size of the FOS, the Rule #1 boundary, and the straightness requirement. Chapter 2 discussed how to check the size and Rule #1 boundary; now we will look at how to inspect the straightness specification.

A straightness control could be inspected as follows:

Establish the first line of the tolerance zone by placing the part surface on a surface plate (see Figure 4-10). The surface plate becomes the true counterpart. Using a gage wire with a diameter equal to the straightness tolerance value, check the distance between the true counterpart and the low points of the line element of the part surface. If the gage wire will not fit between the part and the surface plate, the straightness error of the line element is less than the allowable value. If, at any point along the part, the wire does fit into the space between the part and surface plate, line element straightness is not within its specifications.

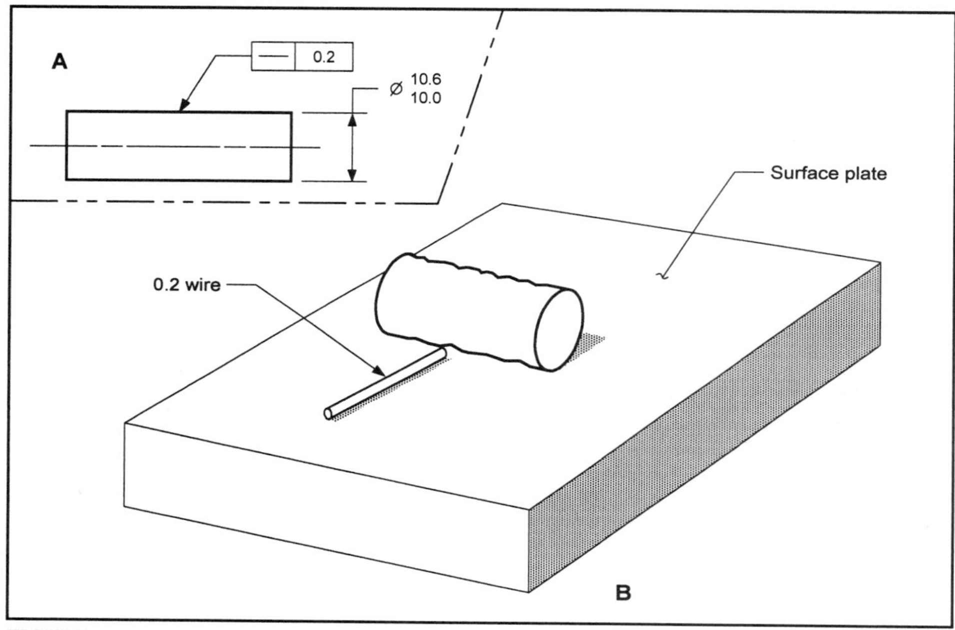

FIGURE 4-10 How Straightness Can be Verified

STRAIGHTNESS AS AN AXIS OR CENTERPLANE CONTROL

How to Determine When a Straightness Control Applies to a FOS

A straightness control is the only form control that can be applied to either a surface or a feature of size. The interpretation of a straightness control applied to a FOS is significantly different from a straightness control applied to a surface. When a straightness control is applied to a FOS, the following conditions apply:

- The tolerance zone applies to the axis or centerplane of the FOS.
- Rule #1 is overridden.
- The virtual condition or outer/inner boundary of the FOS is affected.
- The MMC or LMC modifiers may be used.
- The tolerance value specified may be greater than the size tolerance.

You can tell if a straightness control is applied to a feature or to a FOS by the location of the feature control frame on the drawing. In Figure 4-11*A*, the straightness control is located so that it is directed to the pin surface. The pin surface is a feature; therefore, the symbol applies to the (feature) surface elements of the pin. In Figure 4-11*B*, the straightness control is located so that it is related to a FOS dimension. In this case, the symbol applies to a FOS.

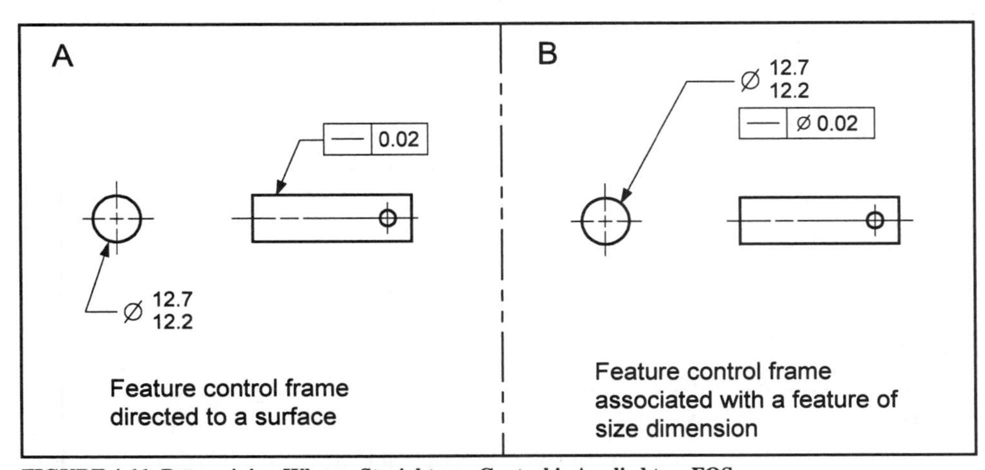

FIGURE 4-11 Determining When a Straightness Control is Applied to a FOS

TECHNOTE 4-5 Straightness of a FOS

Whenever a straightness control is associated with a FOS dimension, it applies to the axis or centerplane of the FOS.

Author's Comment
When straightness is applied to a FOS, it is most often used with the MMC modifier. Therefore, all the examples in this section will use the MMC modifier.

Definition of Straightness as an Axis/Centerplane Control

Straightness of an axis is the condition where an axis is a straight line. ***Straightness of a centerplane*** is the condition where each line element is a straight line. A ***straightness control applied to a FOS*** is a geometric tolerance that limits the amount of straightness error allowed in the axis or centerplane. When a straightness control is applied to a diameter, a diameter symbol modifier is shown in the tolerance portion of the feature control frame, and the tolerance zone is a cylinder. The diameter of the cylinder is equal to the straightness tolerance value. The axis of the FOS must lie within the cylindrical tolerance zone. When a straightness control is applied to a planar FOS, the tolerance zone is two parallel planes. Each line element of the centerplane must lie within the tolerance zone. For an example, see figure 3-16 on page 64.

When a straightness control is applied to a FOS, it can be specified at RFS (by default), at MMC, or at LMC. Remember, RFS is automatic when no modifier is shown.

TECHNOTE 4-6 Straightness of a FOS

Whenever a straightness control is associated with the size dimension of a FOS, the following conditions apply:

- The tolerance zone applies to the axis or centerplane.
- Rule #1 is overridden.
- The virtual condition (outer or inner boundary) is affected.
- MMC or LMC modifiers may be specified.
- The tolerance value may be greater than the specified size tolerance.

Rule #1's Effects on Straightness of a FOS

Whenever Rule #1 applies to a FOS, an automatic straightness control exists for the axis (or centerplane) of the FOS. This automatic control is a result of the interrelationship between Rule #1 and the size dimension. When the FOS is at MMC, the axis (or centerplane) must be perfectly straight. As the FOS departs from MMC, a straightness error equal to the amount of the departure is allowed. Figure 4-12 shows an example of Rule #1's effects on the axis of a FOS. If the straightness provided by Rule #1 is sufficient for the application, there is no need to add a straightness control.

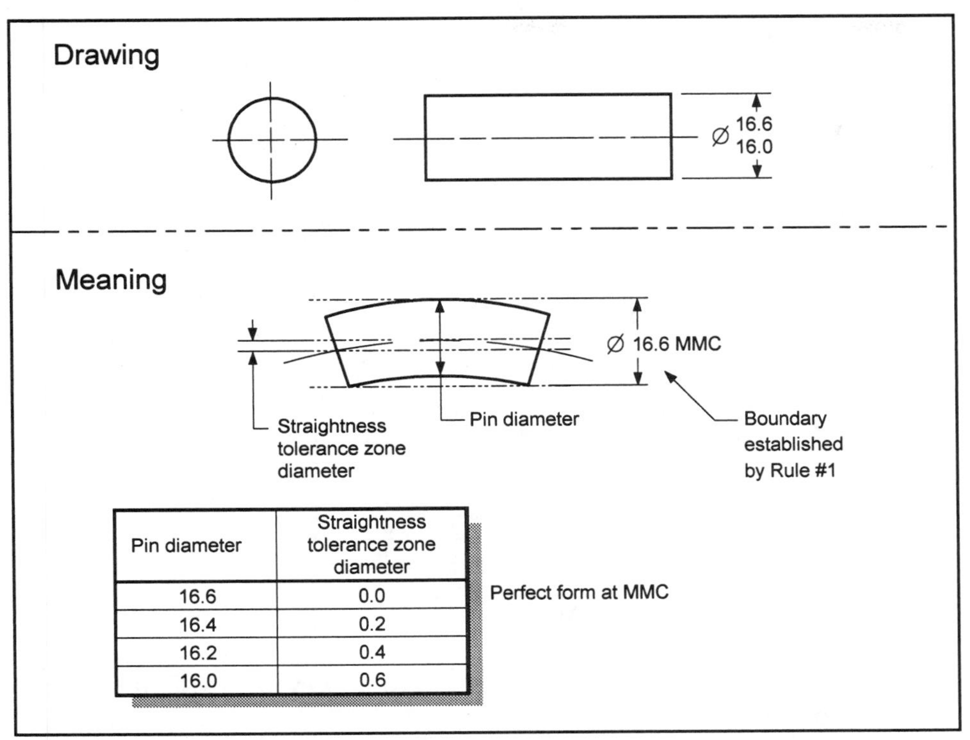

Pin diameter	Straightness tolerance zone diameter
16.6	0.0
16.4	0.2
16.2	0.4
16.0	0.6

FIGURE 4-12 Rule #1's Effects on Axis Straightness

TECHNOTE 4-7 Rule #1's Effects on Straightness of a FOS

Whenever Rule #1 applies to a FOS. . .
- It provides an automatic straightness control for the axis or centerplane.

Straightness at MMC Application

A common reason for applying a straightness control at MMC to a FOS on a drawing is to insure the function of assembly. Whenever the MMC modifier is used in a straightness control, it means the stated tolerance applies when the FOS is produced at MMC. An important benefit becomes available when straightness is applied at MMC: extra tolerance is permissible. As the FOS departs from MMC towards LMC, a bonus tolerance becomes available. An example is shown in Figure 4-13.

For more info. . .
Y14.5 provides a method of adding a note to override Rule #1. See Paragraph 2.7.2.

TECHNOTE 4-8 Straightness Effects of Rule #1

Straightness is the only geometric control that can override Rule #1.

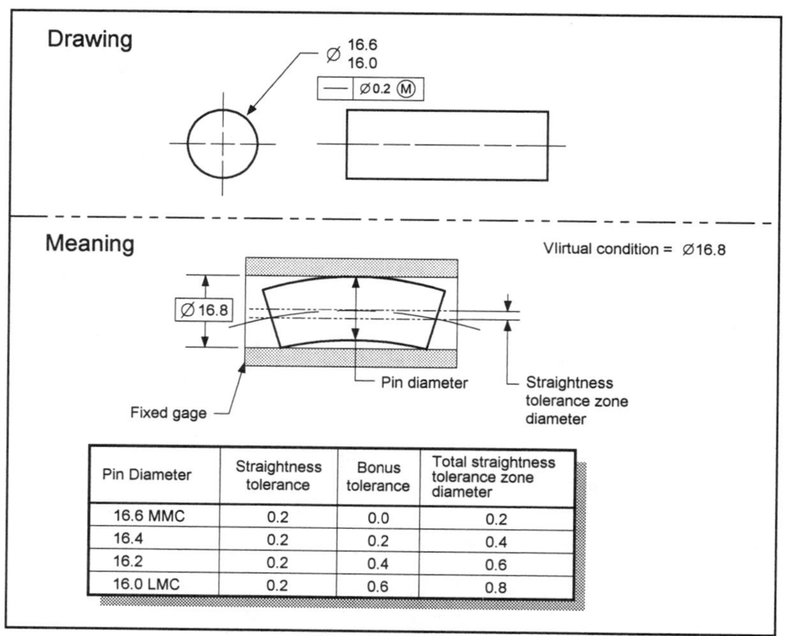

Pin Diameter	Straightness tolerance	Bonus tolerance	Total straightness tolerance zone diameter
16.6 MMC	0.2	0.0	0.2
16.4	0.2	0.2	0.4
16.2	0.2	0.4	0.6
16.0 LMC	0.2	0.6	0.8

FIGURE 4-13 Straightness at MMC Application

Whenever a straightness control is applied to a FOS at MMC, the following conditions apply:

- The FOS must also be within its size tolerance.
- The straightness control specifies a tolerance zone within which the axis or centerplane must lie.
- Rule #1 is overridden.
- A bonus tolerance is permissible.
- The virtual condition of the FOS is affected.
- A fixed gage may be used to verify the straightness.

An example of straightness applied to a centerplane is shown in Figure 3-16.

TECHNOTE 4-9 Straightness of a FOS at MMC

Whenever straightness is applied to a FOS at MMC, a bonus tolerance is permissible.

Indirect Straightness Controls

There are several geometric controls which can indirectly affect the straightness of an axis or centerplane. They are cylindricity, total runout, and in some cases, profile of a surface. When these controls are used, they may affect the straightness of an axis or centerplane of a FOS.

Legal Specification Test for Straightness Applied to a FOS

For a straightness control applied to a FOS to be a legal specification, it must satisfy the following conditions:

- No datum references can be specified in the feature control frame.
- The control must be associated with a feature of size dimension.
- If applied to a cylindrical FOS, a diameter symbol should be specified in the tolerance portion of the feature control frame.
- The control can not contain the projected tolerance zone or tangent plane modifier.
- The tolerance value should be a refinement of other geometric tolerances that control the straightness of the feature.

Figure 4-14 shows a legal specification flowchart for straightness applied to a FOS.

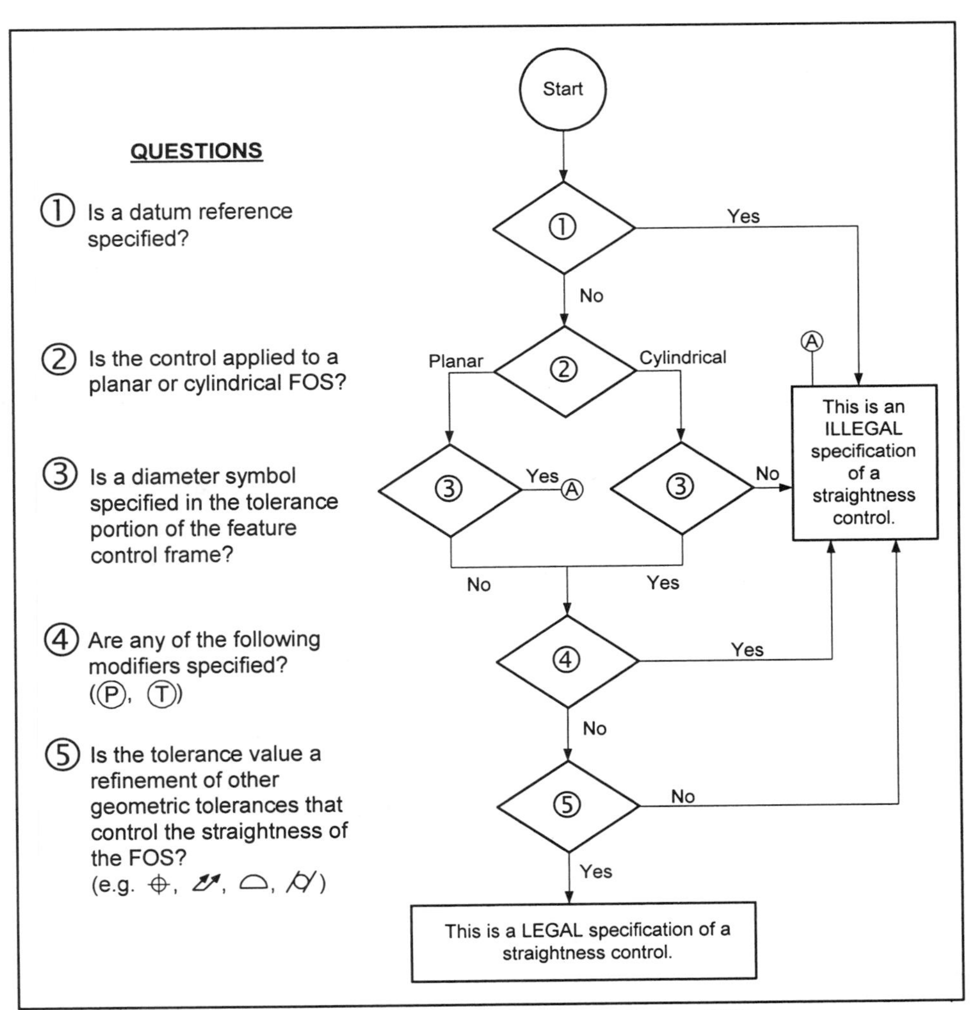

FIGURE 4-14 Legal Specification Flowchart for Straightness Applied to a FOS

Inspecting a Straightness Control (Applied to a FOS at MMC)

In Figure 4-15, a straightness control is applied to the diameter. When inspecting this diameter, two separate checks should be made: the size of the FOS and the straightness of the FOS. Chapter 2 discussed how to check the size of a FOS; now we will look at how to interpret the straightness specification.

The straightness control could be inspected as follows:

Since the straightness control contains an MMC modifier, a fixed gage could be used. The gage would have a hole equal to the virtual condition of the diameter, and the gage would be at least as long as the FOS it is verifying. The FOS would have to pass through the gage to meet the straightness control. The size tolerance of the pin would be checked separately.

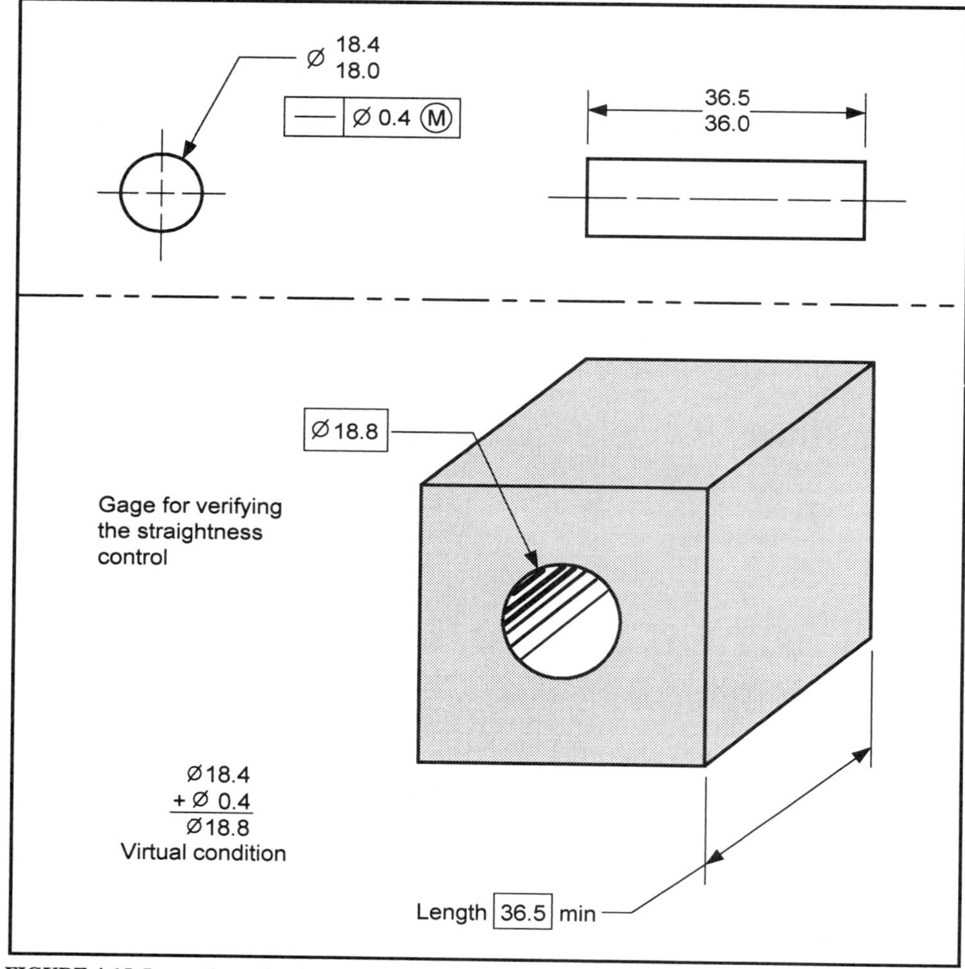

FIGURE 4-15 Inspecting a Straightness Control Applied to a FOS at MMC

CIRCULARITY CONTROL

Definition

Circularity is a condition where all points of a surface of revolution, at any section perpendicular to a common axis, are equidistant from that axis. Circularity can be applied to any part feature with a diametrical (round) cross section.

A ***circularity control*** is a geometric tolerance that limits the amount of circularity on a part surface. It specifies that each circular element of a feature's surface must lie within a tolerance zone of two coaxial circles. It also applies independently at each cross section element and at a right angle to the feature axis. The radial distance between the circles is equal to the circularity control tolerance value. See Figure 4-16.

Author's Comment
The information regarding circularity controls applies to rigid parts only. For parts subject to free-state variation, see Y14.5, Section 6.8.

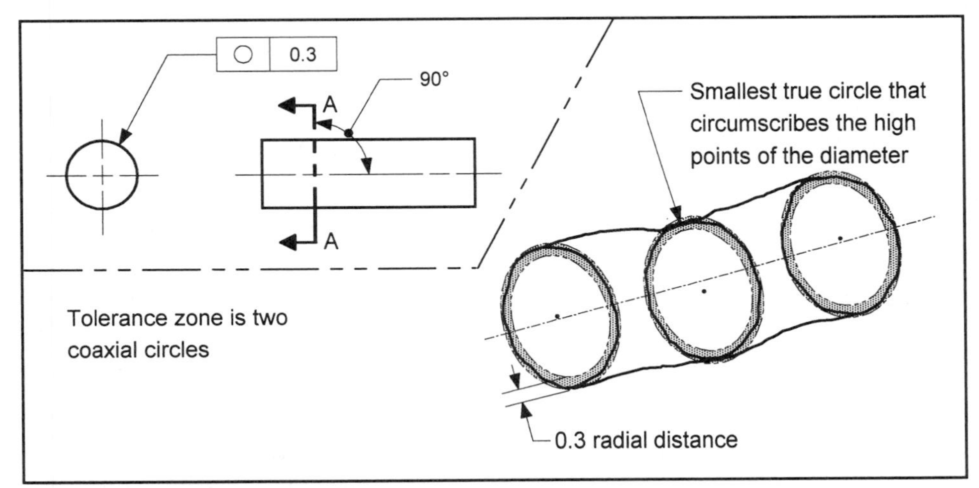

FIGURE 4-16 Circularity Control Example

For more info. . .
See Paragraph 6.4.3 of Y14.5.

A circularity control can only be applied to a surface; therefore, MMC, LMC, diameter, projected tolerance zone, or tangent plane modifiers are not used.

Rule #1's Effects on Circularity

Design Tip

Rule #1 is an indirect form control. The circularity effects of Rule #1 are not inspected; they are a result of the boundary and size limitations. If it is desired to have the circularity of a surface inspected, a circularity control should be specified.

Whenever Rule #1 applies to a FOS with a diametrical cross section, an automatic indirect circularity control exists for its surface. This indirect control is the result of the interrelationship between Rule #1 and the size dimension. When a diameter is at MMC, its cross section elements must be perfectly circular. As a diameter departs from MMC, a circularity error is permissible. Figure 4-17 illustrates an example of how Rule #1 indirectly affects circularity.

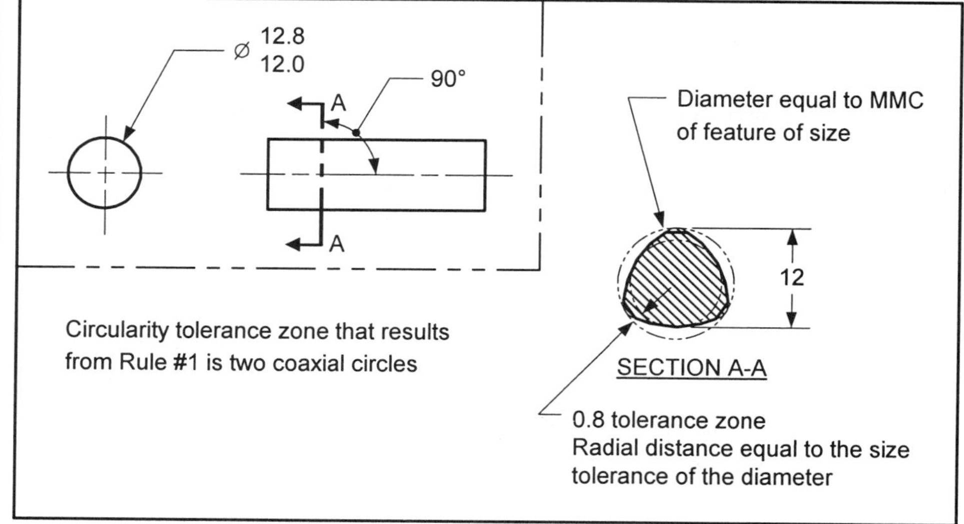

FIGURE 4-17 Rule #1's Effects on Circularity

Figure 4-17 illustrates that whenever a diameter is controlled by Rule #1, its cross section elements must lie between two coaxial circles, one equal to the MMC of the diameter, the second radially smaller by the size tolerance. Therefore, a diametrical dimension automatically restricts the circularity of a diameter to be equal to its size tolerance.

TECHNOTE 4-10 Rule #1 as a Circularity Control

Whenever Rule #1 applies to a FOS with a diametrical cross section, its circularity is automatically restricted to be equal to its size tolerance.

Circularity Application

A common reason for using a circularity control on a drawing is to limit the lobing (out of round) of a shaft diameter. In certain cases, lobing of a shaft diameter will cause bearings or bushings to fail prematurely. In Figure 4-18, the circularity control limits the maximum allowable amount of circularity error of the shaft diameter. In this application, the following statements apply:

- The diameter must be within its size tolerance.
- The circularity control does not override Rule #1.
- The circularity control tolerance must be less than the size tolerance.
- The circularity control does not affect the outer boundary of the FOS.

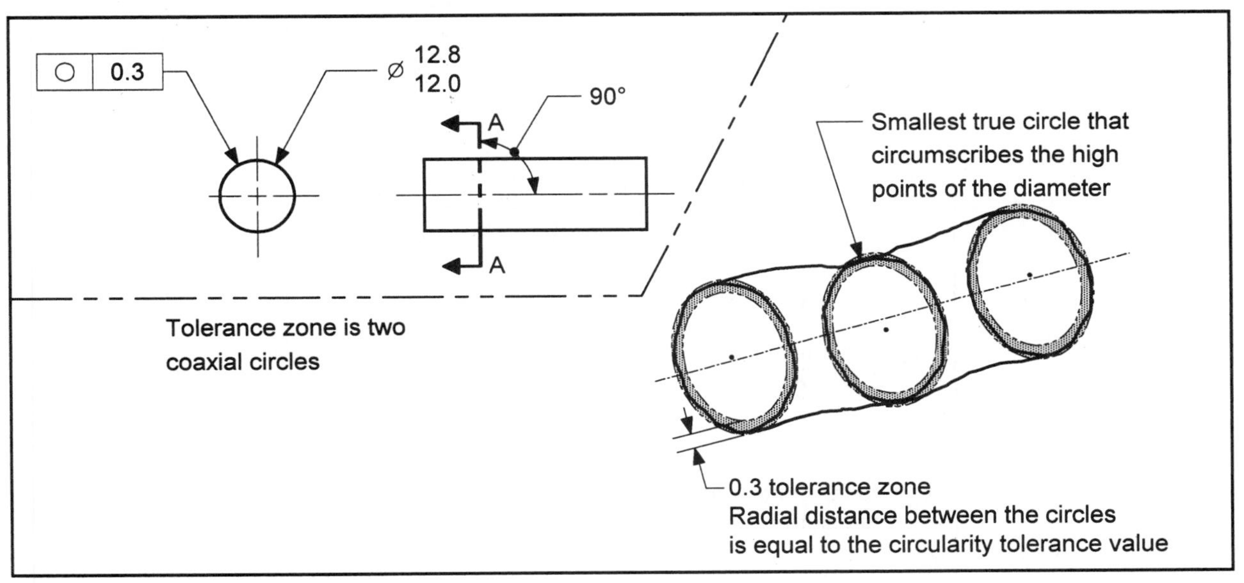

FIGURE 4-18 Circularity Application

Indirect Circularity Controls

There are several geometric controls that can indirectly affect the circularity of a diameter; they are Rule #1, cylindricity, profile, and runout. When any of these controls are used on a diameter, they also limit the circularity error of the diameter. However, indirect circularity controls are not inspected. If it is desired to have the circularity of a diameter inspected, a circularity control should be specified. If a circularity control is specified, its tolerance value must be less than the tolerance value of any indirect circularity control that affects the diameter.

Legal Specification Test for a Circularity Control

For a circularity control to be a legal specification, it must satisfy the following conditions:

- No datum references can be specified in the feature control frame.
- No modifiers can be specified in the feature control frame.
- The control must be applied to a diametrical feature.
- The circularity control tolerance value must be less than any other geometric control that limits the circularity of the feature.

Figure 4-19 shows a legal specification flowchart for a circularity control.

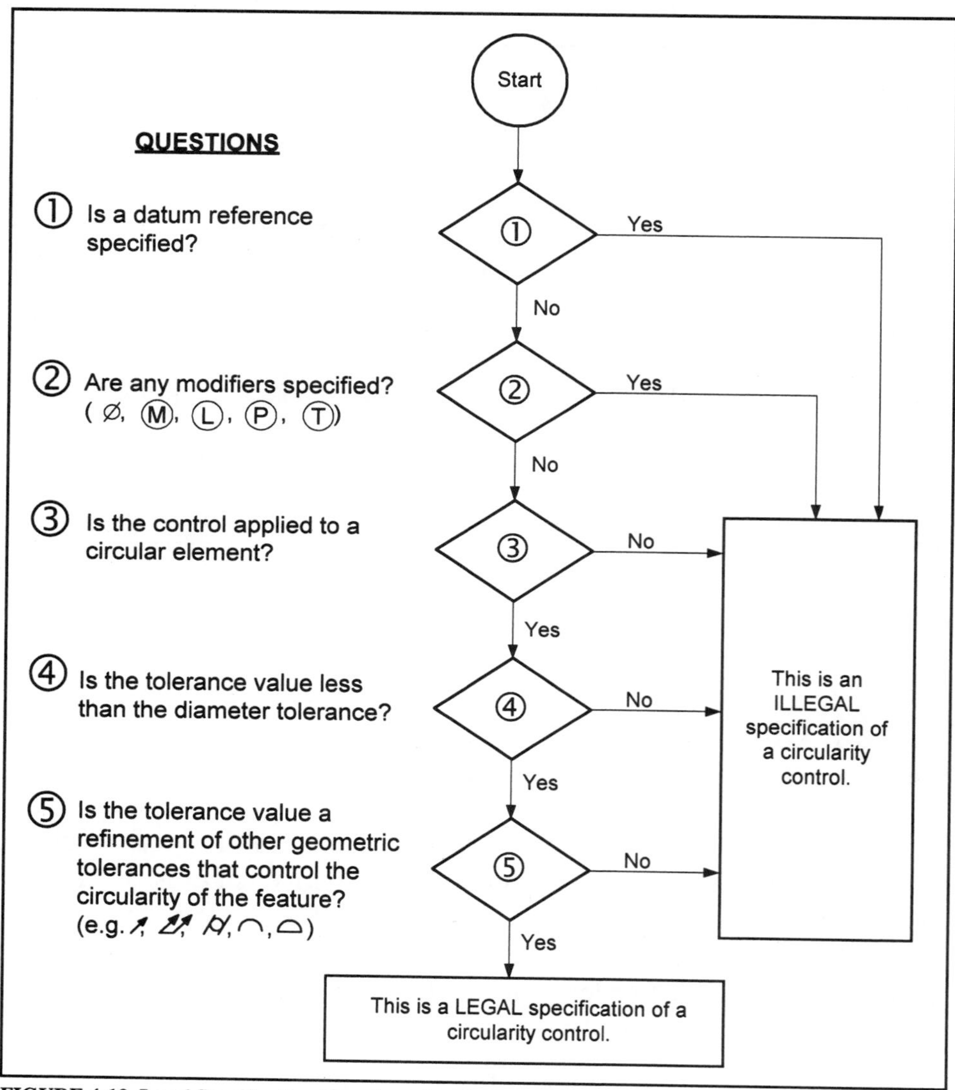

FIGURE 4-19 Legal Specification Flowchart for Circularity

Inspecting Circularity

Figure 4-18 shows a part with a circularity specification. When inspecting this part, the size of the FOS, the Rule #1 boundary, and the circularity requirement must be checked. Chapter 2 discussed how to check the size and Rule #1 boundary; now we will look at how to inspect the circularity specification.

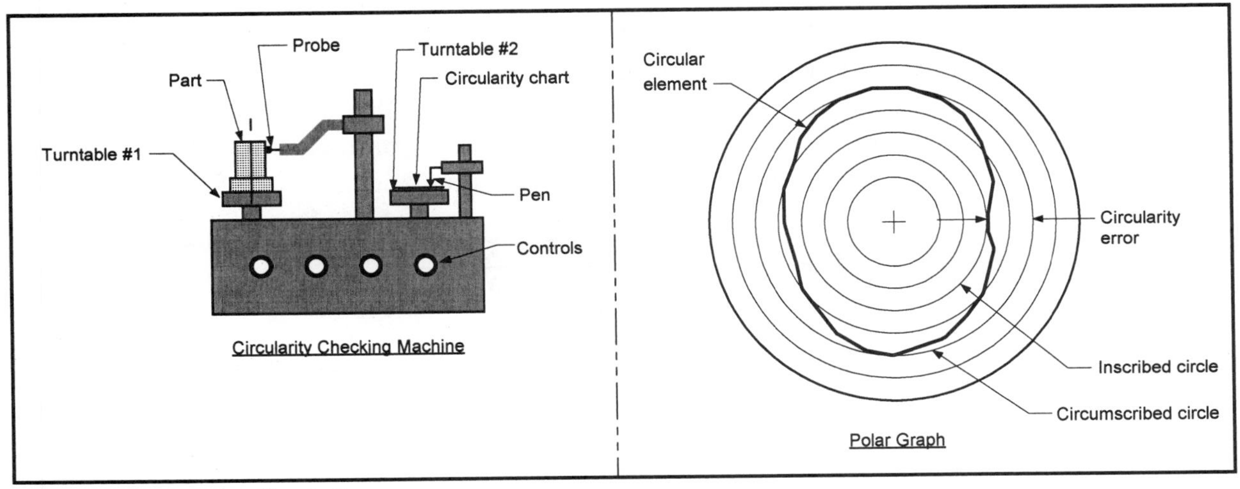

FIGURE 4-20 Inspecting Circularity

A circularity control could be inspected with a special roundness checking device like the one shown in Figure 4-20. A general description of an inspection using this device is:

There are two small turntables. The part being inspected is placed on one turntable. A round paper chart is placed on the other turntable. As both turntables rotate, a probe traces a circular element of the part surface. The electronic componentry of the device magnifies the surface conditions.

Next, a pen draws a circular chart of the surface outline, comparing the radial distance between a circle that circumscribes the high points of the outline drawn on the chart and a coaxial circle that inscribes the low points of the outline drawn on the chart. By multiplying this distance with the magnification scale chosen for the part measurement, a circularity value for the circular element is determined. This process is repeated at several locations until the inspector is confident that the part meets circularity requirements.

CYLINDRICITY CONTROL

For more info. . .
See Paragraph 6.4.4 of Y14.5.

Definition

Cylindricity is a condition of a surface of revolution in which all points of the surface are equidistant from a common axis. A ***cylindricity control*** is a geometric tolerance that limits the amount of cylindricity error permitted on a part surface. It specifies a tolerance zone of two coaxial cylinders within which all points of the surface must lie. A cylindricity control applies simultaneously to the entire surface. The radial distance between the two coaxial cylinders is equal to the cylindricity control tolerance value. A cylindricity control is a composite control that limits the circularity, straightness, and taper of a diameter simultaneously. See Figure 4-21.

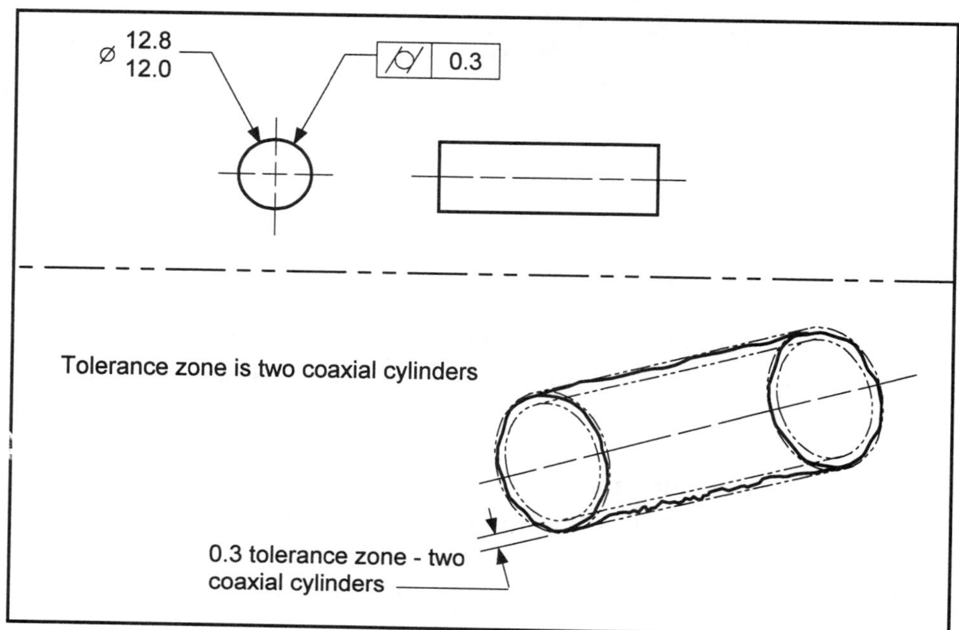

FIGURE 4-21 Cylindricity Control Example

A cylindricity control can only be applied to a surface; therefore, the MMC, LMC, diameter, projected tolerance zone, or tangent plane modifiers are not used.

Rule #1's Effects on Cylindricity

Whenever Rule #1 applies to a cylindrical FOS, an automatic indirect cylindricity control exists for its surface. This indirect control is the result of the interrelationship between Rule #1 and the size dimension. When the diameter is at MMC, its surface must be perfectly cylindrical. As the diameter departs from MMC, a cylindricity error is permissible. Figure 4-22 illustrates an example of how Rule #1 indirectly affects cylindricity.

Design Tip

Rule #1 is an indirect form control. The cylindricity effects of Rule #1 are not inspected; they are a result of the boundary and size limitations. If it is desired to have the cylindricity of a surface inspected, a cylindricity control should be specified.

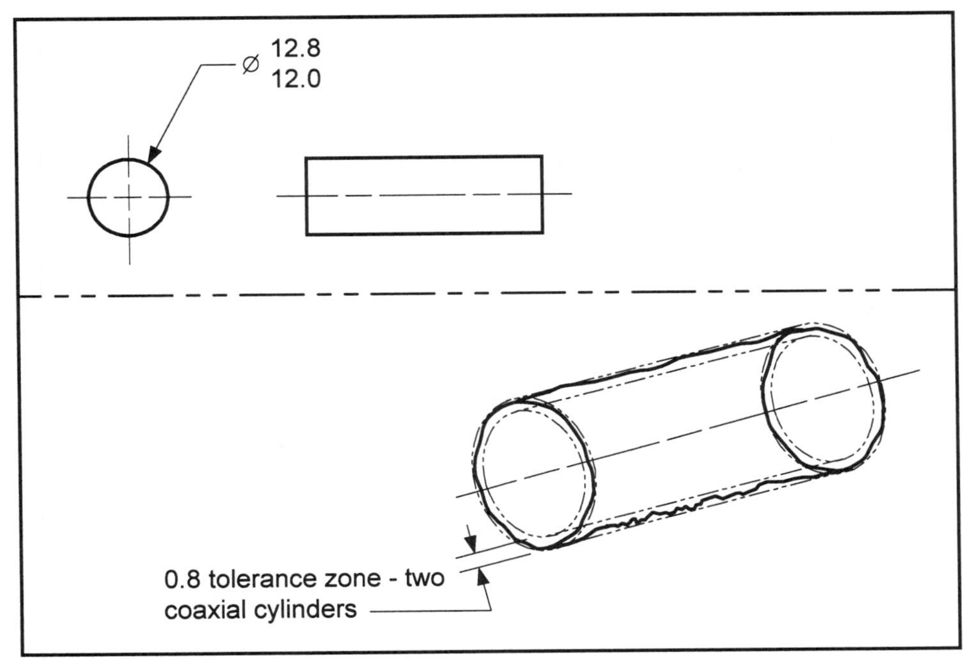

FIGURE 4-22 Rule #1's Effects on Cylindricity

Figure 4-22 illustrates that whenever a diameter is controlled by Rule #1, its surface must lie between two coaxial cylinders, one equal to the MMC of the diameter and the second radially smaller by the size tolerance. Therefore, a diametrical dimension automatically restricts the cylindricity of a diameter to be equal to its size tolerance.

TECHNOTE 4-11 Rule #1 as a Cylindricity Control

Whenever Rule #1 applies to a cylindrical FOS, its cylindricity is automatically restricted to be equal to its size tolerance.

Cylindricity Application

A common reason for a cylindricity control to be used on a drawing is to limit the surface conditions (out of round, taper, and straightness) of a shaft diameter. In certain cases, surface conditions of a shaft diameter will cause bearings or bushings to fail prematurely. In Figure 4-23, the cylindricity control limits the maximum allowable cylindricity error of the shaft diameter. In this application, the following statements apply:

- The diameter must also be within its size tolerance.
- The cylindricity control does not override Rule #1.
- The cylindricity control tolerance must be less than the total size tolerance.
- The cylindricity control does not affect the outer boundary of the FOS.

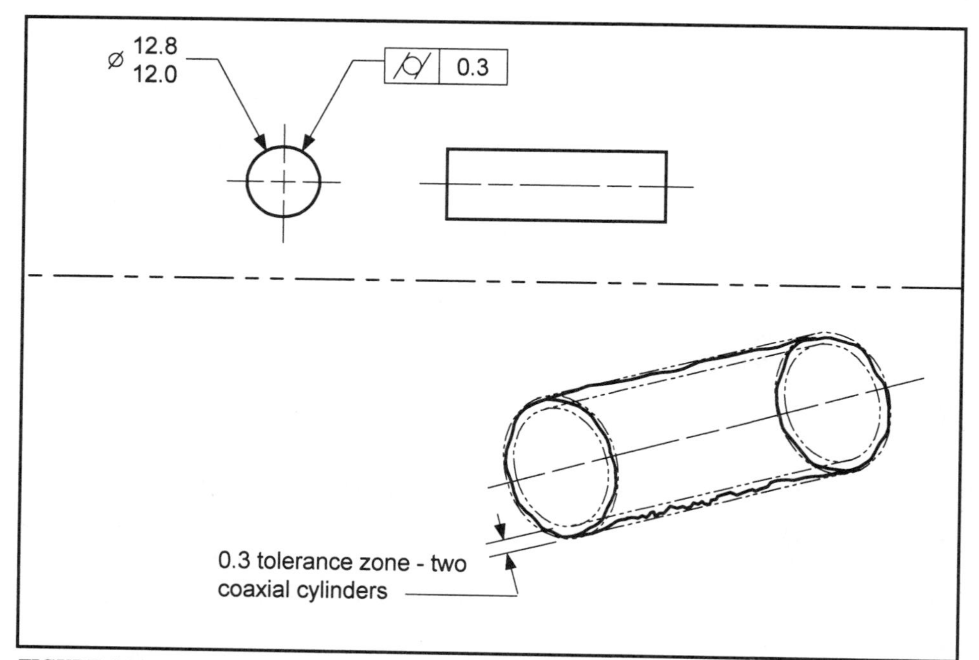

FIGURE 4-23 Cylindricity Application

Indirect Cylindricity Controls

There are several geometric controls that can indirectly affect the cylindricity of a diameter; they are Rule #1, profile of a surface, and total runout. When any of these controls are used on a diameter, they also limit the cylindricity of the diameter. However, indirect cylindricity controls are not inspected. If it is desired to have the cylindricity of a diameter inspected, a cylindricity control should be specified. If a cylindricity control is specified, its tolerance value must be less than the tolerance value of any indirect cylindricity controls that affect the diameter.

Legal Specification Test for a Cylindricity Control

For a cylindricity control to be a legal specification, it must satisfy the following conditions:

- No datum references can be specified in the feature control frame.
- No modifiers can be specified in the feature control frame.
- The control must be applied to a cylindrical feature.
- The cylindricity control tolerance value must be less than any other geometric control that limits the cylindricity of the feature.

Figure 4-24 shows a legal specification flowchart for a cylindricity control.

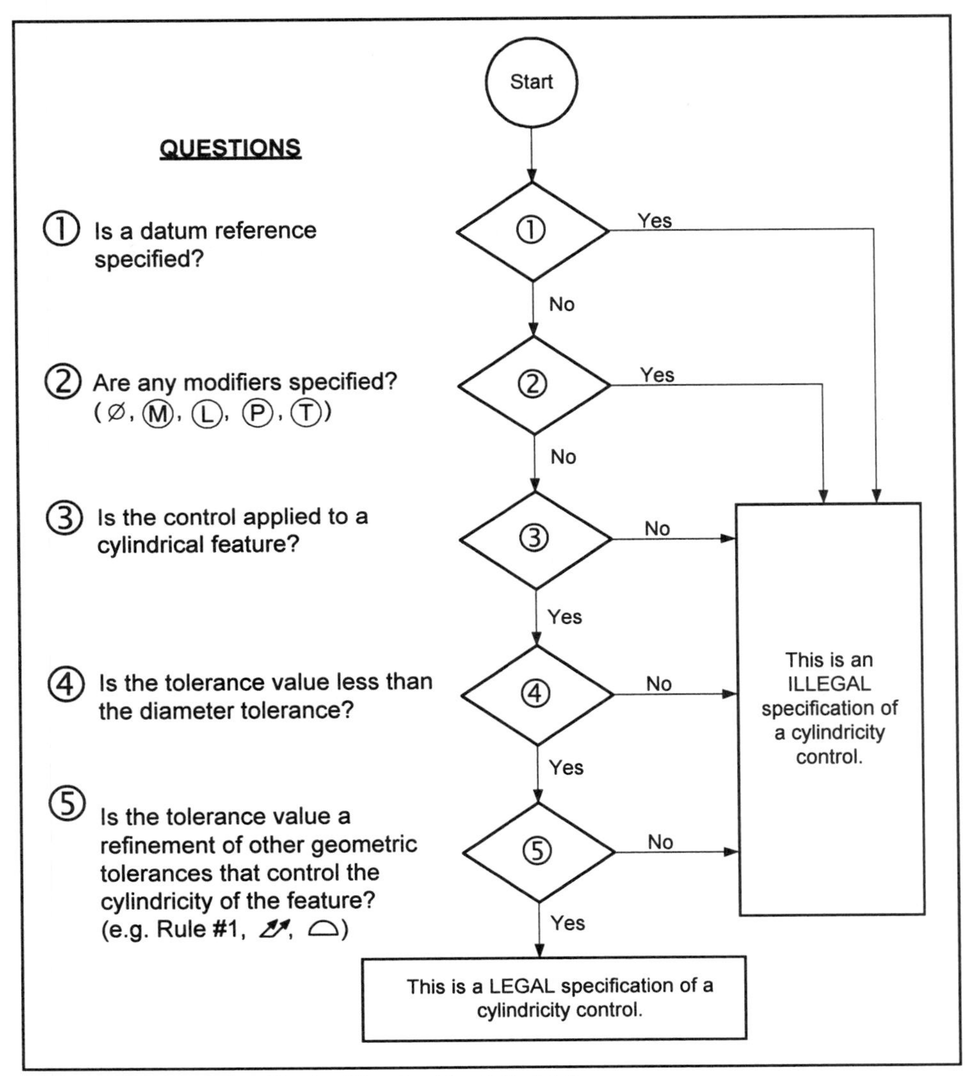

FIGURE 4-24 Legal Specification Flowchart for Cylindricity

Inspecting Cylindricity

When inspecting a part with a cylindricity specification, several separate parameters must be checked; these include the size of the FOS, the Rule #1 boundary, and the cylindricity requirement. Chapter 2 discussed how to check the size and Rule #1 boundary; now we will look at how to inspect the cylindricity specification.

A cylindricity control can be inspected in a manner similar to the inspection for circularity. The difference is that for cylindricity, a spiral path must be traced on the part surface. The surface points would then be compared to the two coaxial cylinders of the cylindricity tolerance zone.

Summary

A summarization of form control information is shown in Figure 4-25.

Symbol	Datum reference required	Can be applied to a		Can affect WCB	Can use Ⓜ or Ⓛ modifier	Can override Rule #1
		Surface	FOS			
⏥	No	Yes	No	No	No	No
⏤	No	Yes	Yes	Yes*	Yes*	Yes*
○	No	Yes	No	No	No	No
⌭	No	Yes	No	No	No	No

* When applied to a FOS

FIGURE 4-25 Summarization of Form Controls

VOCABULARY LIST

New Terms Introduced in this Chapter

Circularity
Circularity control
Cylindricity
Cylindricity control
Flatness
Flatness control
Straightness (axis or centerplane)
Straightness (line element)
Straightness control (FOS)
Straightness control (surface)

Study Tip
Read each term. If you don't recall the meaning of a term, look it up in the chapter.

ADDITIONAL RELATED TOPICS

Topic	ASME Y14.5M-1994 Reference
• Straightness controls applied on a per unit basis	Paragraph 6.4.1.1.4
• Flatness controls applied on a per unit basis	Paragraph 6.4.2.1.1
• Circularity applied to non-rigid parts	Paragraph 6.8

Author's Comment
These topics, plus advanced coverage of many of the topics introduced in this text, will be covered in my new book on advanced GD&T concepts.

QUESTIONS AND PROBLEMS

1. Describe what flatness is. ______________________________

2. Describe the tolerance zone for a flatness control.
____

3. Describe how a flatness tolerance zone is located. ______________________________

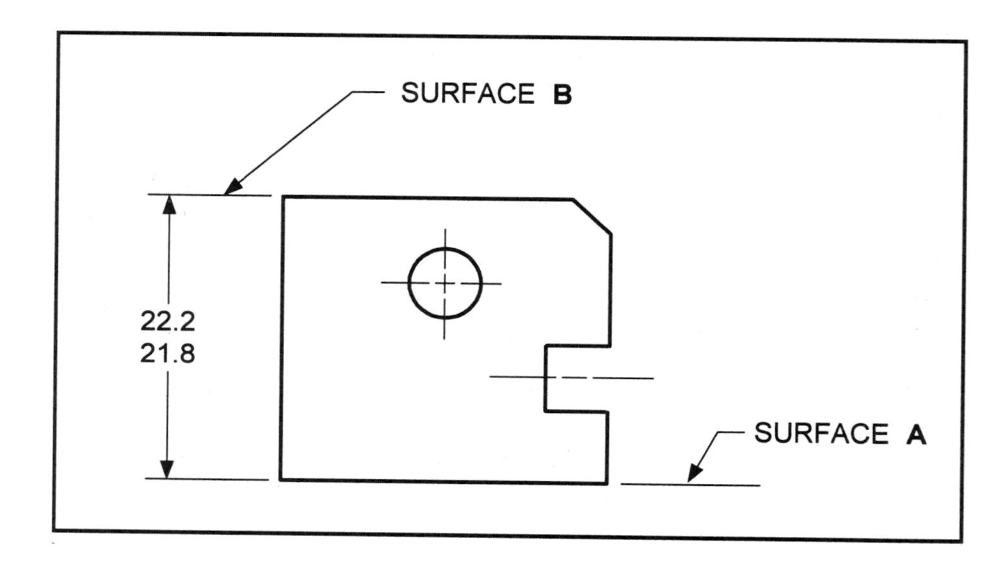

4. On the drawing above, what is the maximum allowable flatness error of surface *A*?________

5. On the drawing above, what is the maximum allowable flatness error of surface *B*?________

6. For each flatness control shown below, indicate if it is a legal specification. If a control is illegal, explain why.

   A. | ⏥ | 0.1 | ______________________________

   B. | ⏥ | 0.1 Ⓜ | ______________________________

   C. | ⏥ | 0.1 | A | ______________________________

   D. | ⏥ | ⌀ 0.1 | ______________________________

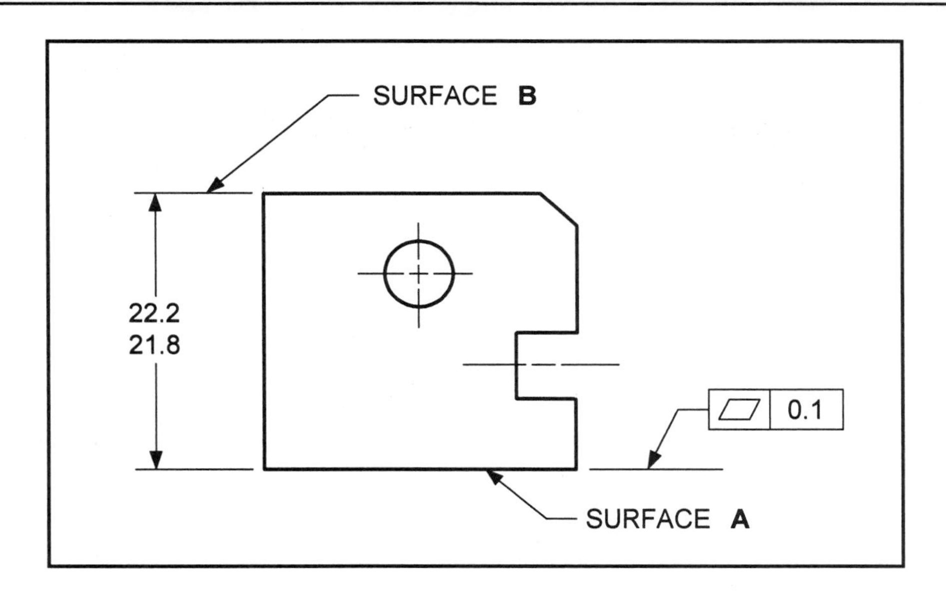

Questions 7-12 refer to the drawing above.

7. What is the maximum permissible flatness error of surface *A*?____________________
   Surface *B*?______________

8. What is the outer boundary of the 21.8 - 22.2 dimension? ____________________

9. Could the flatness control tolerance value be increased to 0.5? ___________ Explain why or why not.

   ____

10. If the 21.8 - 22.2 dimension was increased to 21.6 - 22.4, would this change the flatness tolerance zone on surface *A*?____________ Explain why or why not.

    ____

11.

If the part was . . .	The flatness error of surface *B* would be limited to . . .	The flatness error of surface *A* would be limited to . . .
At MMC		
At LMC		
At 22.0		

12. Can a flatness control be applied to a FOS?

    ____

13. Describe how a flatness control can be inspected.

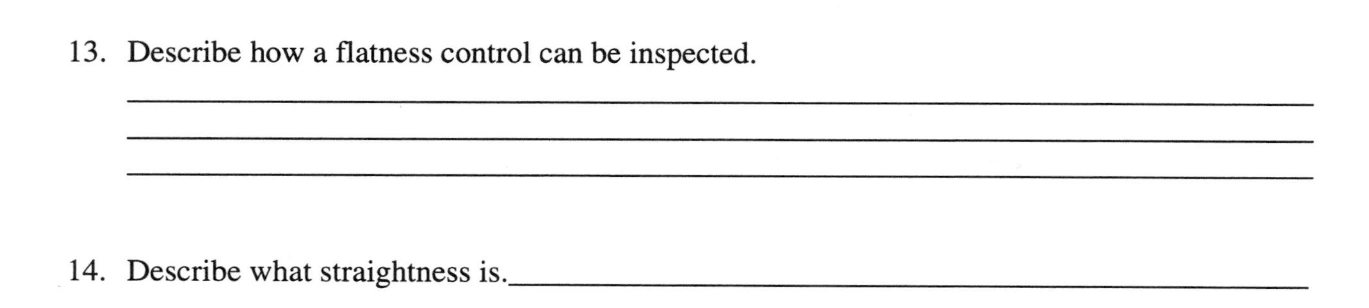

14. Describe what straightness is.

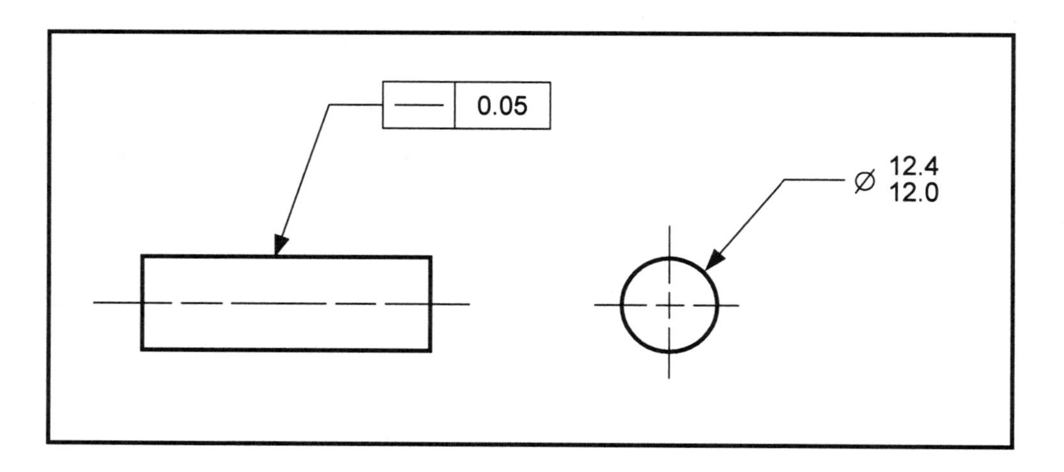

Questions 15 - 18 refer to the drawing above.

15. Describe the tolerance zone for the straightness callout.

16. The outer boundary of the pin is________________.

17. If the straightness control was removed, what would control the straightness of the surface elements?

18. Describe how the straightness control could be inspected.

19. Each straightness control shown below is applied to a surface. Indicate if it is a legal specification. If a control is illegal, explain why.

A. | — | 10 | ____

B. | — | 0.1 Ⓜ | ____

C. | — | 0.1 | A | ____

D. | — | Ø 0.1 | ____

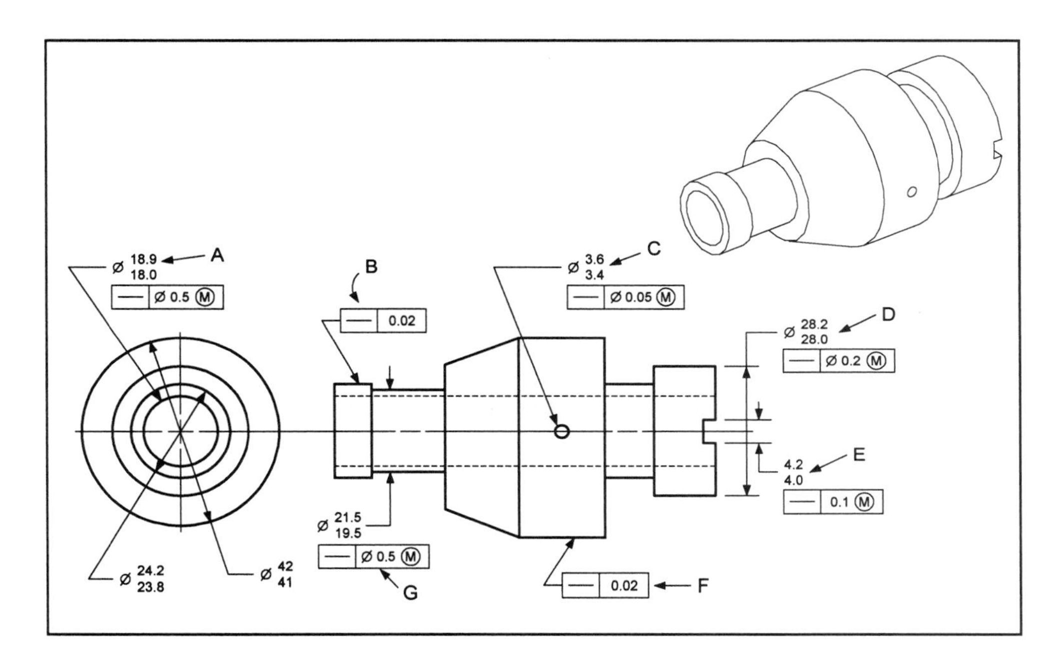

20. Use the drawing above to fill in the chart below. (Use N/A if not applicable.)

Dimension at letter	The straightness control shown is applied to a Surface	or FOS	The VC, OB, or IB of the FOS is	Does Rule #1 apply to the FOS?
A				
B				
C				
D				
E				
F				
G				

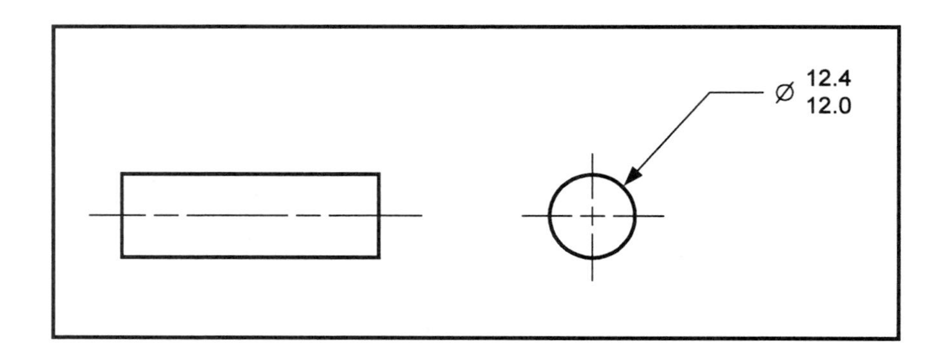

Questions 21 and 22 refer to the drawing above.

21. What limits the straightness of the diameter of the pin?

____

22. The outer boundary of the pin is ____________________.

23. Circle the letter for each of the four conditions that exist when a straightness control (with an MMC modifier) is applied to a feature of size.

A. The FOS size tolerance (of the FOS) may be violated.

B. Rule #1 is overridden.

C. The virtual condition of the FOS is equal to zero.

D. The MMC modifier cannot be used.

E. The straightness control affects the virtual condition.

F. The flatness of the surface is also controlled.

G. A fixed gage can be used to verify the straightness.

H. Rule #1 applies.

I. A bonus tolerance is permissible.

J. A datum must be referenced.

K. A variable gage must be used.

L. No bonus is allowed.

24. In each case, draw and dimension the gage for the straightness control and calculate the maximum amount of bonus tolerance possible and the maximum total allowable tolerance.

Part	Gage	Max. Bonus / Total Allowable Tolerance
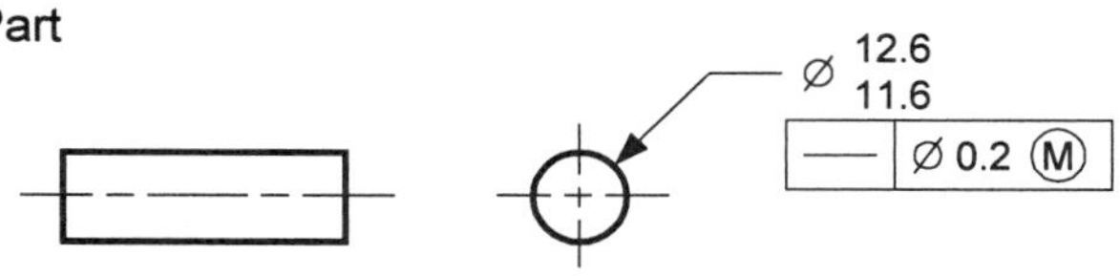 A Part — ⌀ 12.6 / 11.6; — ⌀ 0.2 (M)	Gage	Max. Bonus Total Allowable Tolerance
B Part 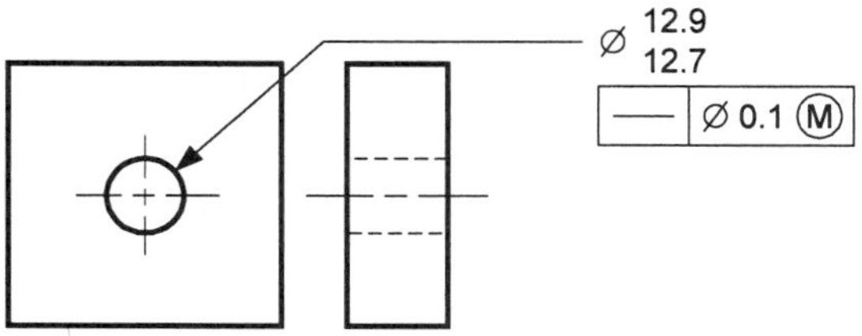 — ⌀ 12.9 / 12.7; — ⌀ 0.1 (M)	Gage	Max. Bonus Total Allowable Tolerance
C Part 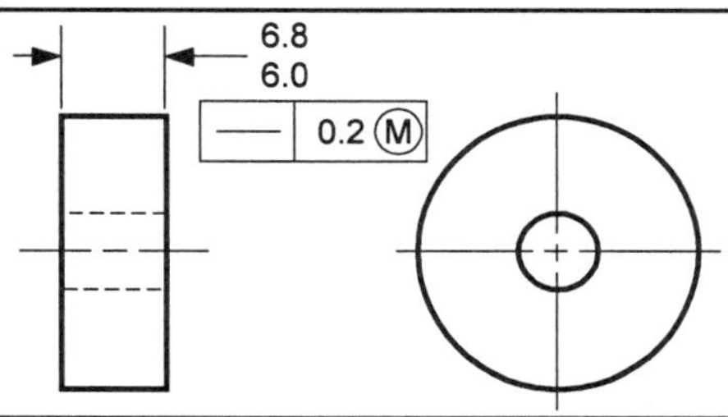 — 6.8 / 6.0; — 0.2 (M)	Gage	Max. Bonus Total Allowable Tolerance
D Part 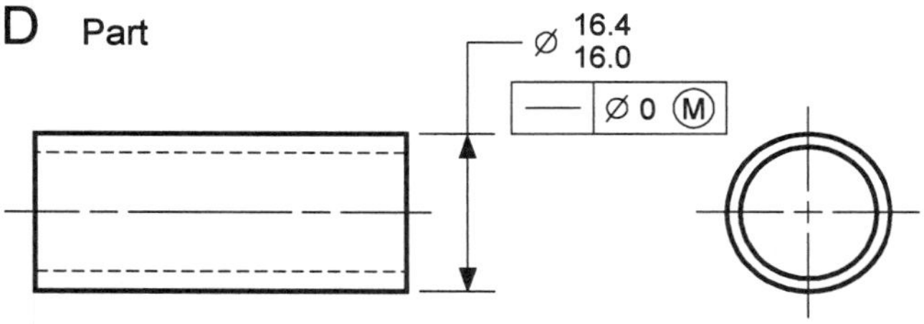 — ⌀ 16.4 / 16.0; — ⌀ 0 (M)	Gage	Max. Bonus Total Allowable Tolerance
E Part 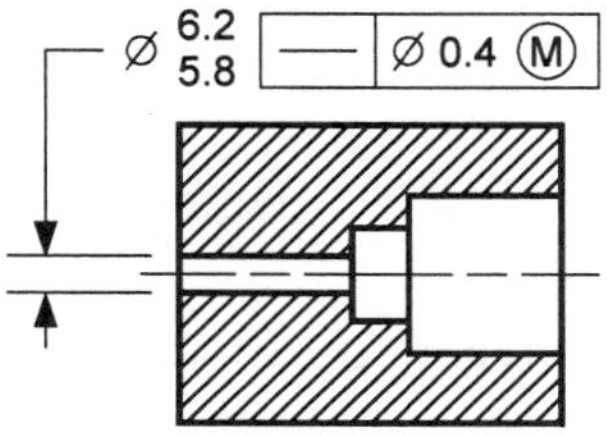 — ⌀ 6.2 / 5.8; — ⌀ 0.4 (M)	Gage	Max. Bonus Total Allowable Tolerance
F Part 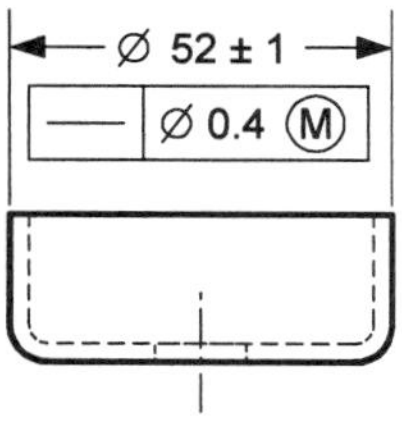 — ⌀ 52 ± 1; — ⌀ 0.4 (M)	Gage	Max. Bonus Total Allowable Tolerance

25. Describe circularity. ____

26. Describe the tolerance zone for a circularity control. ____

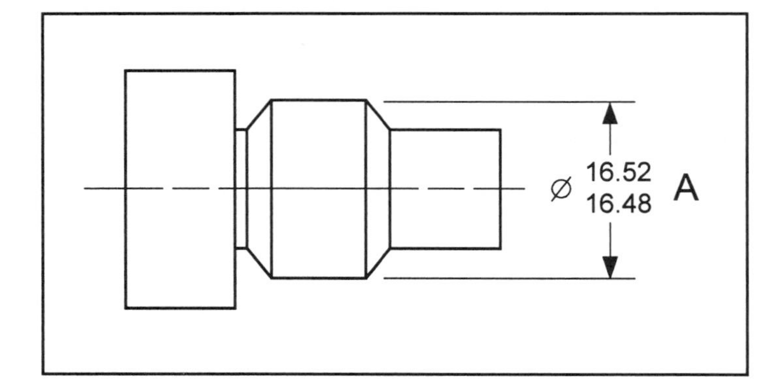

Questions 27 and 28 refer to the drawing above.

27. What controls the circularity of diameter *A*? ______________________

28. The maximum circularity error possible for diameter *A* is ______________________

29. Circle the letter for each of the three conditions that exist when a circularity control is applied to a diameter.

    A. The diameter must be within its size tolerance.
    B. Rule #1 applies.
    C. Rule #1 does not apply.
    D. The virtual condition is affected.
    E. A bonus tolerance is allowable.
    F. The circularity control tolerance value must be less than the size tolerance.

30. For each circularity control shown below, indicate if it is a legal specification. If the control is illegal, explain why.

    A. 
     ____

    B. ○ | 0.2 Ⓢ ______________________

    C. ○ | 0.1 ______________________

    D. ○ | 0.1 | A ______________________

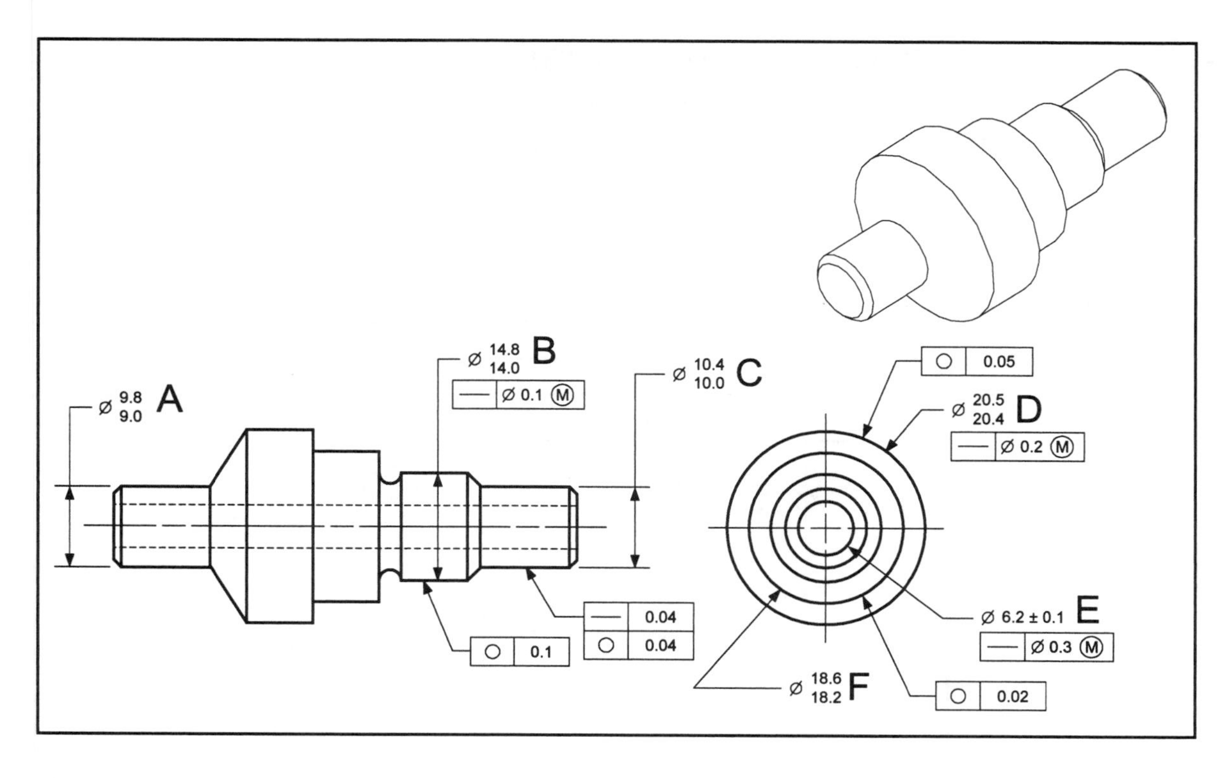

31. Using the drawing above, fill in the chart below.

Diameter	WCB	Max circularity error possible	Max straightness of axis error possible	Max straightness of line element error possible	Rule #1 applies (YES/NO)
A					
B					
C					
D					
E					
F					

32. Describe how a circularity control can be inspected.

____

____

33. Describe what cylindricity is. ____

____

34. Describe the tolerance zone for cylindricity. ____

____

____

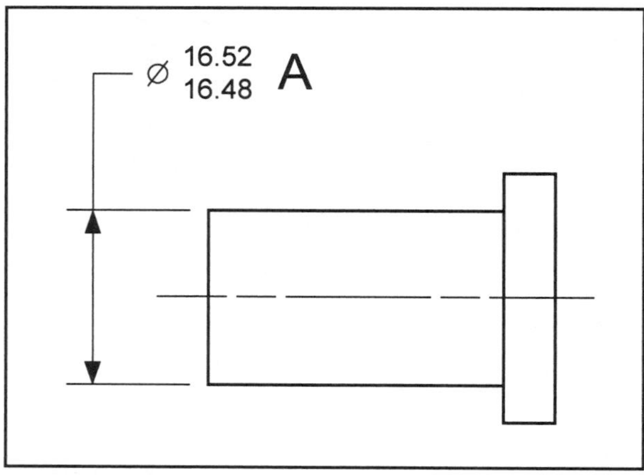

Questions 35 and 36 refer to the drawing above.

35. What controls the cylindricity of diameter *A*? _______________________

36. The maximum cylindricity error possible for diameter *A* is _______________________.

37. Circle the letter for each of the three conditions that exist when a cylindricity control is applied to a diameter.

    A. Rule #1 applies.
    B. Rule #1 does not apply.
    C. The worst-case boundary is not affected.
    D. A bonus tolerance is available.
    E. It limits the size tolerance of the diameter.
    F. The diameter must also be within its size tolerance.

38. Describe how a cylindricity control can be verified. _______________________

____

____

____

39. For each cylindricity control shown below, indicate if it is a legal specification. If a control is illegal, explain why.

A. | ⌭ | 0.02 | A | ______________________________

B. | ⌭ | 0.02 | ______________________________

C. | ⌭ | 0.02 Ⓛ | ______________________________

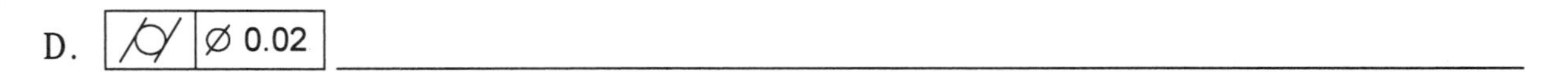

D. | ⌭ | Ø 0.02 | ______________________________

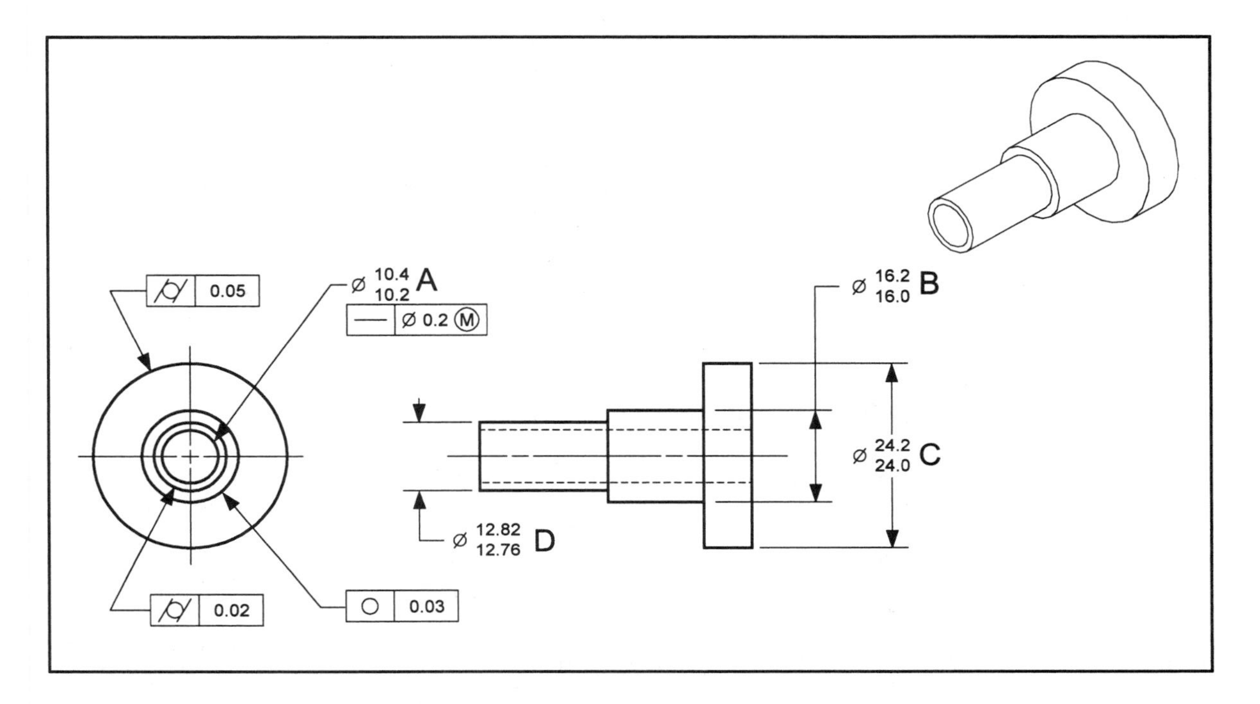

40. Using the drawing above, fill in the chart below.

Diameter	Rule #1 applies (YES/NO)	WCB	Max straightness of axis error possible	Max circularity error possible	Max cylindricity error possible
A					
B					
C					
D					

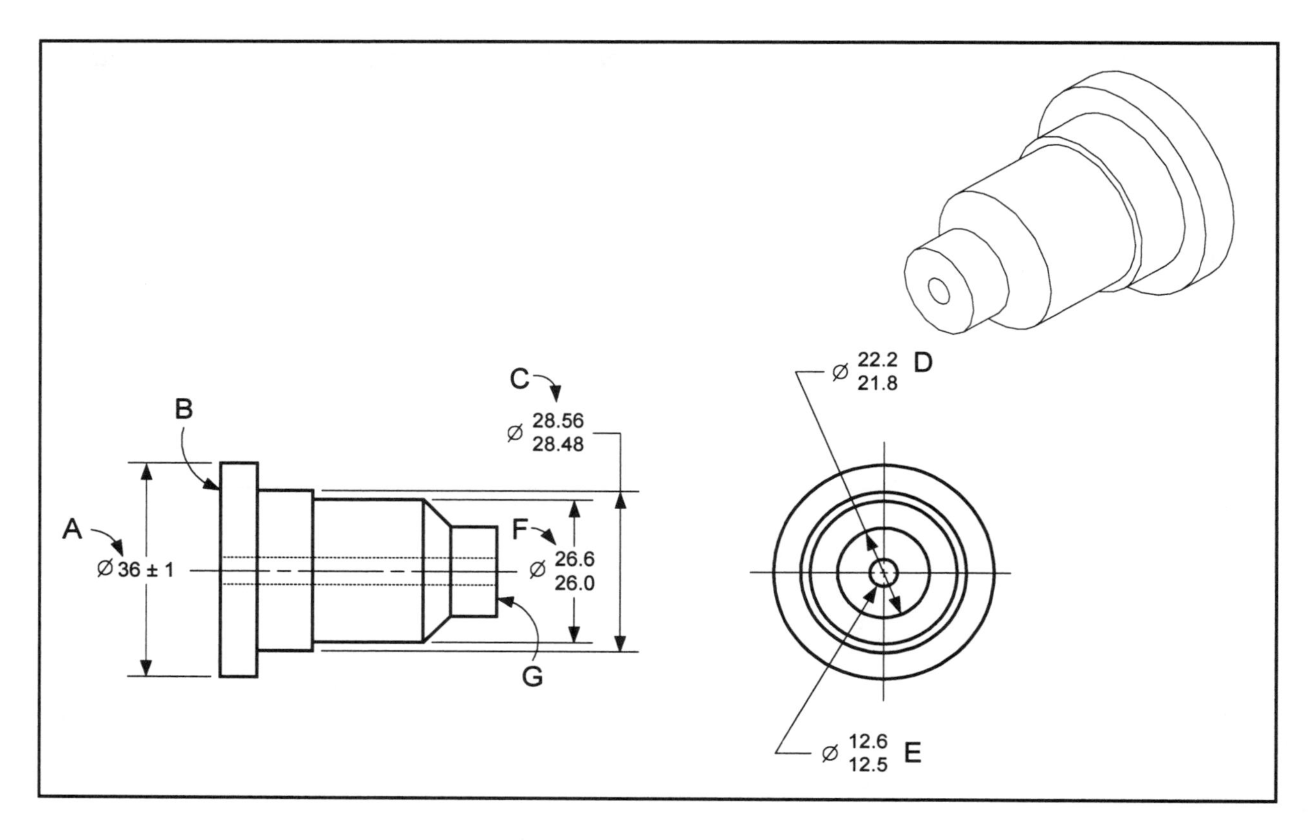

41. Use the instructions below to complete the drawing above.

   a. Add a form control to diameter *A* that will allow the virtual condition of the diameter to be 37.6. The control should allow a bonus tolerance.

   b. Add a form control to surface *B*. The control will allow a maximum flatness error of 0.1.

   c. Add a form control to limit the cylindricity error of diameter *C* to 0.02 maximum.

   d. Add a form control to diameter *D* to limit the straightness of the surface elements to 0.05 maximum.

   e. Add a form control to diameter *E* that will allow the virtual condition of the diameter to be 12.0. The control should allow a bonus tolerance.

   f. Add a form control to diameter *F* to limit the straightness of the surface elements to 0.08 maximum.

   g. Add a form control to surface *G* to allow a maximum flatness error of 0.2.

Chapter 5

Datums (Planar)

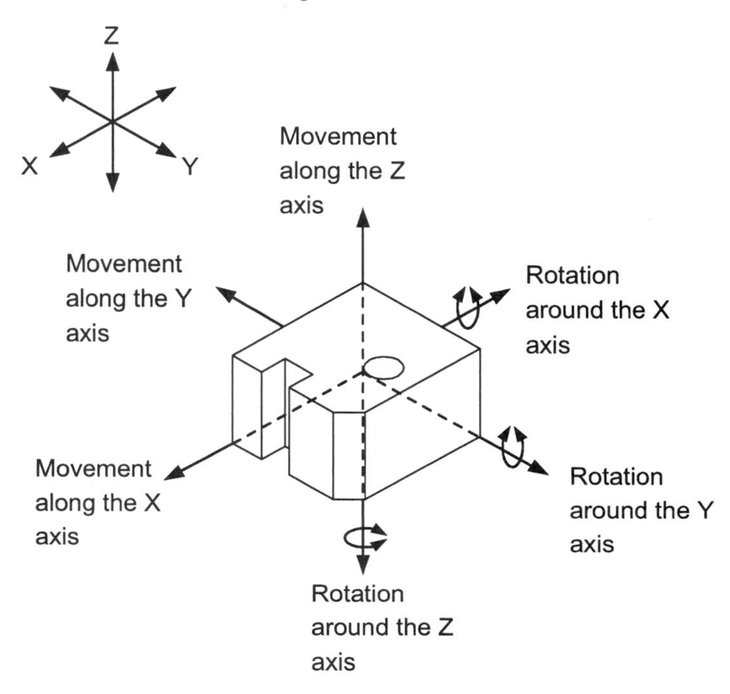

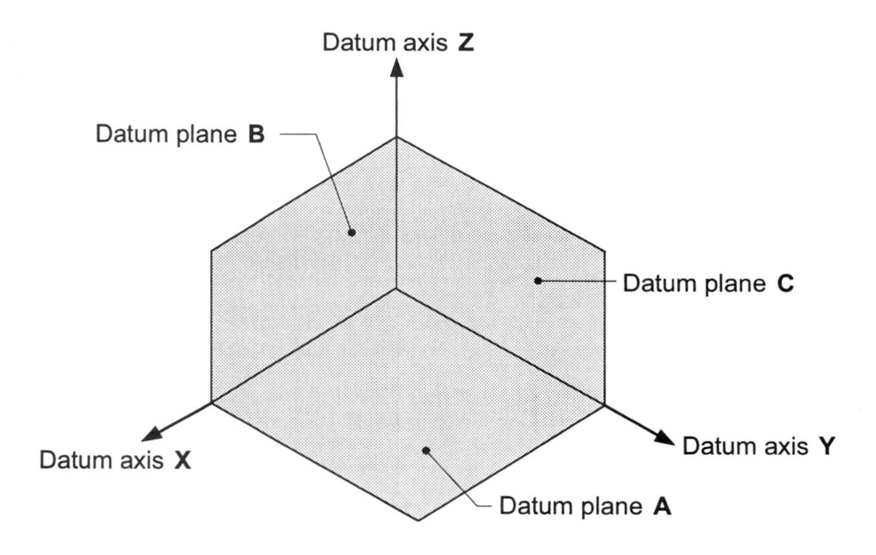

INTRODUCTION

This chapter is designed to help you read and understand the information related to datums on drawings. The datum system is an important part of the language of geometric tolerancing. The ***datum system*** is a set of symbols and rules that communicates to the drawing user how dimensional measurements are to be made. The datum system is used for two reasons: First, the datum system allows the designer to specify in which sequence the part is to contact the inspection equipment for the measurement of a dimension. Second, it allows the designer to specify which part surfaces are to contact the inspection equipment for the measurement of a dimension.

The datum system provides three important benefits:

- It aids in making repeatable dimensional measurements.
- It aids in communicating part functional relationships.
- It aids in making the dimensional measurement as intended by the designer.

CHAPTER GOALS AND OBJECTIVES

There are Two Goals in this Chapter:

5-1. Understand the datum system (planar datums).
5-2. Interpret datum targets.

Study Tip
Take a few minutes to fully understand these objectives. When reading this chapter, look for information to help you master these objectives.

Performance Objectives that Demonstrate Mastery of These Goals

Upon completion of this chapter, each student should be able to:

Goal 5-1 (pp. 114-126)

- Describe the datum system.
- List three benefits of the datum system.
- Describe an implied datum.
- List two shortcomings of implied datums.
- List two consequences of implied datums.
- Define a datum.
- Define a datum feature.

- Define a true geometric counterpart.
- Describe a datum feature simulator.
- Describe a simulated datum.
- Draw the datum feature symbol.
- Describe four ways to specify a planar datum.
- Explain the basis for choosing datum features.
- Describe a datum reference frame.
- Describe what controls the orientation of datum features.
- List the six degrees of part freedom in space.
- Explain the 3-2-1 Rule.
- Explain the difference between a datum-related dimension and a FOS (non datum-related) dimension.
- Describe the datum reference frame for a part with inclined datum features.
- Describe coplanar datum features.

Goal 5-2 (pp. 126-133)

- Explain datum targets.
- List two situations where datum targets should be used.
- Recognize the datum target symbol.
- State when a datum target specification is on the front or back surface in a view on a drawing.
- Describe why basic dimensions are used to locate datum targets.
- List three requirements of datum target applications.
- Draw the symbol for point, line, and area datum targets.
- Draw a simulated gage for a point datum target application.
- Draw a simulated gage for a line datum target application.
- Draw a simulated gage for an area datum target application.
- Specify datum targets on a drawing.

IMPLIED DATUMS

Definition

An ***implied datum*** is an assumed plane, axis, or point from which a dimensional measurement is made. Implied datums are an old concept from coordinate tolerancing. An example of implied datums is shown in Figure 5-1. In this figure, the bottom and left sides of the block are considered implied datums.

Shortcomings of Implied Datums

Implied datums have two major shortcomings. First, they do not clearly communicate to the drawing user which surfaces should contact the inspection equipment. When the drawing does not clearly specify which surfaces are to contact the inspection equipment, the inspector must make an assumption. Second, implied datums do not communicate to the drawing user in which sequence the part should be brought into contact with the inspection equipment. If the order is not clearly specified, each inspector could assume a different sequence. Each sequence would then produce different results for the part measurements. Figure 5-1 illustrates this problem.

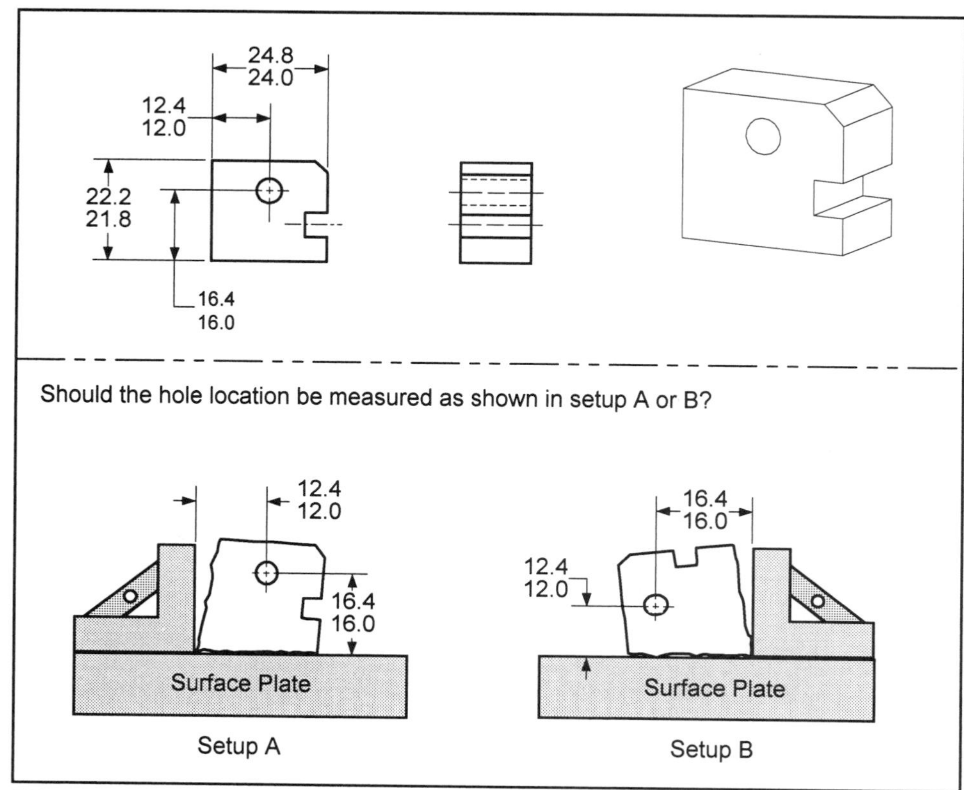

FIGURE 5-1 Implied Datums

Consequences of Implied Datums

The use of implied datums results in two consequences:

- Good parts are rejected.
- Bad parts are accepted.

The use of implied datums requires the inspector to assume which part surfaces should contact the inspection equipment and in what sequence. A part that is to specification when measured properly may be rejected when measured from the wrong surfaces or when using the wrong sequence. Also, a part that would be out of specification when measured properly, may pass inspection when measured from the wrong surfaces or using the wrong sequence.

Design Tip
Whenever dimensioning the location of a FOS on a part, do not use implied datums.

PLANAR DATUMS

General Information on Datums

A ***datum*** is a theoretically exact plane, point, or axis from which a dimensional measurement is made. A ***datum feature*** is a part feature that contacts a datum. A ***planar datum*** is the true geometric counterpart of a planar datum feature. A ***true geometric counterpart*** is the theoretical perfect boundary or best fit tangent plane of a specified datum feature.

For more info. . .
See Paragraph 4.2 of Y14.5.

Depending upon the type of datum feature, a true geometric counterpart may be:

- A tangent plane contacting the high points of a surface.
- A maximum material condition boundary.
- A least material condition boundary.
- A virtual condition boundary.
- An actual mating envelope.
- A mathematically defined contour.
- A worst-case boundary.

Design Tip
Whenever a surface is used as a primary datum feature, a flatness control should be applied to the surface. This will improve the stability of the part on its datum feature simulator.

Since a true geometric counterpart is theoretical, the datum is assumed to exist in, and be simulated by, the associated inspection (or processing) equipment. The inspection equipment (or gage surfaces) used to establish a datum is called the ***datum feature simulator***. For example, surface plates and gage surfaces—though not perfect planes—are of such quality that they are assumed to be perfect and are used as simulated datums. A ***simulated datum*** is the plane (or axis) established by the datum feature simulator. For practical purposes in industry, a simulated datum is used as the datum. Figure 5-2 shows a datum, true geometric counterpart, datum feature simulator, and simulated datum plane.

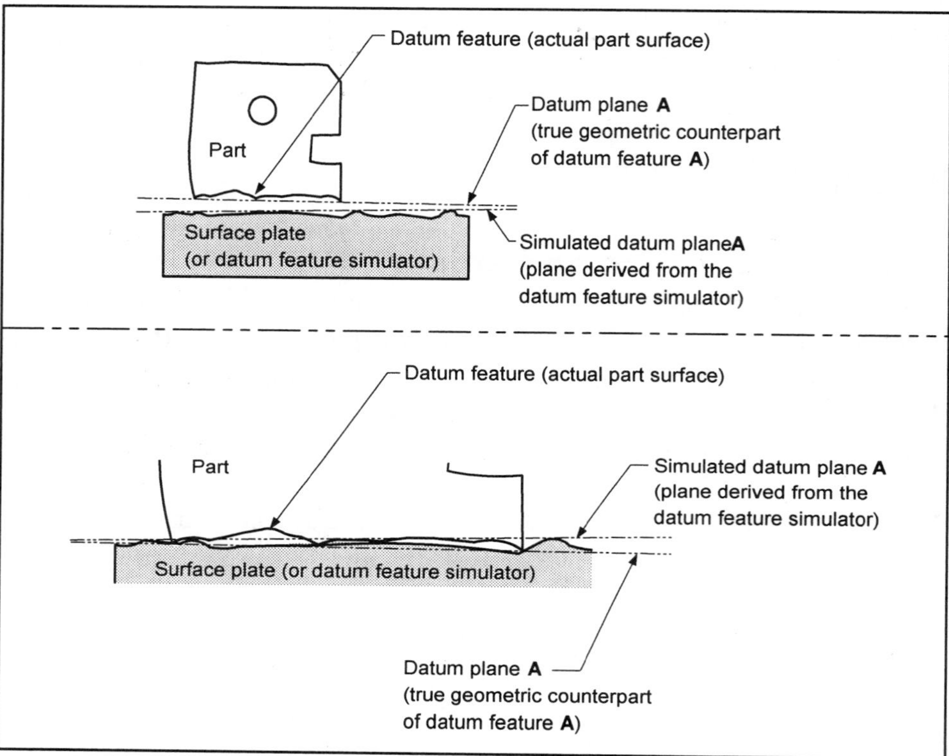

FIGURE 5-2 Planar Datums

TECHNOTE 5-1 Datum Features and Datums

- Datum features are part features and they exist on the part.
- A datum feature simulator is the inspection equipment that includes the gage elements used to establish a simulated datum.
- Datums are theoretical reference planes, points, or axes and are simulated by the inspection equipment.
- For practical purposes, a simulated datum is considered a datum.

Datum Feature Symbol

The symbol used to specify a datum feature on a drawing is shown in Figure 5-3. It is called the datum feature symbol. The method of attaching this symbol to a part feature determines if it designates a planar datum or a FOS datum. Figure 5-3 shows four ways of displaying the datum feature symbol to denote a planar datum.

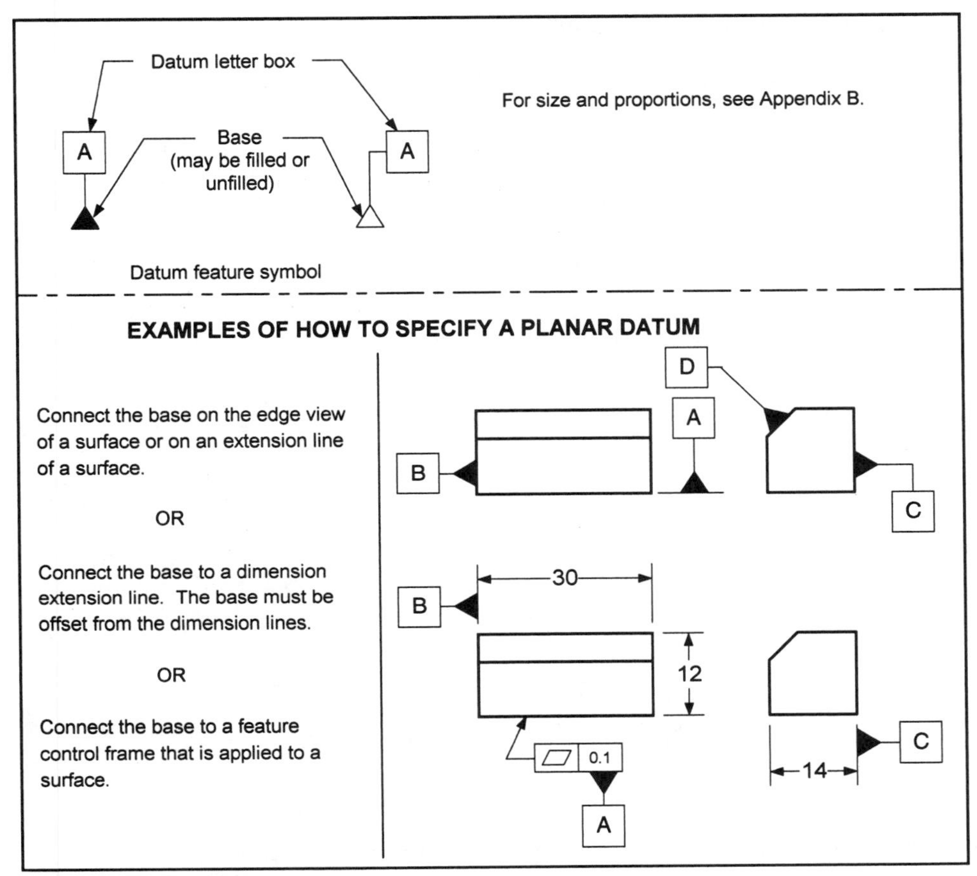

FIGURE 5-3 Datum Feature Symbol

For more info. . .
See Paragraphs 4.3 and 4.5.11.1 of Y14.5.

Datum Selection

Datum features are selected on the basis of part function and assembly requirements. The datum features are often the features that orient (stabilize) and locate the part in its assembly. For example, the part in Figure 5-4 mounts on surface *A* and is located by diameter *B*. For assembly, the holes need to be located relative to the features that mount and locate the part to the mating part. Therefore, surface *A* and diameter *B* are designated as datum features.

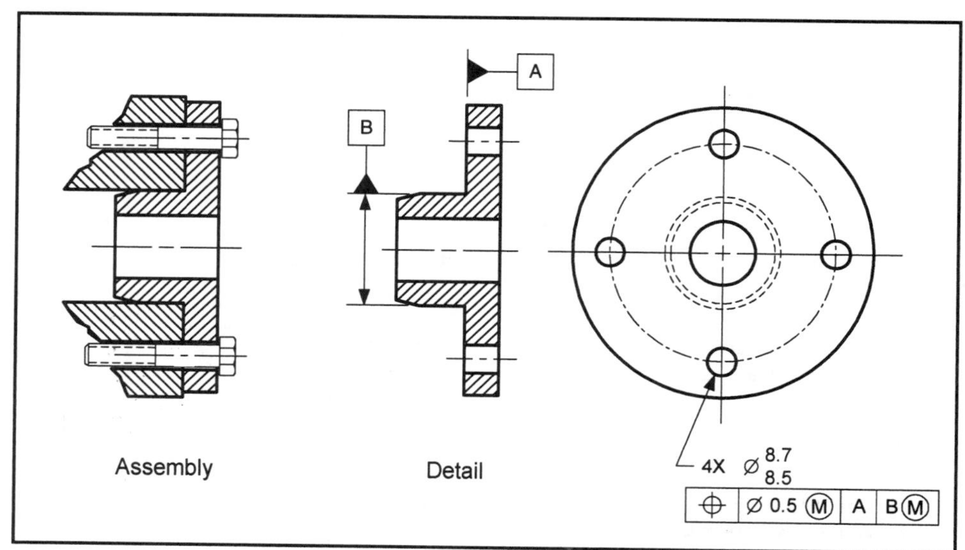

FIGURE 5-4 Datum Selection

TECHNOTE 5-2 Datum Selection

Datum features are selected on the basis of part function and assembly requirements; they are often the features that orient (stabilize) and locate the part in its assembly.

Referencing Datums in Feature Control Frames

After datums are specified, the drawing must also communicate when and how the datums should be used. This is typically done through the use of feature control frames. When feature control frames are specified, they reference the datums to be used for their measurement. For example, in Figure 5-4, the bolt holes are toleranced relative to the datums through the use of a geometric tolerance. Since the part is clamped against surface *A*, surface *A* will establish the orientation of the part in space and is referenced as the primary datum. The part is located by the pilot diameter; therefore, it is referenced as the secondary datum.

When feature control frames reference datums, they also specify the sequence for contacting the part to the datums referenced. The sequence is determined by reading the feature control frame from the left. The first compartment that contains a datum reference denotes the datum feature that is to contact the inspection equipment first. The second compartment that contains a datum reference denotes the datum feature that is to contact the inspection equipment second. The third compartment that contains a datum reference denotes the datum feature that is to contact the inspection equipment third. Figure 5-5 shows an example of how to interpret the datum sequence in a feature control frame.

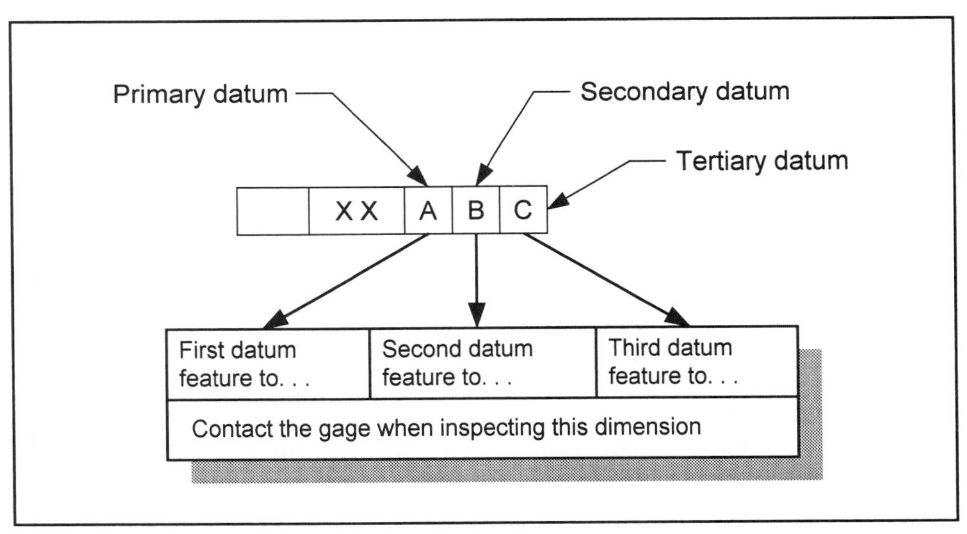

FIGURE 5-5 Referencing Datums in Feature Control Frames

Datum Reference Frame

One of the purposes of the datum system is to limit the movement of a part so that repeatable measurements can be made during inspection. When a part is free to move in space, it has six degrees of freedom. The six degrees of freedom are rotation around the *x, y*, or *z* axis and movement along the *x, y*, or *z* axis. They are shown in Figure 5-6. In order to restrict the six degrees of freedom on a part with planar datums, it takes the use of three datum planes. When three datum planes are used, they are considered to be a datum reference frame. A ***datum reference frame*** is a set of three mutually perpendicular datum planes. The datum reference frame provides direction as well as an origin for dimensional measurements.

Author's Comment
When three datums are referenced on a part with all planar datums, all six degrees of freedom are restricted.

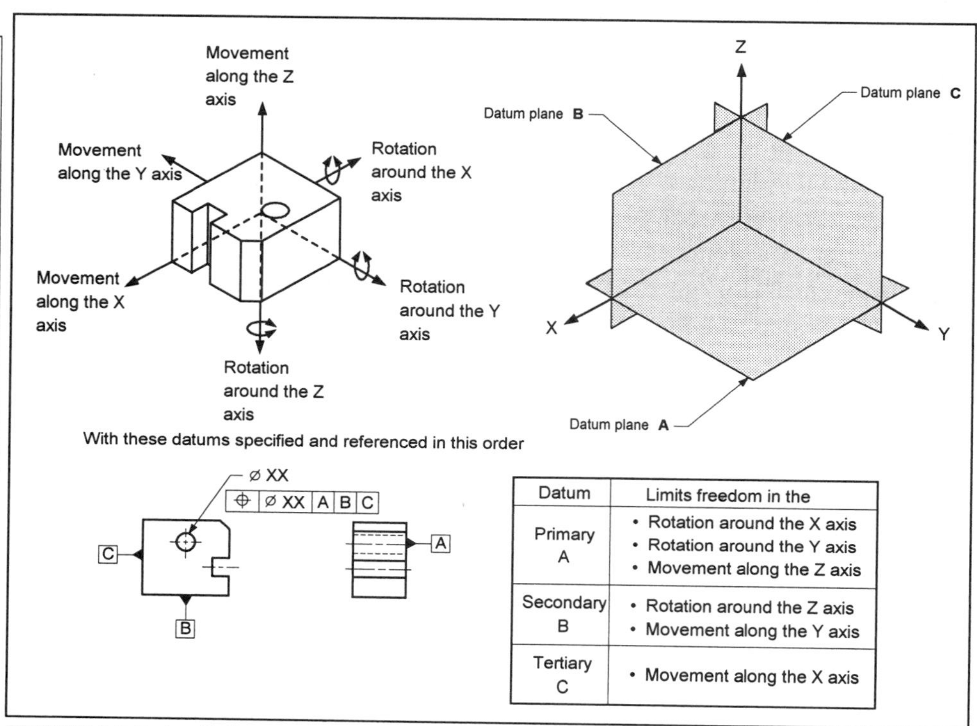

Datum	Limits freedom in the
Primary A	• Rotation around the X axis • Rotation around the Y axis • Movement along the Z axis
Secondary B	• Rotation around the Z axis • Movement along the Y axis
Tertiary C	• Movement along the X axis

FIGURE 5-6 Datum Reference Frame

The planes of a datum reference frame have zero perpendicularity tolerance to each other by definition. Measurements are taken perpendicular from the datum planes. Figure 5-7 shows a datum reference frame for the part from Figure 5-6.

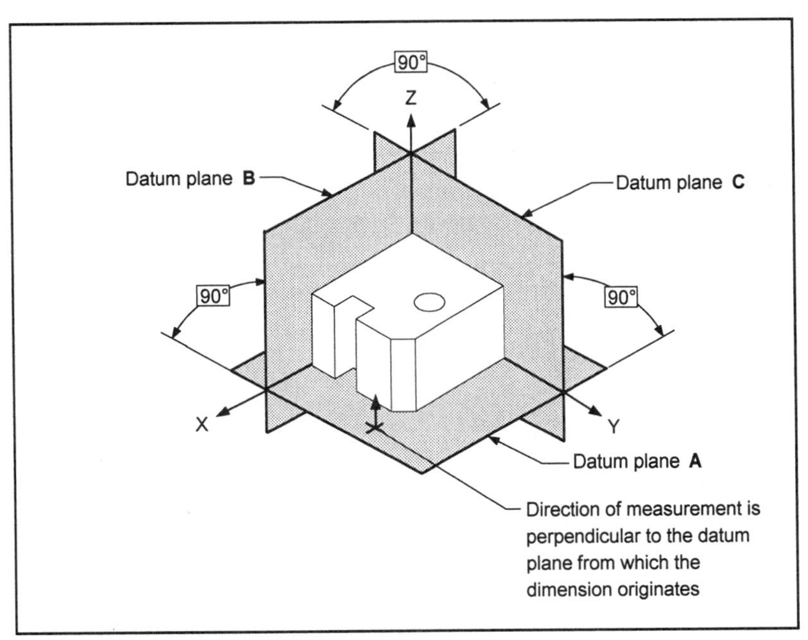

FIGURE 5-7 Datum Reference Frame

The 90° angles between datum planes are basic. The 90° angles between actual part surfaces have tolerance. The angular tolerance of the part surfaces may be specified on the drawing or in a general note. Figure 5-8 shows a part where the part surfaces (datum features) are not exactly 90° to each other. A measurement on this part will produce different results based on which part surface touches the datum reference frame first, second, and third.

When making a location measurement on a part feature, the six degrees of freedom are restricted by using a datum reference frame. The method of bringing a part into contact with the planes of the datum reference frame has a significant impact on the measurement of the part dimensions.

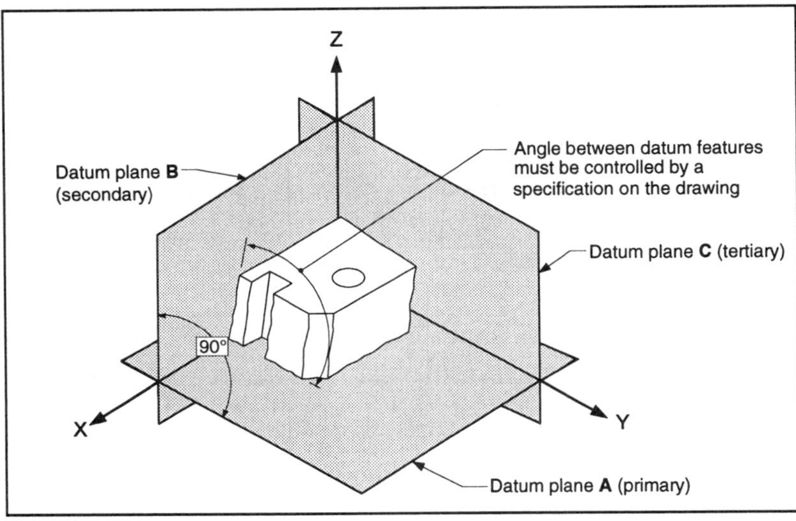

FIGURE 5-8 Datum Features

TECHNOTE 5-3 Datum Feature Relationships

The tolerance between part datum features that are shown at 90° must be specified in a general note, in titleblock tolerances, or by other means.

For more info. . .
See Paragraph 4.4.1 of Y14.5.

The feature control frames on a drawing specify the order in which the part surfaces are to contact the datum reference frame for a dimensional measurement. Refer to Figure 5-9 while reading this explanation.

A feature control frame is read from left to right. In this case, datum *A* is referenced first, datum *B* is referenced second, and datum *C* is referenced third.

The feature control frame indicates that the part should contact datum *A* first. In a dimensional measurement, the first datum plane that the part contacts is called the ***primary datum***. The primary datum establishes the orientation of the part (stabilizes the part) to the datum reference frame. The part contacts the datum plane with at least three points of contact. The primary datum restricts three degrees of freedom: movement along the *Z* axis, rotation around the *X* axis, and rotation around the *Y* axis. Three degrees of freedom remain unrestricted.

The feature control frame indicates that the part should contact datum *B* second. In a dimensional measurement, the second datum plane that the part contacts is called the ***secondary datum***. The secondary datum locates the part (restricts part movement) within the datum reference frame. The part may have a line contact with the secondary datum; therefore, it requires a minimum of two points of contact with the secondary datum plane. The secondary datum restricts two additional degrees of freedom: rotation around the *Z* axis and movement along the *Y* axis. One degree of freedom remains unrestricted.

The feature control frame indicates that the part should contact datum *C* third. In a dimensional measurement, the third datum plane that the part contacts is called the ***tertiary datum***. The tertiary datum locates the part (restricts part movement) within the datum reference frame. The part may have a single point of contact with the tertiary datum; therefore, it requires a minimum of one point of contact with the tertiary datum plane. The tertiary datum restricts the last remaining degree of freedom: movement along the *X* axis.

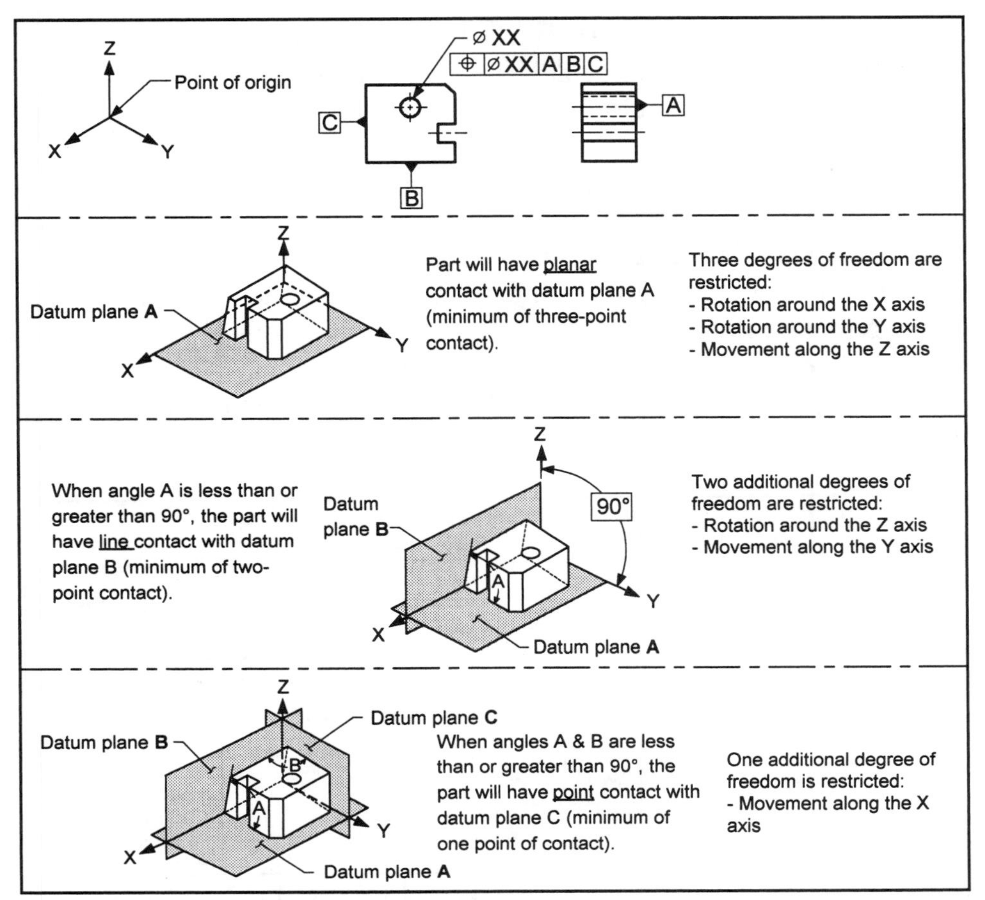

FIGURE 5-9 Primary, Secondary, and Tertiary Datums

The 3-2-1 Rule

The ***3-2-1 Rule*** defines the minimum number of points of contact required for a part datum feature with its primary, secondary, and tertiary datum planes. The 3-2-1 Rule only applies on a part with all planar datums. The primary datum feature has at least three points of contact with its datum plane. The secondary datum feature has at least two points of contact with its datum plane. The tertiary datum feature has at least one point of contact with its datum plane. The 3-2-1 Rule applies to planar datum features only.

Author's Comment

The three-point contact is the ideal condition. Form error on a planar datum feature can cause one-point contact or "rocking" of a part on a primary datum plane. In these cases, the part must be stabilized on its primary datum plane. See Y14.5, Paragraph 4.5.1

TECHNOTE 5-4 3-2-1 Rule

The 3-2-1 Rule defines the minimum points of contact with the primary datum as 3, the secondary datum as 2, and the tertiary datum as 1.

Datum-Related Versus FOS Dimensions

Only dimensions that are related to a datum reference frame through geometric tolerances should be measured in a datum reference frame. If a dimension is not associated to a datum reference frame with a geometric tolerance, then there is no specification on how to locate the part in the datum reference frame. In Figure 5-10, the hole locations are related to the datum reference frame, *D* primary, *E* secondary, and *F* tertiary. During inspection of the hole location dimensions, the part should be mounted in datum reference frame *D-E-F*, but the overall dimensions are not related to the datum reference frame. The overall dimensions are FOS dimensions. During inspection of the overall dimensions, the part should not be mounted in the datum reference frame.

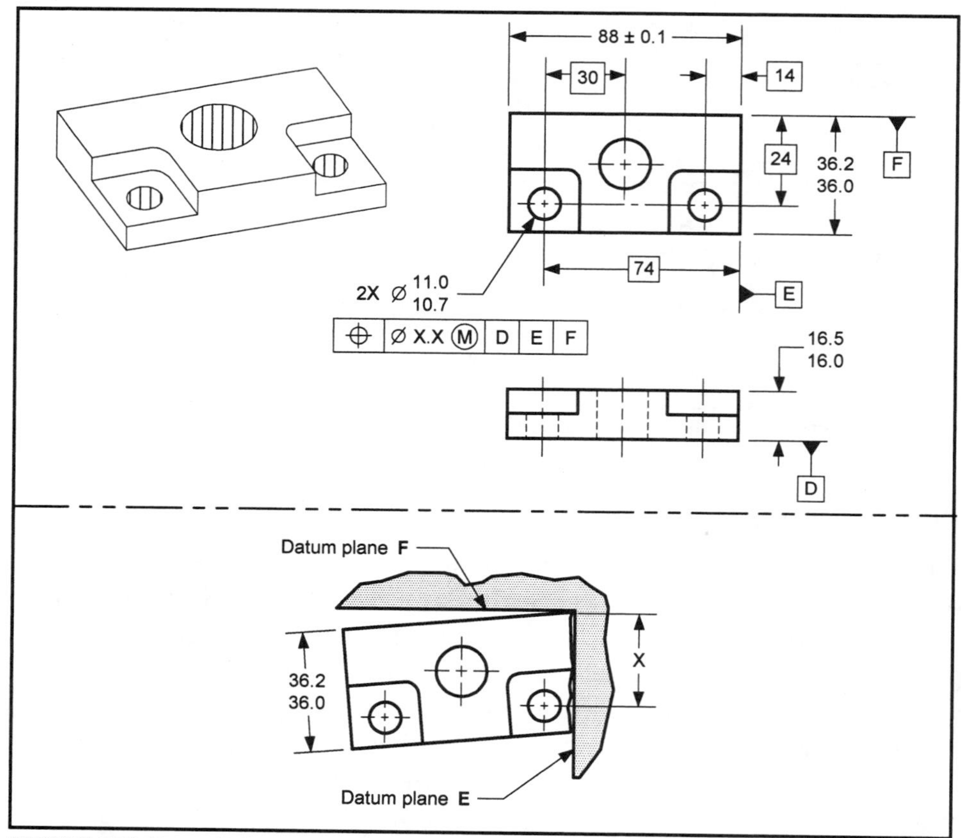

FIGURE 5-10 Datum-Related Versus Feature of Size Dimensions

> **TECHNOTE 5-5 Datum-Related Dimensions**
>
> Only dimensions related to a datum reference frame through the use of geometric tolerances are to be measured in a datum reference frame.

Inclined Datum Features

An ***inclined datum feature*** is a datum feature that is at an angle other than 90°, relative to the other datum features. On parts with datum features (surfaces) at angles other than 90°, the datum reference frame will contain planes at the basic angle of the part surface. The part shown in Figure 5-11 has a surface 60° from datum feature *A*; this surface is designated as datum feature *C*. The datum reference frame would have three perpendicular planes. However, the inclined datum feature would have its datum feature simulator oriented at the basic angles shown on the drawing. For this type of datum reference frame, the angle of the surface must be specified as basic, and the surface is controlled by a geometric tolerance.

For more info. . .
See Paragraph 4.4.1.1 of Y14.5.

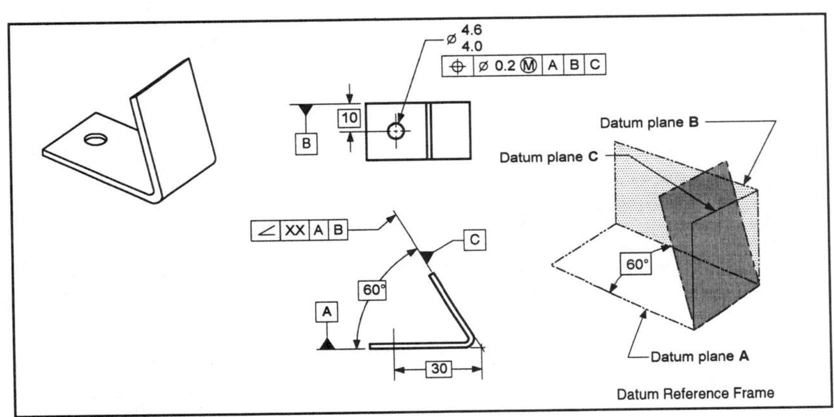

FIGURE 5-11 Inclined Datum Features

Multiple Datum Reference Frames

In certain cases, the functional requirements of a part call for the part to contain more than one datum reference frame. A part may have as many datum reference frames as needed to define its functional relationships. The datum reference frames may be at right angles or at angles other than 90°. Also, a datum plane may be used in more than one datum reference frame. Figure 5-12 shows a part with three datum reference frames.

Author's Comment
On complex parts, it is common to have multiple datum reference frames. I have seen drawings with as many as thirty datum reference frames.

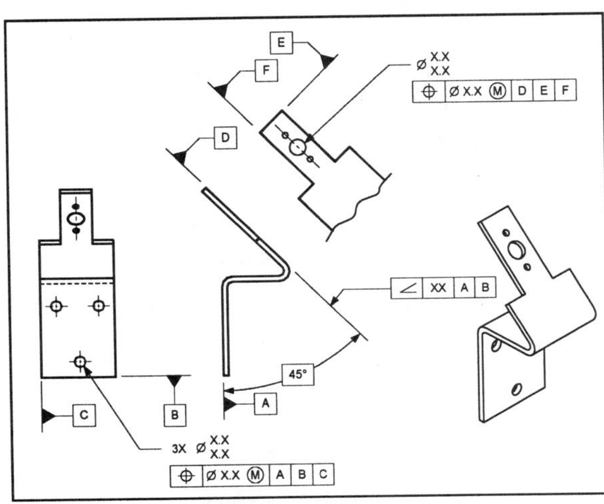

FIGURE 5-12 Datum Reference Frames

For more info. . .
The profile control is explained in Chapter 12.

Coplanar Datum Features

Coplanar surfaces are two or more surfaces that are on the same plane. ***Coplanar datum features*** are two or more datum features that are on the same plane. A single datum plane can be established from multiple surfaces. In this case, a datum feature symbol is attached to a profile control. The profile control limits the flatness and coplanarity of the surfaces. The note following this profile control—"two surfaces"—also denotes that datum feature *A* is comprised of two surfaces. Figure 5-13 illustrates coplanar datum features.

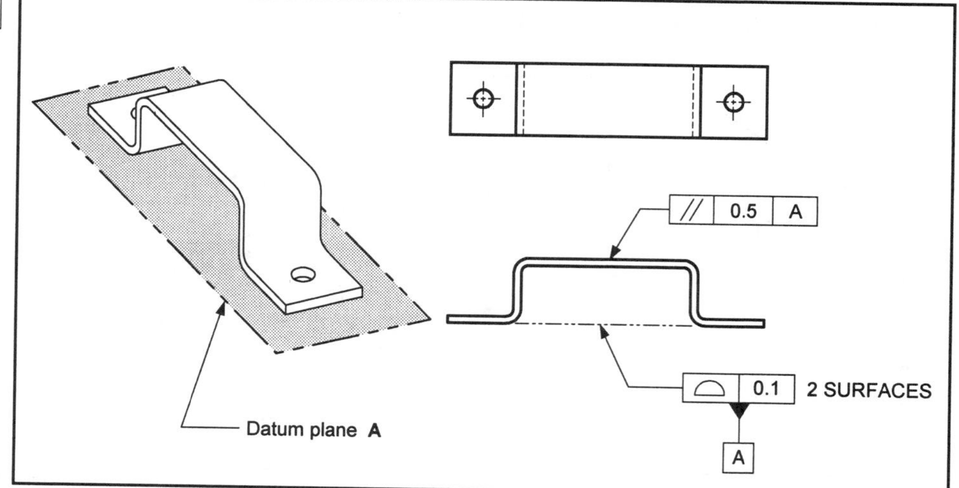

FIGURE 5-13 Coplanar Datum Features

For more info. . .
See Paragraph 4.5 of Y14.5.

DATUM TARGETS

General Information on Datum Targets

Datum targets are symbols that describe the shape, size, and location of gage elements that are used to establish datum planes or axes. Datum targets are shown on the part surfaces on a drawing, but they actually do not exist on a part. Datum targets describe gage elements. The gage elements only contact a portion of the part surface. Datum targets can be specified to simulate a point, line or area contact on a part. The use of datum targets allows a stable and repeatable relationship for a part with its gage.

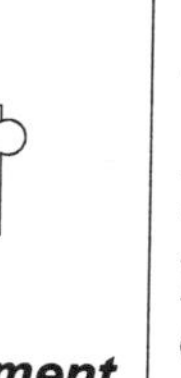

Author's Comment
Datum targets may also be used to establish a datum axis or centerplane.

Datum targets should be specified on parts where it is not practical (or possible) to use an entire surface as a datum feature. A few examples are castings, forgings, irregularly shaped parts, plastic parts, and weldments. These types of parts often do not have a planar datum feature, or the datum feature is likely to be warped or bowed; this results in an unsuitable contact with a full datum plane. Often the part will rock, wobble, or not rest in the same position on a full datum plane.

TECHNOTE 5-6 Datum Target Definition

Datum targets are symbols that describe the shape, size and location of gage elements that are used to establish datum planes, axes, and points.

Datum Target Symbols

A datum target application uses two types of symbols: a datum target identification symbol and symbols that denote which type of gage elements are to be used. The datum target identification symbol is shown in Figure 5-14*A*. The symbol is divided into two parts with a horizontal line. The bottom half denotes the datum letter and the target number associated with that datum; the top half contains gage element size information when applicable. The leader line from the symbol specifies whether the datum target exists on the surface shown or on the hidden surface side of the part. If the leader line is solid, the datum target exists on the surface shown (datum targets *B1* and *B2* in Figure 5-14). If the leader line is dashed, it denotes that the datum target exists on the hidden surface of the part (datum targets *A1*, *A2*, and *A3* in Figure 5-14). When a datum target symbol is used on a drawing, it is often accompanied by the datum feature symbol.

Author's Comment
Using datum targets often improves measurement repeatability and does not increase part cost.

Three symbols used to denote the type of gage element in a datum target application are the symbols for a target point, a target line, and a target area. A datum target application is shown in Figure 5-14*B*.

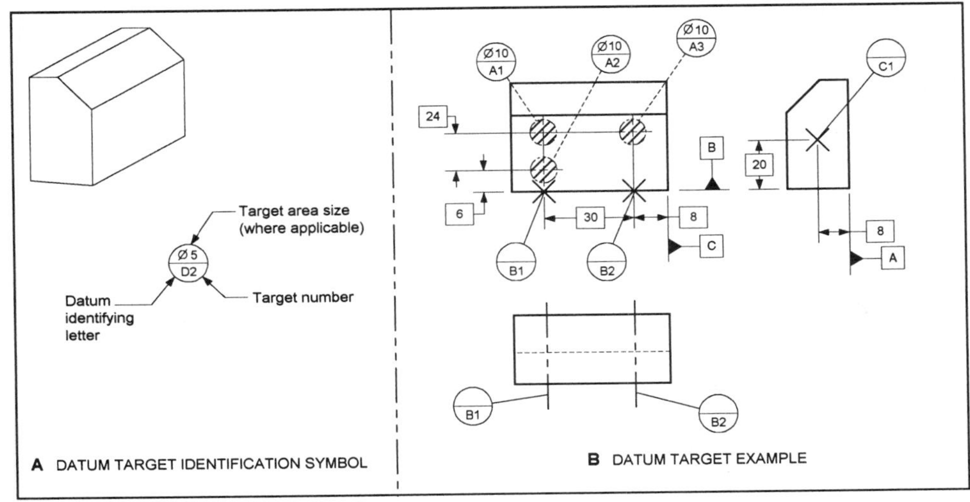

FIGURE 5-14 Datum Target Symbols

TECHNOTE 5-7 When to Use Datum Targets

Datum targets should be used whenever...

- It is not practical to use the entire surface as a datum plane.
- The designer suspects the part may rock or wobble when the datum feature contacts the datum plane.
- Only a portion of the feature is used in the function of the part.

Design Tip
When using three datum target points to establish a primary datum plane, the points should be as far apart as possible and should not be in a straight line.

A datum target point is specified by an *X*-shaped symbol, consisting of a pair of lines intersecting at 90°. The symbol is shown and dimensioned on the plan view of the surface to which it is being applied (see Figure 5-15*A*). Where this type of view is not available, the symbol can be shown and dimensioned in two adjacent views (see Figure 5-15*B*). Basic dimensions should be used to locate datum target points relative to each other and the other datums on the part. Basic dimensions are used to describe the location of datum targets, which assures that there will be minimum variation between gages. A datum target point is often simulated in a gage with a spherical-tipped gage pin. The gage pin for a target point is shown in Figure 5-16.

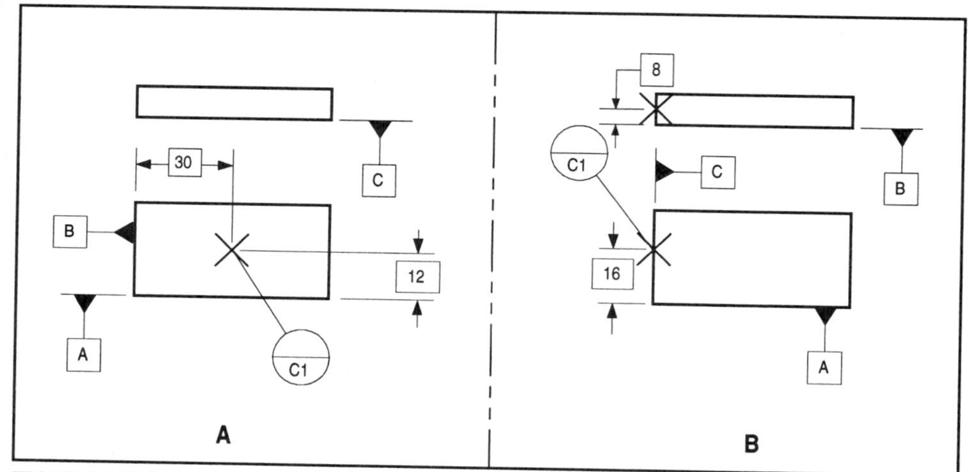

FIGURE 5-15 Datum Target Points

TECHNOTE 5-8 Datum Targets and Basic Dimensions

Basic dimensions should be used to describe the location of datum targets, which assures that there will be minimum variation between gages.

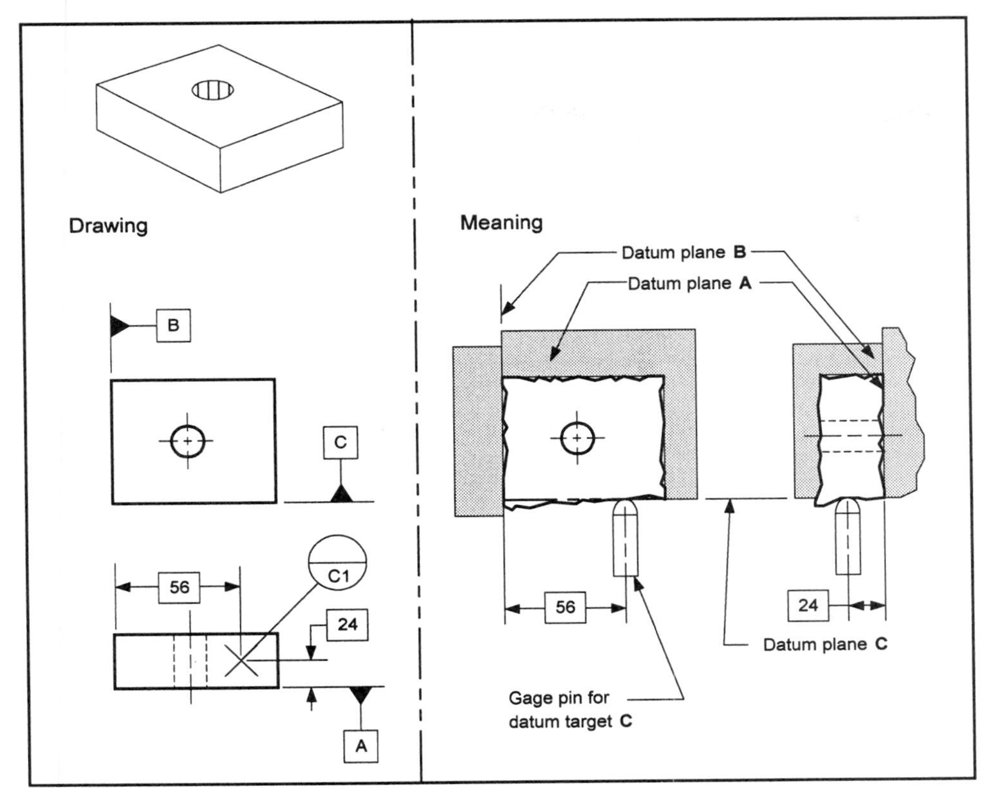

FIGURE 5-16 Datum Target Points

Author's Comment
When basic dimensions are used to define datum targets, gage tolerances apply to the basic dimensions.

Notice that datum plane *C* in Figure 5-16 is across the top of the spherical-tipped pin and is perpendicular to both datum planes *A* and *B*. The actual part surface may be above or below the datum plane.

Figure 5-17 illustrates the three ways to specify a datum target line: a phantom line on the plan view of a surface, an *X* on the edge view of a surface, and a combination of a phantom line and an *X*. Basic dimensions should be used to locate the datum targets relative to other targets and datums.

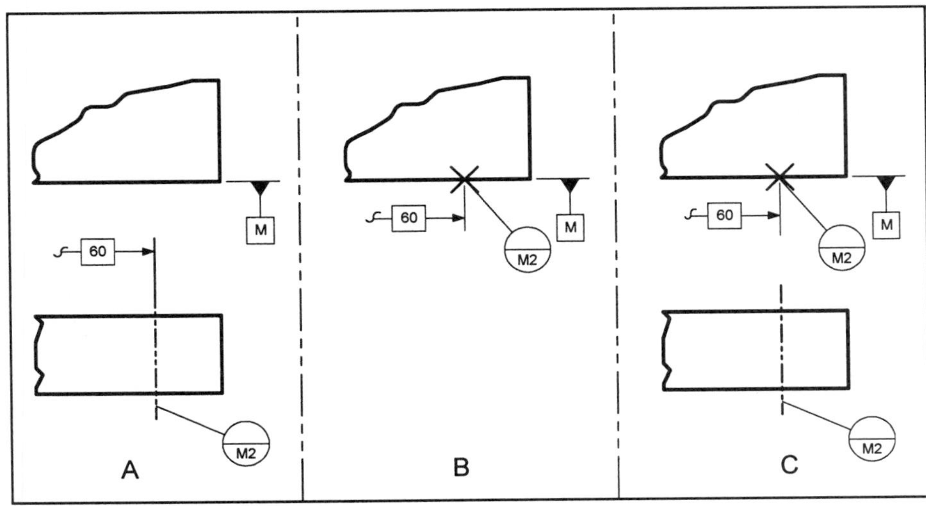

FIGURE 5-17 Symbols for Datum Target Lines

A datum target line is often simulated in a gage by the side of a cylindrical gage pin. Figure 5-18 illustrates a gage for simulating a datum target line.

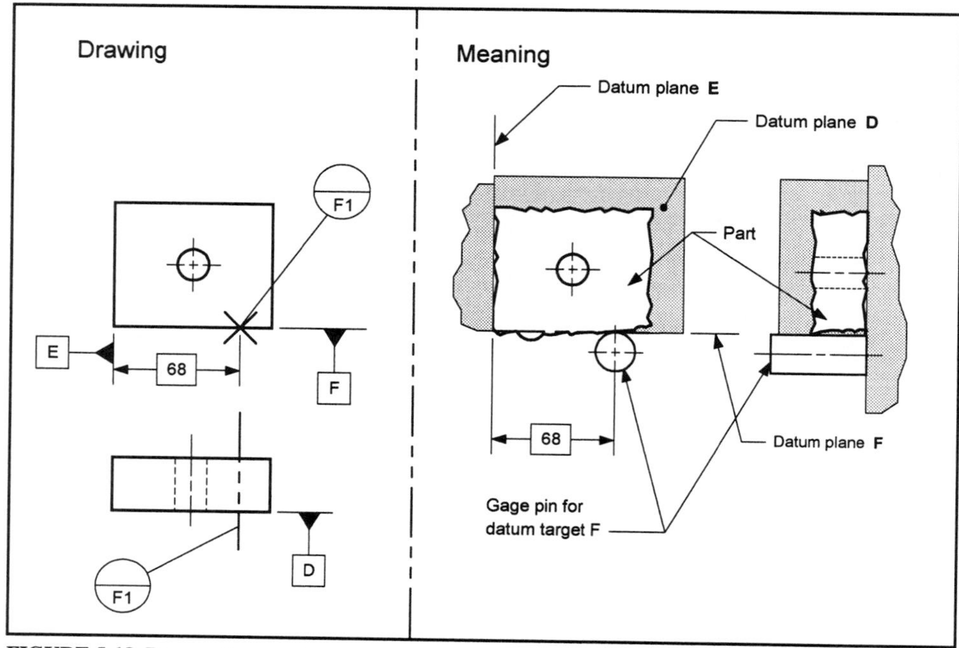

FIGURE 5-18 Datum Target Lines

Where it is preferred to use a gage element that represents an area of contact on a part surface, a datum target area is specified. There are three ways to specify a datum target area. A datum target area is designated by drawing the outline of the gage for simulating the datum target area on the part surface. The outline is shown in phantom lines, and the shape is cross-hatched as shown in Figure 5-19*A*. The shape, size, and location of the area are described with basic dimensions. If the target area is circular, the diameter may be specified in the upper half of the datum target symbol as shown in Figure 5-19*B*.

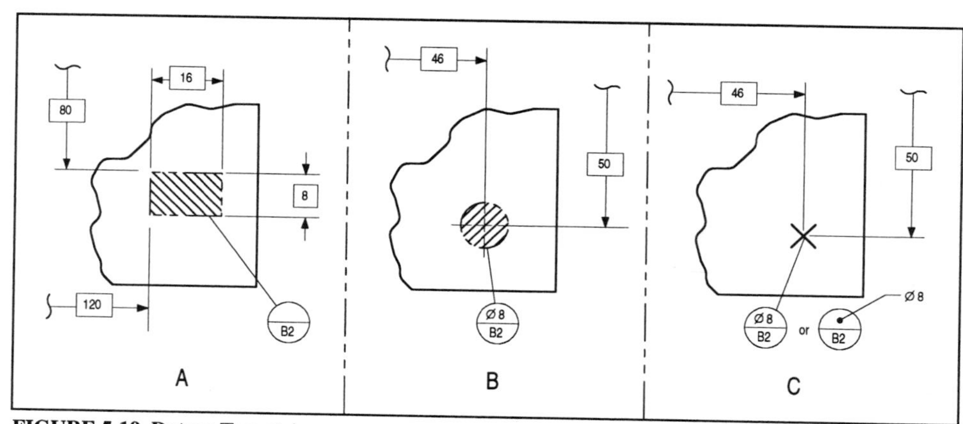

FIGURE 5-19 Datum Target Areas

Where it is impractical to show the circular target area, the method in Figure 5-19*C* may be used. Figure 5-20 shows an application with datum targets. Note that flat-tipped gage pins are used to simulate the area datum targets specified on the drawing. When the leader line between the datum target symbol and the datum target is dashed (like datum target *A1, A2,* and *A3*), the gage pin contacts the back surface of the part in the view shown.

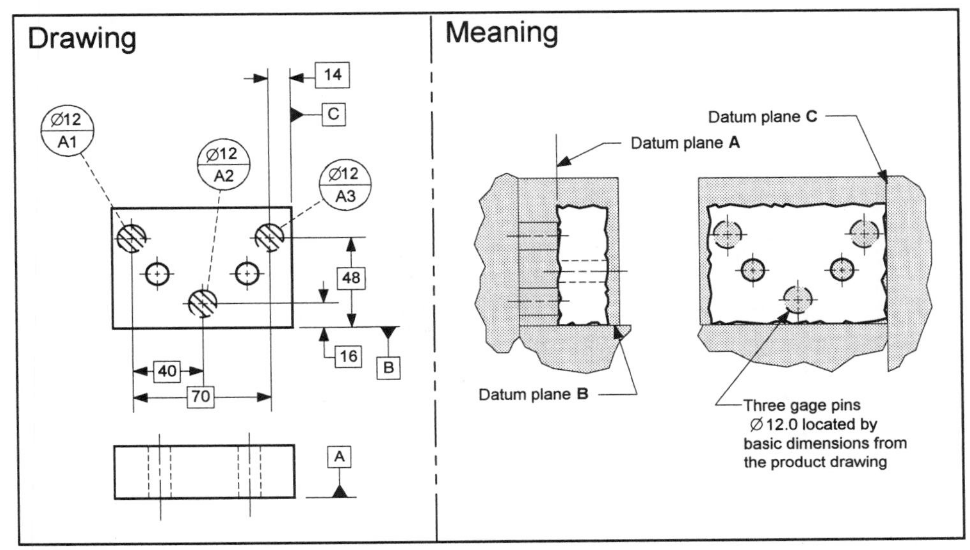

FIGURE 5-20 Datum Target Applications

Datum Target Applications

Although there are only three datum target symbols, they can be used in a variety of ways and on a number of different types of parts. This section illustrates and explains three different datum target applications:

1. Creating a partial reference frame from partial surfaces
2. Creating a partial datum reference frame from offset surfaces
3. Creating a complete datum reference frame from irregular surfaces

When using datum targets to establish a complete datum reference frame, three requirements should be met:

1. Basic dimensions should be used to define and locate the datum targets.
2. The datum reference frame must restrain the part in all six degrees of freedom.
3. The part dimensioning must ensure that the part will rest in the gage in only one orientation and location.

For more info. . .
See Paragraph 4.6.4.1 of Y14.5.

In certain applications, it may be desired to create a datum reference frame by using a partial surface of the datum feature. Figure 5-21 shows an application that uses datum targets to define gage elements that contact partial surfaces of the datum feature. In Figure 5-21, datum targets *A1* and *B1* designate gage elements that contain a portion of the surfaces they contact. When a datum target area (or line) is shown on a cylindrical surface like datum target *B1*, it wraps around the diameter of the part.

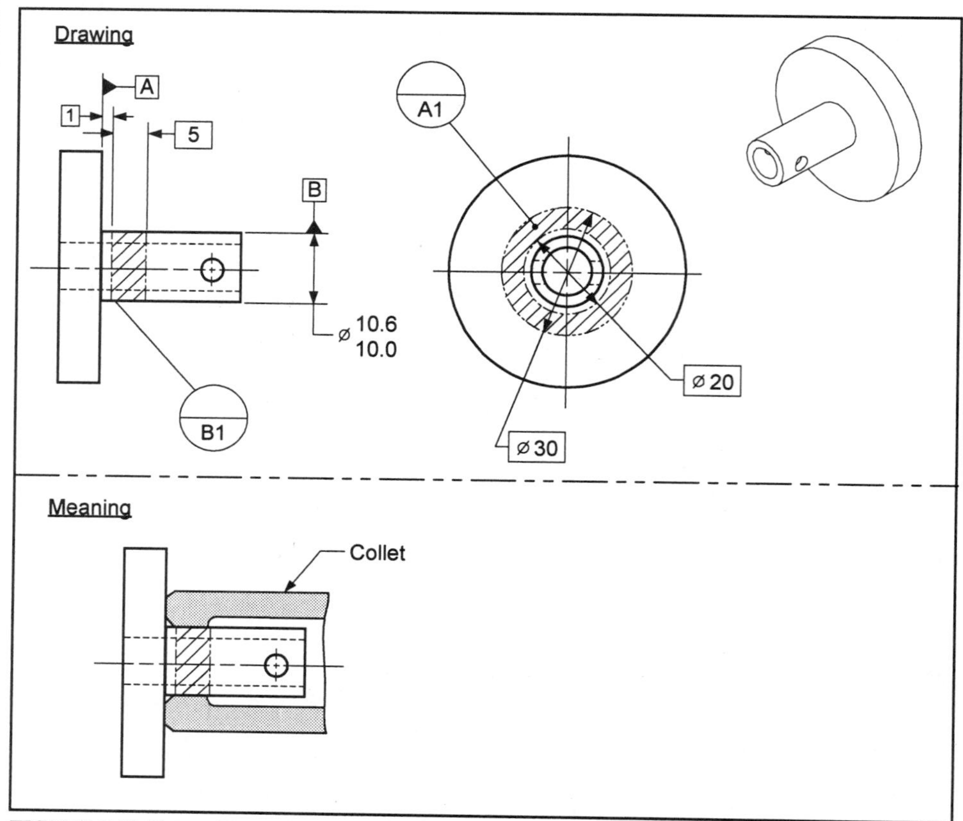

FIGURE 5-21 Datum Target Area Application

In other applications, you may want to create a datum reference frame by using points (or lines, or areas) from offset surfaces of a part. Figure 5-22 shows an application that uses datum targets. Gage elements that contact the offset surfaces are used to establish the datum reference frame. In Figure 5-22, datum target *A1* is on one surface, and datum targets *A2* and *A3* are on a second surface.

With datum targets on offset surfaces, a basic dimension is used to define the offset of the datum targets (gage pins). A profile tolerance may be used to define the offset tolerance for the part surfaces. A datum feature symbol may be attached to either of the surfaces where the datum targets are shown.

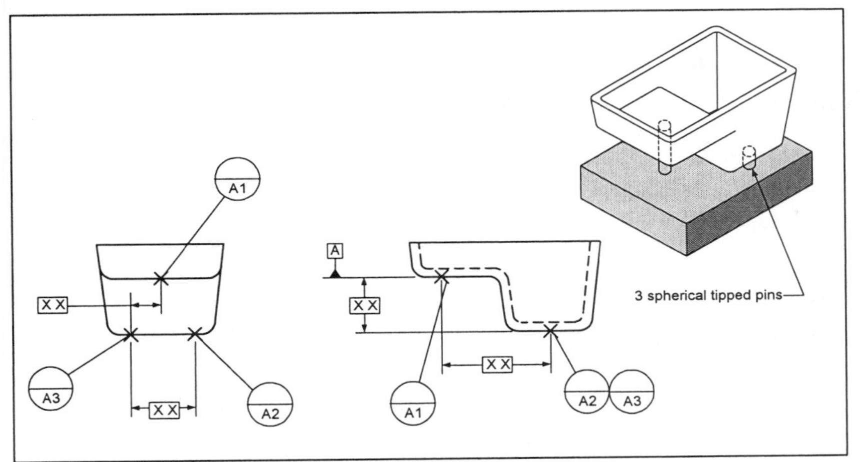

FIGURE 5-22 Datum Target Point Application

Design Tip

A secondary and tertiary datum are required when the primary datum is established with datum targets.

When datum target points, lines, or areas are used to establish a primary datum plane, it is a good practice to specify them as far apart as possible. This will improve the stability of the part on the gage.

In some applications, you may wish to create a datum reference frame on a part with irregular surfaces. Figure 5-23 shows an application that uses datum targets. Gage elements that contact the points on the irregular surfaces are used to establish the datum reference frame. In Figure 5-23, datum target *A1* is an area target, and it establishes datum plane *A*. Datum targets *B1* and *B2* are line targets, and they establish datum plane *B*. Datum target *C1* is a line target, and it establishes datum plane *C*. Note the use of basic dimensions to define the size and location of the datum targets. A profile tolerance may be used to define the tolerance for the irregular part surfaces. A datum feature symbol may be attached to either of the planes where the datum targets are shown.

For more info. . .

See Paragraph 4.6.3.1 of Y14.5.

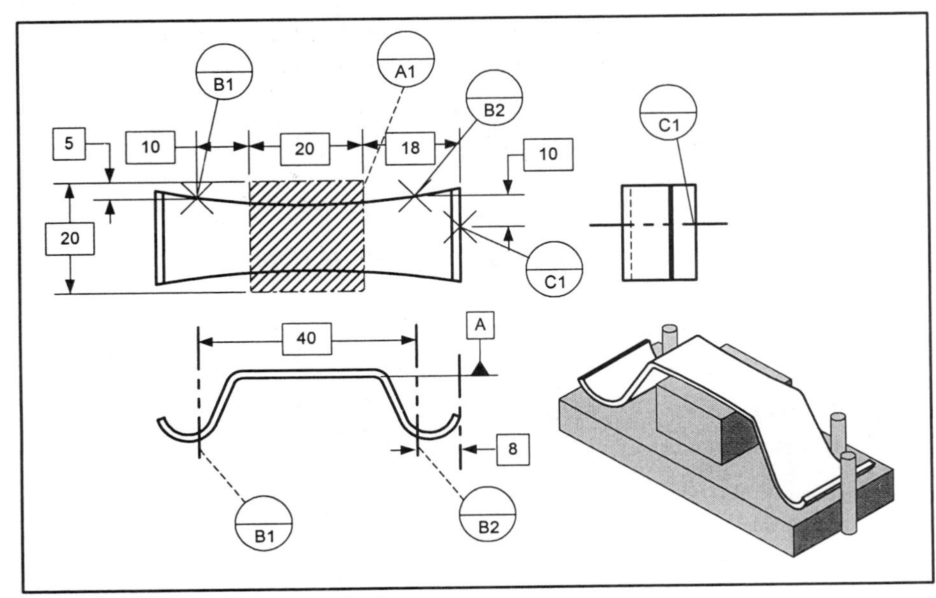

FIGURE 5-23 Datum Target Application

Study Tip
Read each term. If you don't recall its meaning, look it up in the chapter.

VOCABULARY LIST

New Terms Introduced in this Chapter

Coplanar datum features
Coplanar surfaces
Datum
Datum feature
Datum feature simulator
Datum reference frame
Datum system
Datum target
Implied datum
Inclined datum features
Planar datum
Primary datum
Secondary datum
Simulated datum
Tertiary datum
3-2-1 Rule
True geometric counterpart

ADDITIONAL RELATED TOPICS

Author's Comment
These topics, plus advanced coverage of many of the topics introduced in this text, will be covered in my new book on advanced GD&T concepts.

Topic	ASME Y14.5M-1994 Reference
• Temporary and permanent datum features	Paragraph 4.3.1
• Partial surface as datum features	Paragraph 4.5.10
• Mathematically defined surfaces	Paragraph 4.5.10.1
• Datums established from complex or irregular surfaces	Paragraph 4.6.7

QUESTIONS AND PROBLEMS

1. Describe the datum system.

2. List three benefits of the datum system.

3. Describe what an implied datum is.

4. List two shortcomings of implied datums.

5. List two consequences of implied datums.

6. Define a datum.

7. Define a datum feature.

8. Describe the true geometric counterpart of a planar datum feature.

9. Draw the datum feature symbol.

10. Describe two ways to specify a planar datum feature.

11. Describe the basis for choosing datum features.

12. Define the term, "datum reference frame."

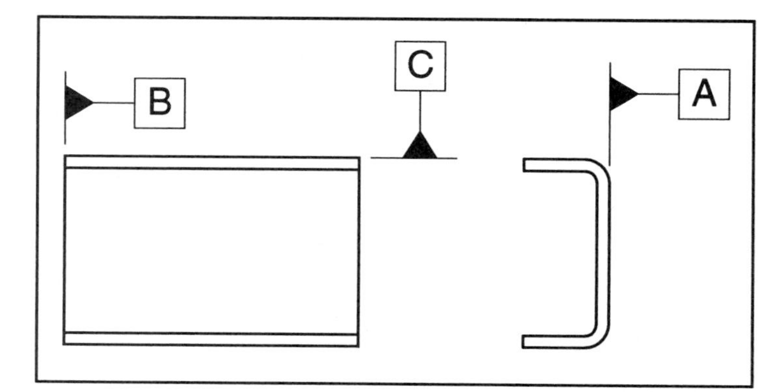

13. On the drawing above, what controls the squareness between datum features *A* & *B*?

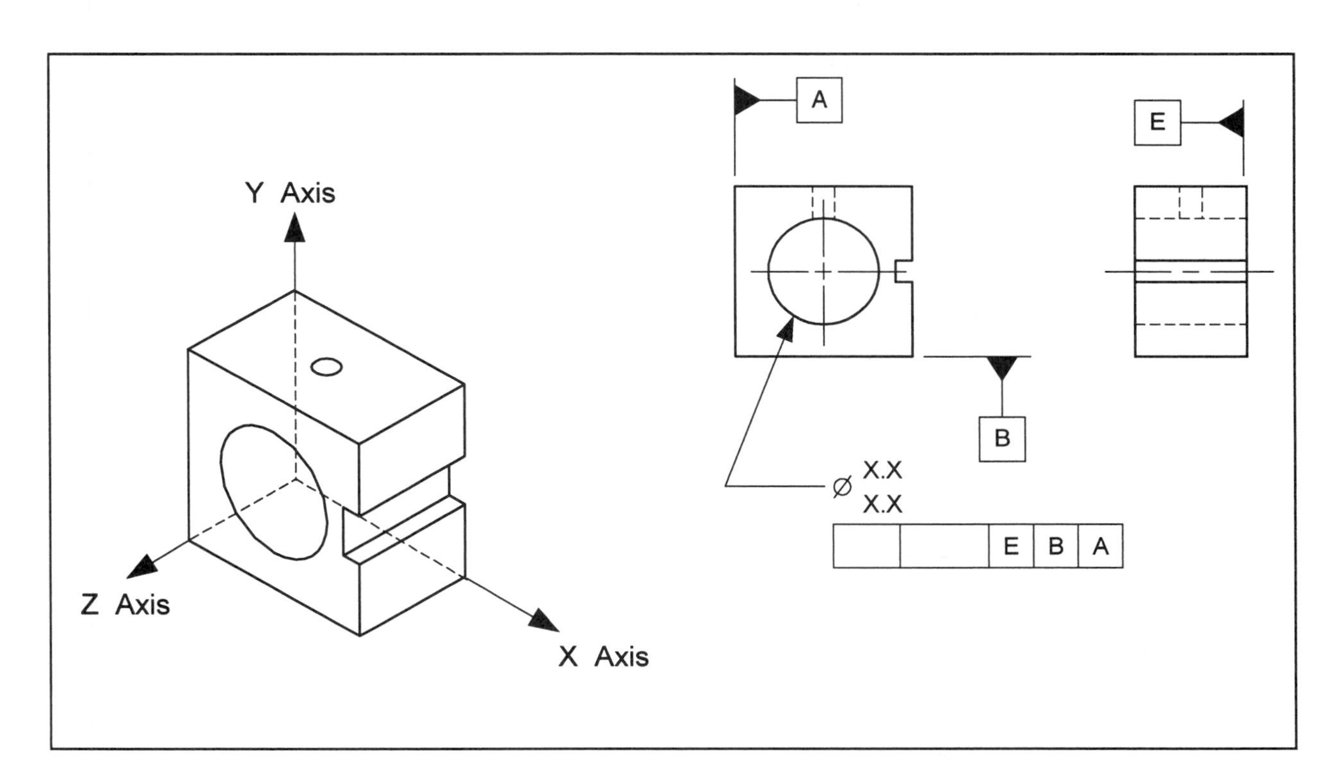

Questions 14-17 refer to the datum sequence and drawing above.

14. List the six degrees of freedom.

______________________________ ______________________________
______________________________ ______________________________
______________________________ ______________________________

15. List the minimum number of points of contact that the part must have with . . . .
    a. Datum plane *E*____________________
    b. Datum plane *B*____________________
    c. Datum plane *A*____________________

16. List the degrees of freedom that contact with datum plane *E* will restrict.

____
____
____

17. List the degrees of freedom that contact with datum plane *B* will restrict (after the part contacts the datum plane *E*).

____
____
____

18. Describe coplanar datum features.____

____

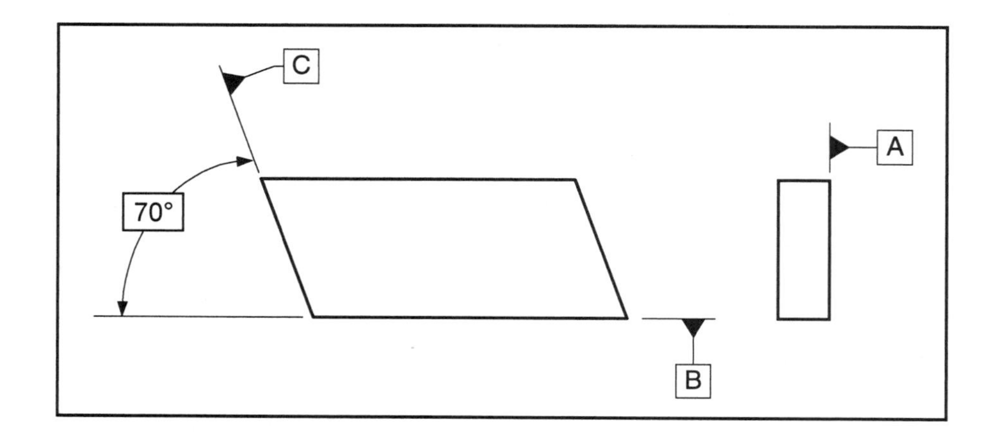

19. Draw the datum reference frame for the part shown above.

20. Explain how a drawing user can determine if a dimension is related to a datum reference frame.

____

____

____

21. Explain the term, "datum target."

____

____

22. List two places where datum targets should be used.

____

____

23. Why should basic dimensions be used to define datum targets?

____

____

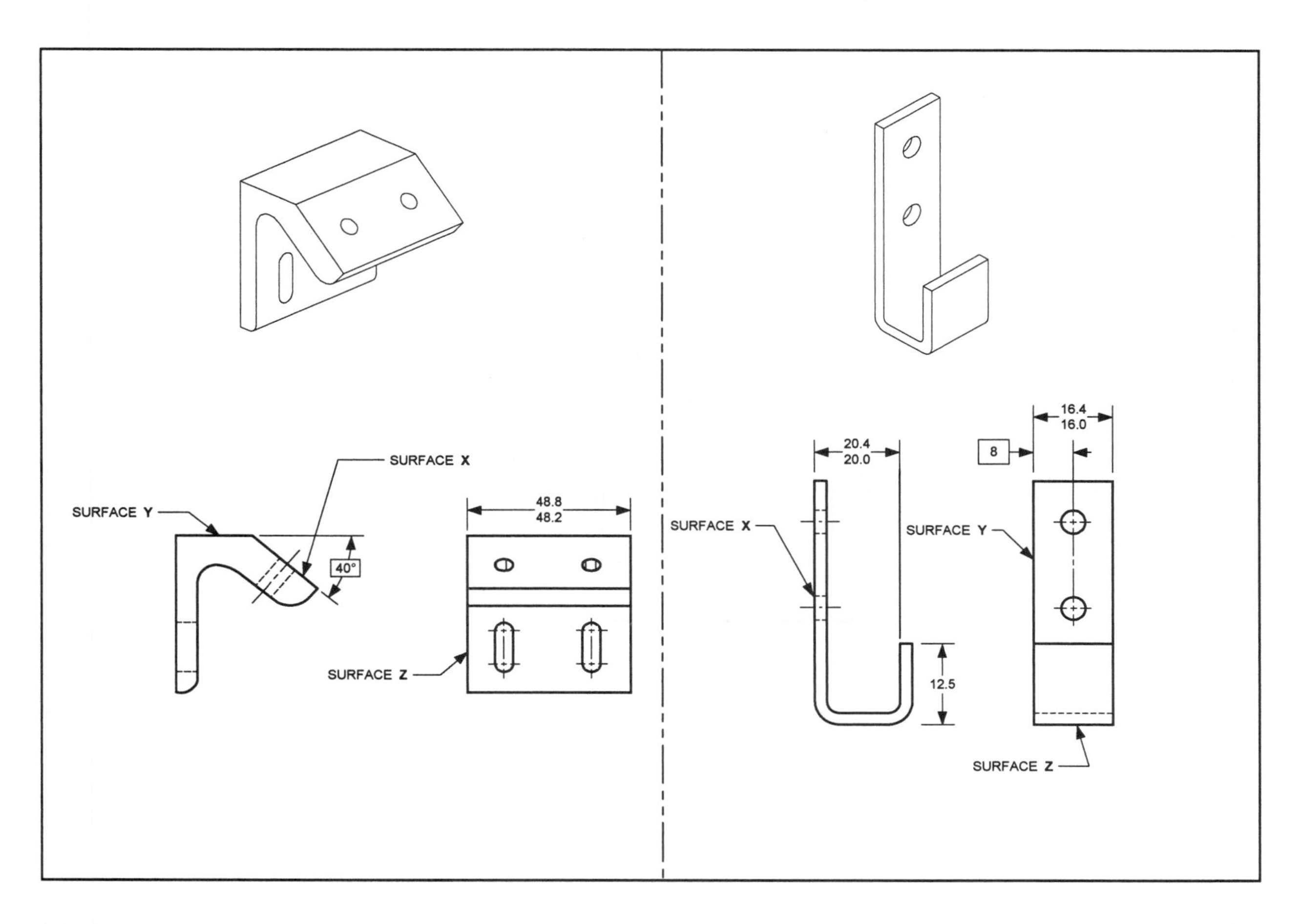

24. On both drawings above, sketch the datum symbols to specify surface *X* as datum feature *B*, surface *Y* as datum feature *C*, surface *Z* as datum feature *D*.

25. List three requirements of datum target applications.
   1. ____
   2. ____
   3. ____

26. (A1) This is called a ______________________ symbol.

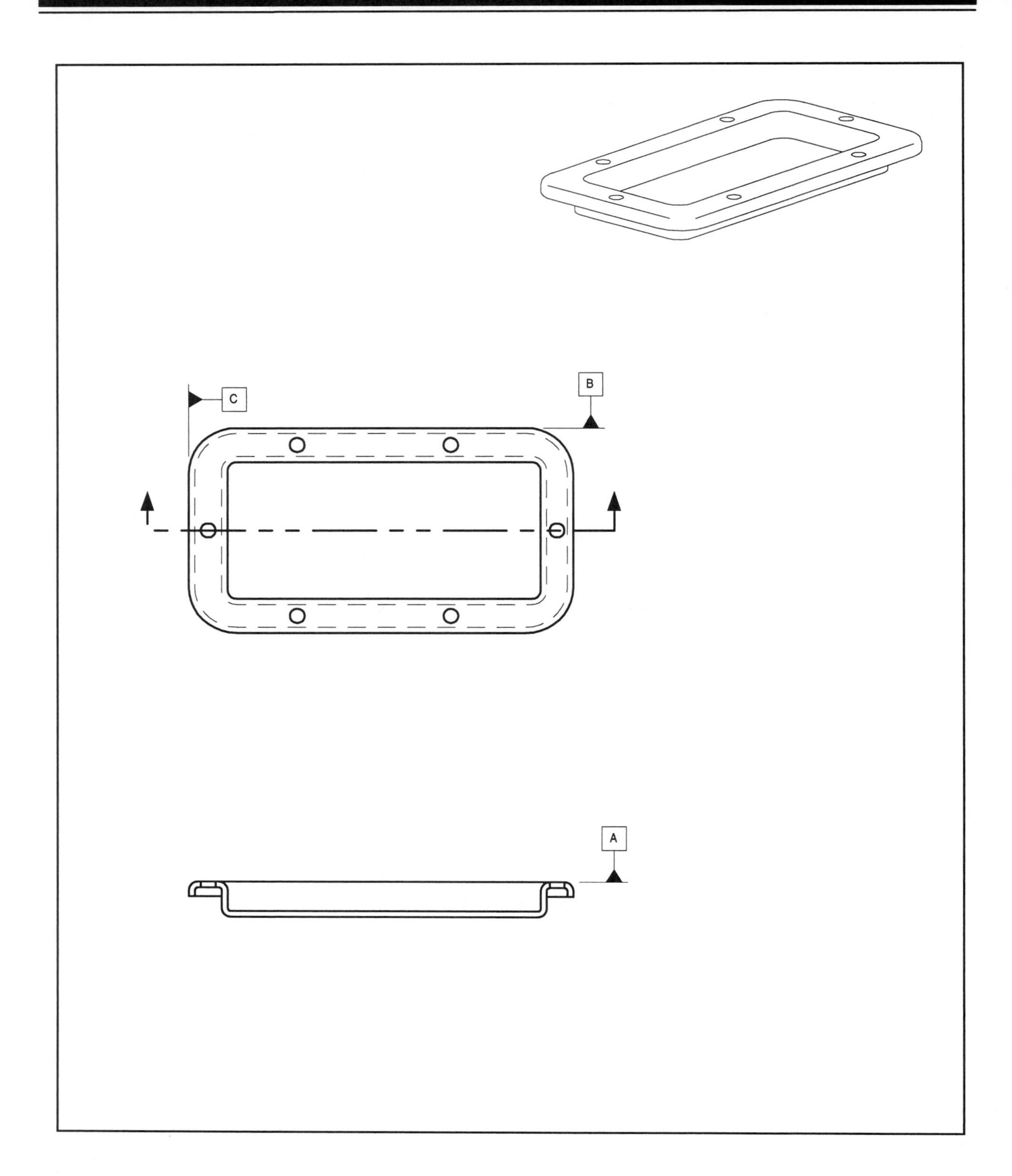
C
B
A

Questions 27 and 28 refer to the drawing on page 140.

27. Draw and dimension the datum target according to the following instructions.
    a. Use three 15 mm. dia. area targets to establish datum plane *A*.
    b. Use two-line datum targets to establish datum plane *B*.
    c. Use a one-line datum target to establish datum plane *C*.
    d. Assign a flatness control to limit the flatness tolerance of datum feature *A* to within 0.1 mm.

28. Are the datum targets used when inspecting the flatness control? Explain why or why not.

    ____
    ____
    ____

29. In the space below, draw the symbol for a datum target point, line, and area.

Point   Line   Area

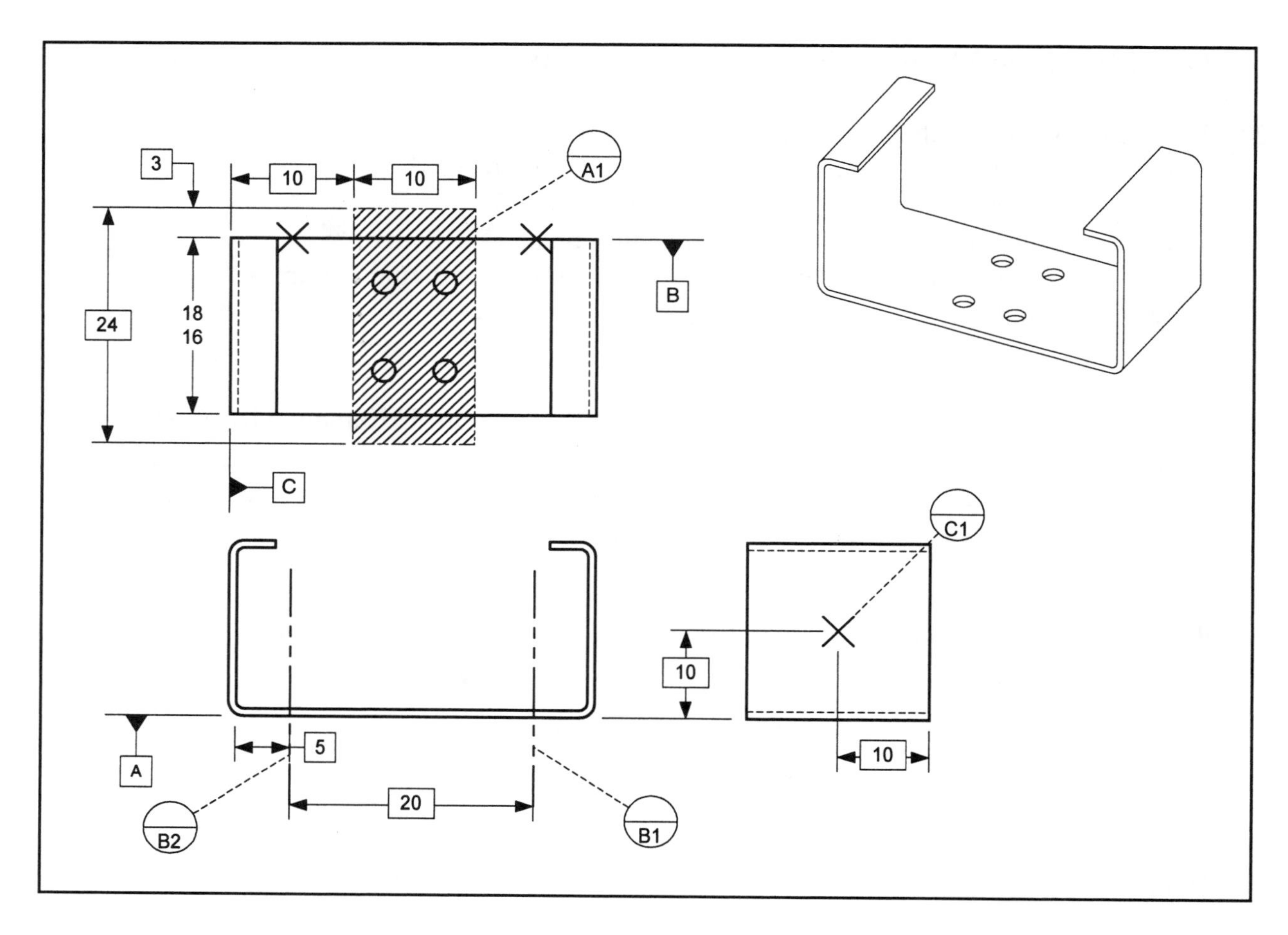

30. Using the figure above, sketch the simulated datums for the datum targets.

Chapter 6

Datums (Axis and Centerplane)

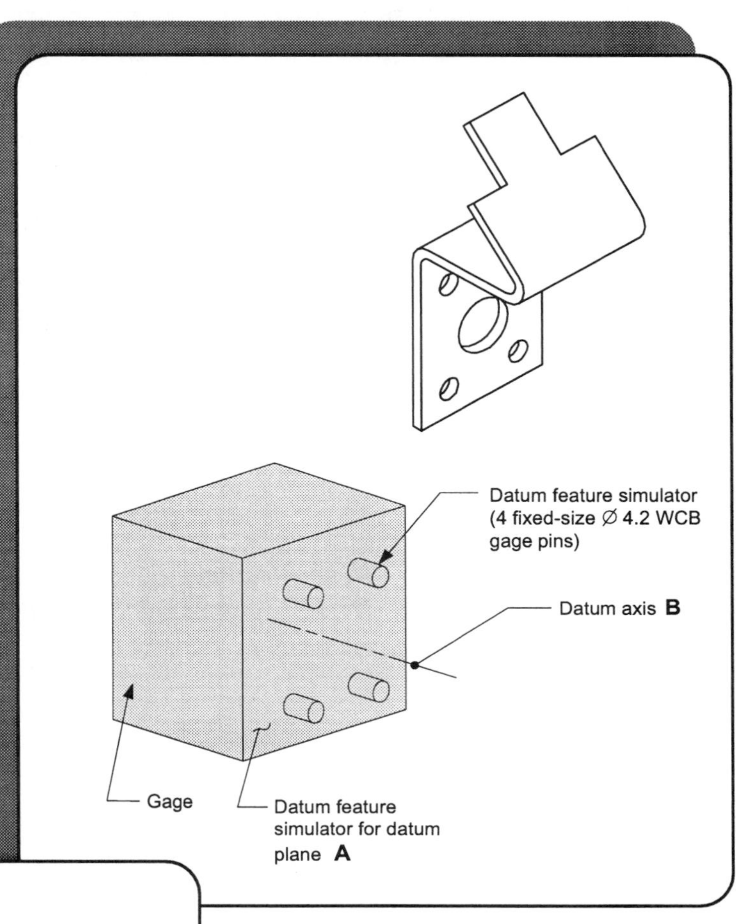

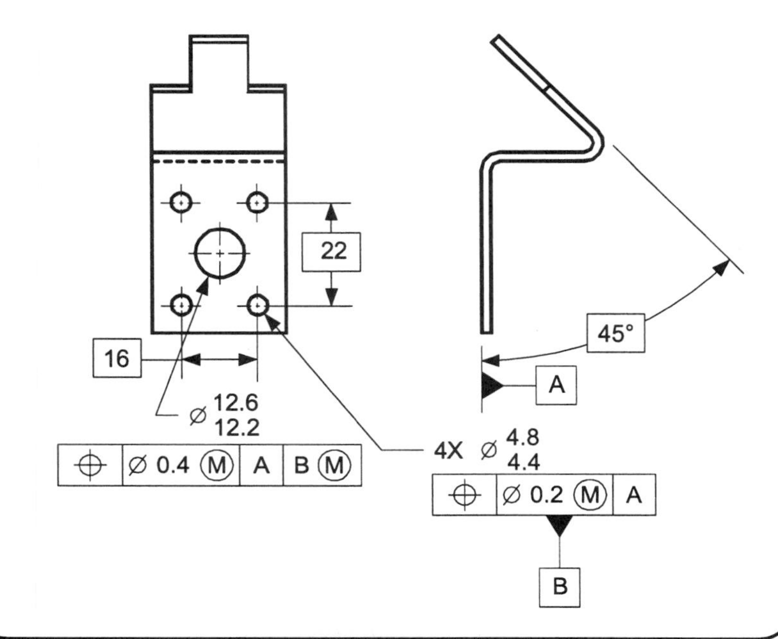

INTRODUCTION

This chapter will help you to read and understand datum reference frames established from feature of size datum features. The information in this chapter is limited to FOS datum features referenced at RFS and MMC.

Author's Comment
This text does not include FOS datums referenced at LMC.

When a FOS is used as a datum feature, it usually results in an axis or a centerplane as the datum. Diametrical and planar features of size may be used as datum features. When a FOS is used as a datum feature, the drawing must indicate the material condition at which the true geometric counterpart applies (MMC or RFS). RFS is the default condition; if MMC is desired, it must be specified. The material condition for a FOS datum feature is specified in the feature control frames that reference the datum.

When a diameter is used as a datum feature, it results in a datum axis. When a planar FOS is used as a datum feature, it results in a datum centerplane.

CHAPTER GOALS AND OBJECTIVES

There are Two Goals in this Chapter:

6-1. Interpret FOS datum specifications (RFS).
6-2. Interpret FOS datum specifications (MMC).

Performance Objectives that Demonstrate Mastery of These Goals

Upon completion of this chapter, each student should be able to:

Study Tip
Take a few minutes to fully understand these objectives. When reading this chapter, look for information to help you master these objectives.

Goal 6-1 (pp. 146-152)

- Describe the datum that results from a FOS datum feature.
- Describe five ways to specify an axis or centerplane as a datum.
- Explain how FOS datum references communicate size condition.
- Draw the datum feature simulator for an external FOS datum feature (RFS primary).
- Draw the datum feature simulator for an internal FOS datum feature (RFS primary).
- Draw the datum feature simulator for an external FOS datum feature (RFS secondary).
- Draw the datum feature simulator for an internal FOS datum feature (RFS tertiary).
- Describe coaxial datum features.

Goal 6-2 (pp. 153-165)

- List two effects of referencing a FOS datum at MMC.
- Describe a special-case FOS datum.
- List two cases where a datum feature referenced at MMC is simulated at a special-case FOS datum condition.
- Explain the concept of datum shift.
- Recognize when datum shift is permissible.
- Calculate the amount of datum shift possible in a datum application.
- Draw the datum feature simulator for an external FOS datum feature (MMC primary).
- Draw the datum feature simulator for an internal FOS datum feature (MMC primary).
- Draw the datum feature simulator for a FOS datum feature (MMC primary with a virtual condition).
- Draw the datum feature simulator for a FOS datum feature (MMC secondary with a virtual condition).
- Draw the datum axis when using a pattern of features of size as a datum feature (MMC secondary).
- Explain how changing the datum reference sequence in a feature control frame affects the part and gage.

FOS DATUM FEATURES

Specifying an Axis or Centerplane as a Datum

A FOS is specified as a datum feature by associating the datum identification symbol with the FOS. When a FOS is specified as a datum feature, it results in an axis or centerplane as a datum. There are five common ways to specify an axis or centerplane as a datum. They are:

Design Tip
To specify an axis or centerplane as a datum, the datum identification symbol must be associated with a FOS.

1. The datum identification symbol can be touching the surface of a diameter to specify the axis as a datum. (See Figure 6-1*A*.)
2. The datum identification symbol can be touching the beginning of a leader line of a FOS to specify an axis datum. (See Figure 6-1*B*.)
3. The datum identification symbol can be touching a feature control frame to specify an axis or centerplane as a datum. (See Figure 6-1*C*.)
4. The datum identification symbol can be in line with a dimension line and touching the extension line on the opposite side of the dimension line arrowhead of a FOS to specify an axis or a centerplane as a datum. (See Figure 6-1*D*.)
5. The datum identification symbol can replace one side of the dimension line and arrowhead. When the dimension line is placed outside the extension lines of a FOS dimension, it specifies an axis or centerplane as a datum. (See Figure 6-1*E*.)

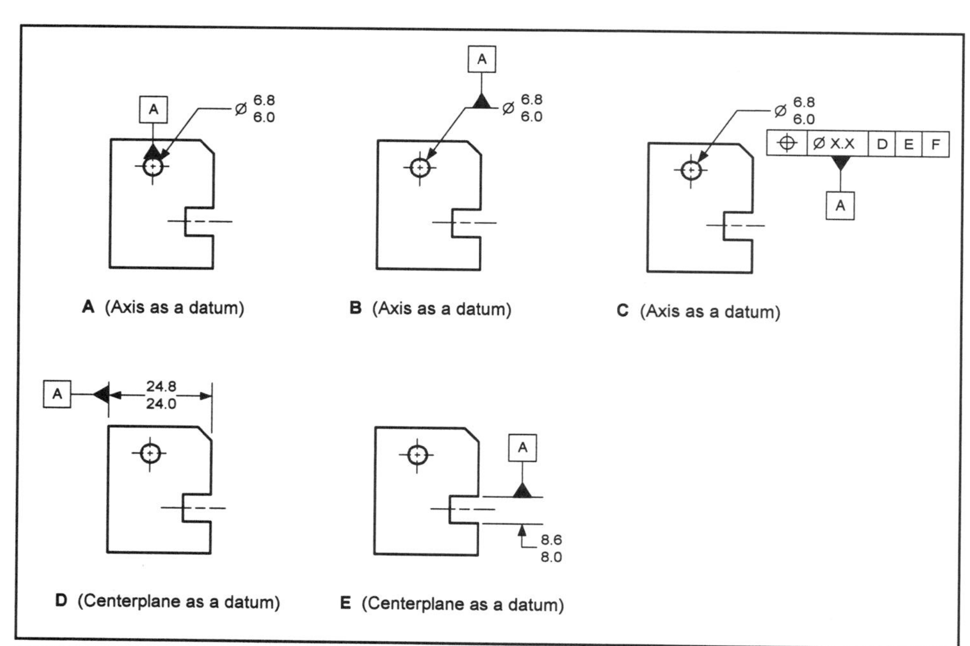

FIGURE 6-1 Specifying an Axis or Centerplane as a Datum

Datum Terminology

When a diameter is designated as a datum feature, the datum axis is derived from placing the part in a datum feature simulator (gage elements). Figure 6-2 shows the terminology used to describe the various elements in a datum axis application. The datum feature is the surface of the part. The gage element that holds the part datum feature is the datum feature simulator; it is considered the true geometric counterpart of the datum feature. The axis of the gage element that holds the part datum feature is the simulated datum axis and is considered to be the datum axis. The datum axis becomes the origin of measurement for dimensions that are related to the datum.

For more info. . .
The terminology used here is defined in Chapter 5, pages 115-16.

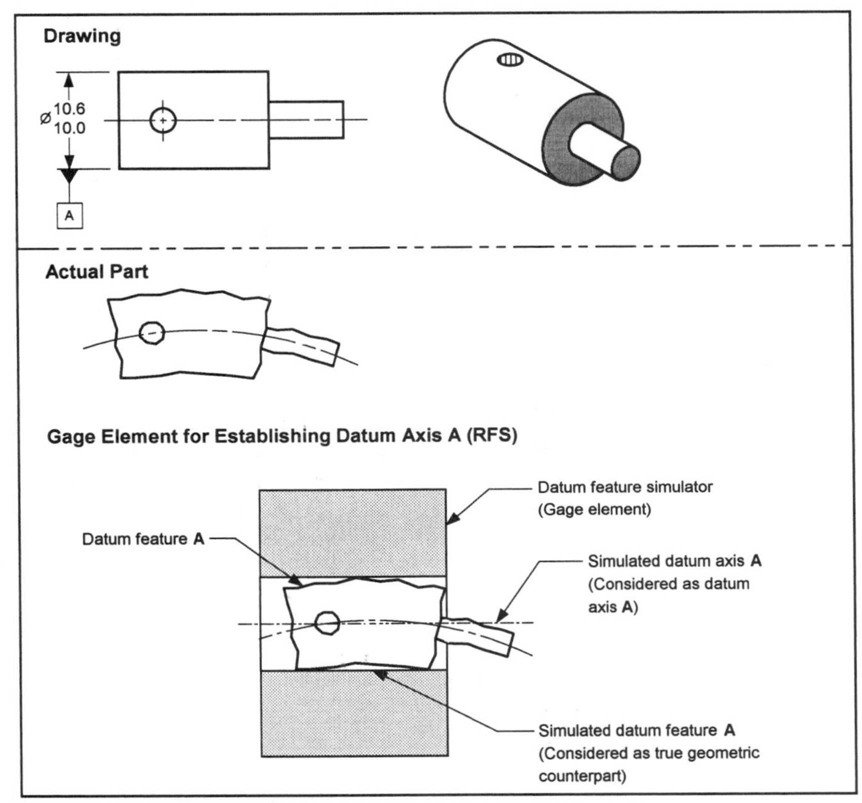

FIGURE 6-2 Datum Terminology

Referencing FOS Datum Features

When using FOS datum features on drawings, the material condition for establishing the datum axis or centerplane must be communicated. This is typically done through feature control frames that reference the FOS datum features. The feature control frame indicates both the material condition (MMC, LMC, or RFS) and the datum sequence (primary, secondary, or tertiary). Whenever a FOS datum feature is referenced in a feature control frame, without showing any modifier, it is automatically referenced in the RFS condition. This condition results from Rule #2.

TECHNOTE 6-1 Referencing FOS Datum Features

When referencing FOS datum features, the following items apply:
- The datum sequence must be specified.
- The material condition (MMC or LMC) must be specified.
- If no material condition is specified, RFS is the default condition.

FOS DATUM FEATURE APPLICATIONS (RFS)

Design Tip
When a diameter (or width) is used to establish a datum axis (or centerplane) RFS primary, its length must be sufficient to establish a repeatable relationship (orientation) with the gage. For example, a hole in a stamping (like the one shown in Figure 5-12) would not have sufficient length to be used as a primary datum.

Datum selection and referencing are directly related to the function of the part. Usually the features that orient and locate the part in its assembly become the datum features for the part.

Datum Axis RFS Primary

Where a diameter is designated as a datum feature and referenced in a feature control frame at RFS, a datum axis is established. The datum axis is established through physical contact between the inspection equipment and the datum feature. A datum feature simulator surrounds (or fills) the diameter. Devices that are adjustable in size—such as a precision chuck, collet, or centering device—are used as the datum feature simulator (true geometric counterpart). The datum feature simulator holds the part securely on the datum feature. The axis of the datum feature simulator becomes the datum axis and establishes the orientation of the part. Figures 6-3*A* and 6-3*B* show examples of an external and internal diameter used to establish a primary datum axis RFS.

For more info. . .
See Y14.5 Paragraph 4.5.3.

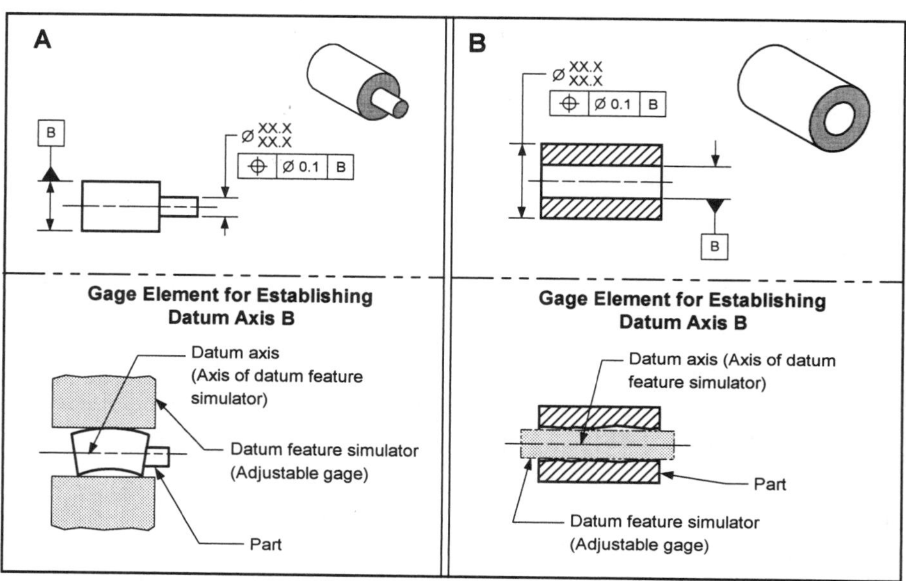

FIGURE 6-3 Primary Datums RFS

Datum Centerplane RFS Primary

When a parallel plane FOS is designated as a datum feature and referenced in a feature control frame at RFS, a datum centerplane is established. The datum centerplane is established through physical contact between the inspection equipment and the datum feature. A datum feature simulator surrounds (or fills) the FOS. Devices that are adjustable in size, such as contracting or expanding parallel plates, are used as the datum feature simulator (true geometric counterpart). The datum feature simulator holds the part securely on the datum feature. The centerplane of the datum feature simulator becomes the datum centerplane and establishes the orientation of the part. Figures 6-4*A* and 6-4*B* show examples of an external and internal FOS used to establish a primary datum centerplane RFS.

Design Tip
Using a FOS as a primary datum feature at RFS is expensive. RFS datum references should only be specified when the function of the part would benefit.

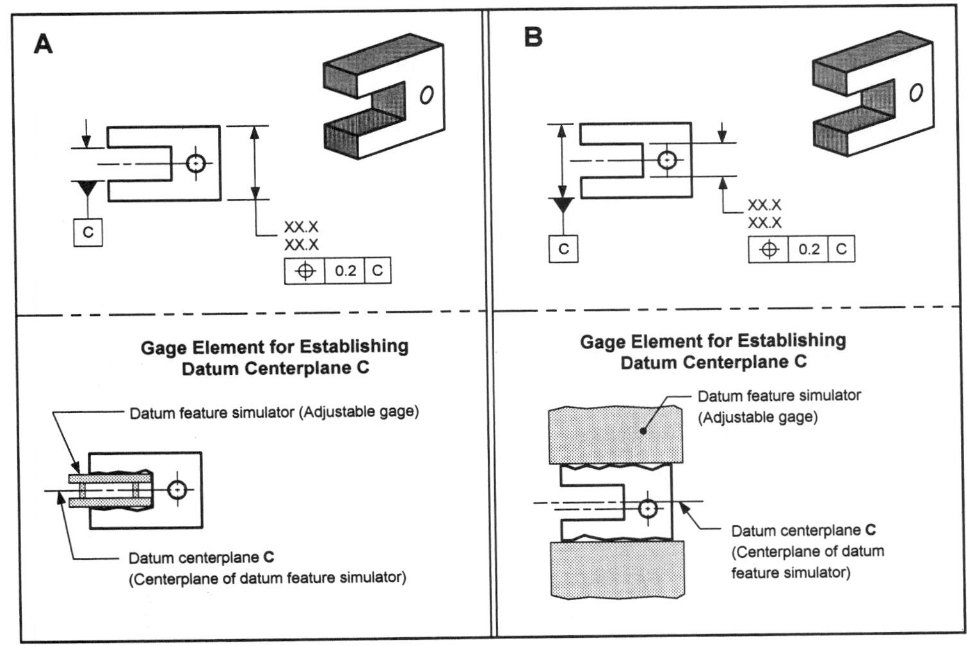

FIGURE 6-4 Primary Datums RFS

TECHNOTE 6-2 Primary Datum RFS

When a FOS is used as a datum feature and referenced as a primary datum RFS, the following applies:

- An adjustable gage element is needed to simulate the true geometric counterpart of a datum feature.
- The part is oriented by the gage.
- The part is held securely in the gage.

Datum Axis RFS Secondary

When a part is oriented by a surface and located by a diameter, it is common to have the surface and diameter designated as datum features. Figure 6-5 shows an example. When referencing the datums with the face primary and the diameter secondary (RFS), the following conditions apply:

- The part will have a minimum of three-point contact with the primary datum plane.
- The secondary datum feature simulator will be adjustable in size.
- The datum axis is the axis of the datum feature simulator.
- The datum axis is perpendicular to the primary datum plane.
- A second and third datum plane will be associated with the datum axis.

For more info. . .
See Paragraph 4.5.3C of Y14.5.

If an angular and/or rotational relationship is important, a tertiary datum is necessary.

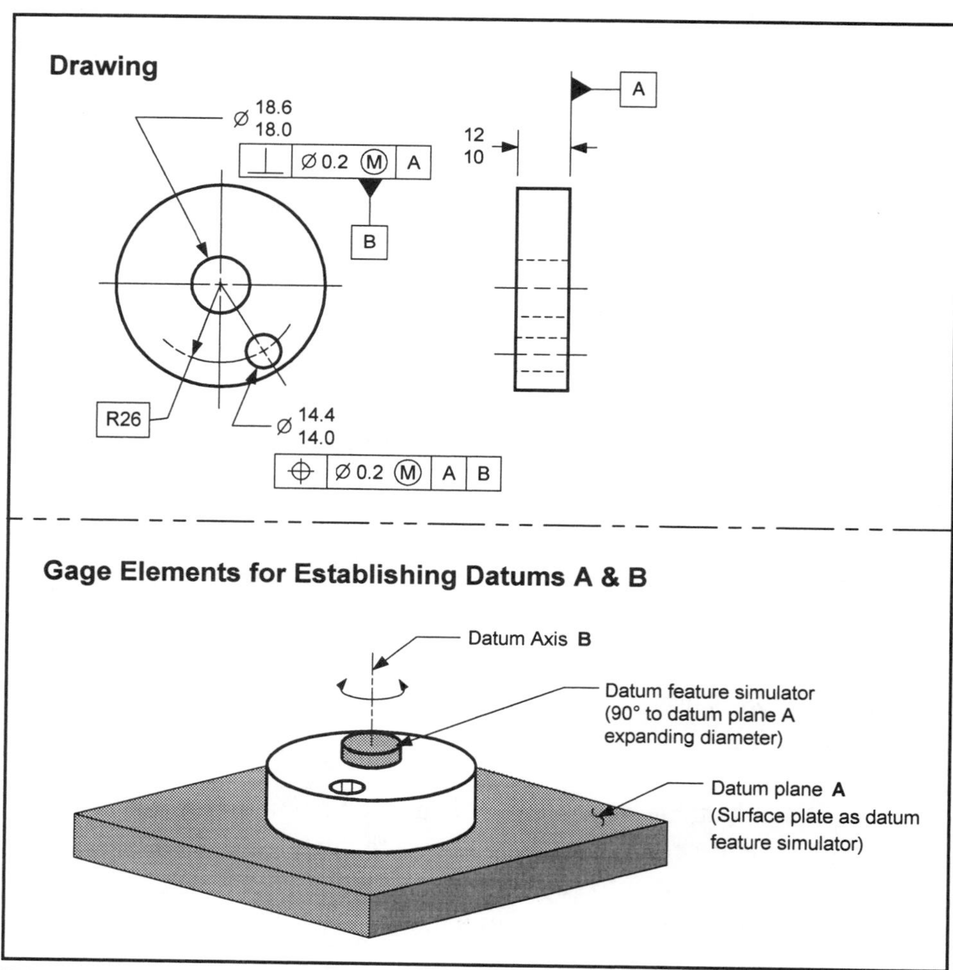

FIGURE 6-5 Datum Axis RFS Secondary

Datum Axis RFS Secondary, Datum Centerplane RFS Tertiary

When a part is oriented by a surface, located by a diameter, and has an angular relationship relative to a FOS, it is common to have the surface, diameter, and FOS designated as datum features. Figure 6-6 shows an example. When referencing the datums with the face primary, diameter secondary (RFS), and slot tertiary (RFS), the following conditions apply:

- The part will have a minimum of three points of contact with the primary datum plane.
- A datum axis perpendicular to the primary datum plane will exist.
- A datum centerplane that will pass through the datum axis and be perpendicular to the primary datum plane will exist.

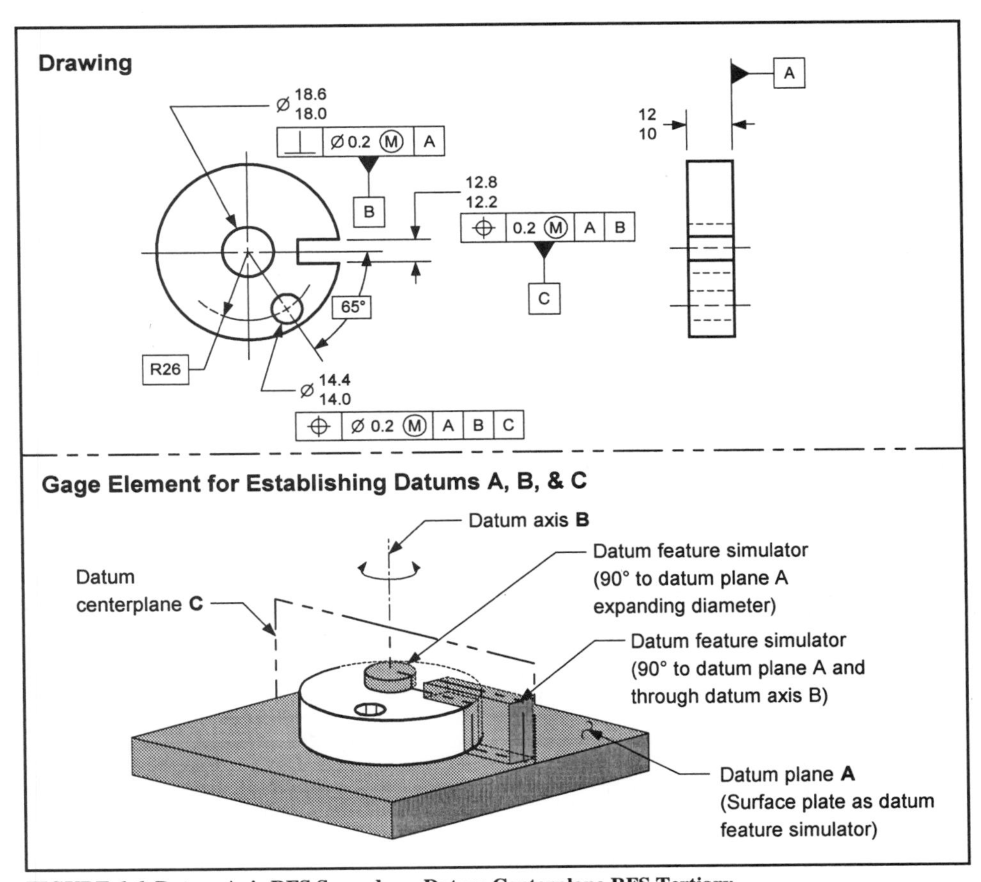

FIGURE 6-6 Datum Axis RFS Secondary, Datum Centerplane RFS Tertiary

Datum Axis from Coaxial Diameters, RFS Primary

For more info. . .
An example of coaxial diameters referenced at MMC is shown in Figure 8-16.

For some parts, it is desired to create a single datum axis from two coaxial diameters. ***Coaxial diameters*** are two (or more) diameters that are shown on the drawing as being on the same centerline (axis). When coaxial diameters are used to establish a datum axis, they are called ***coaxial datum features***.

The method shown in Figure 6-7 may be used to establish a datum axis from coaxial datum features. In this drawing, each coaxial diameter is designated as a datum feature; the datum reference letters (separated by a dash) are then entered in one compartment. The datum axis is established by the two coaxial diameters. The datum axis is simulated by contacting the high points of both datum features simultaneously. Coaxial diameters are often used when both datum features have an equal role in locating the part in its assembly. A single datum axis from two coaxial diameters is normally used as a primary datum.

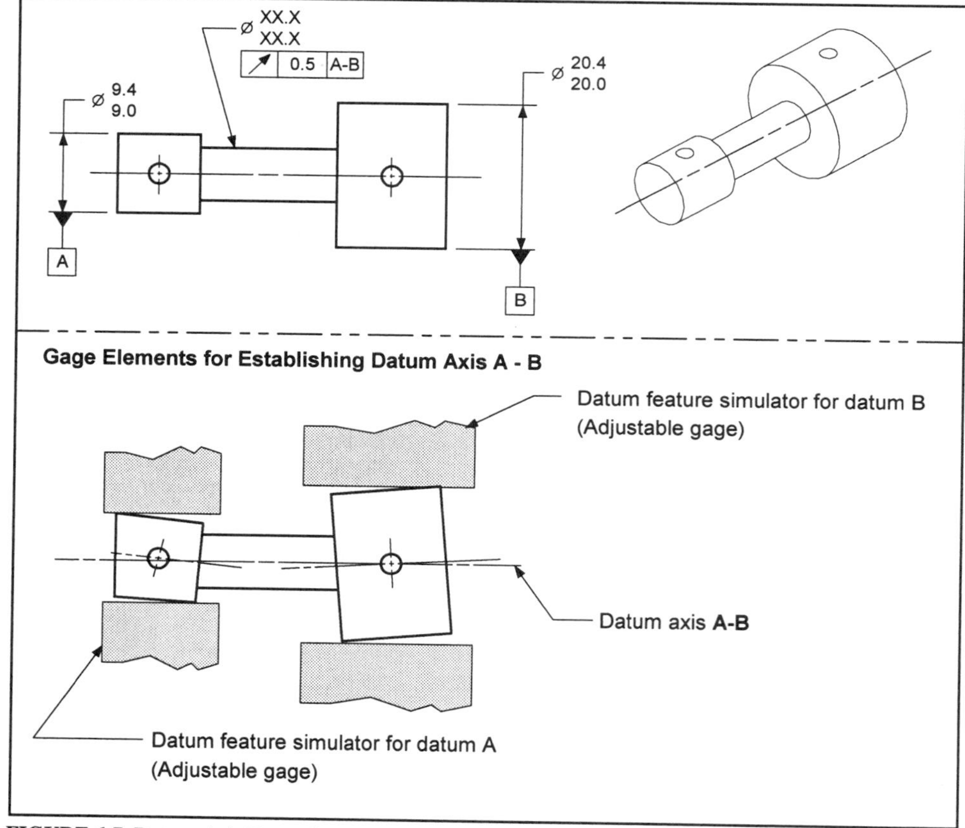

FIGURE 6-7 Datum Axis From Coaxial Diameters RFS Primary

FOS DATUM FEATURE REFERENCED AT MMC

Effects of Specifying the MMC Modifier

Where a FOS datum is referenced at MMC, the gaging equipment that serves as the datum feature simulator is a fixed size. The datum axis or centerplane is the axis or centerplane of the gage element. The size of the true geometric counterpart of the datum feature is determined by the specified MMC limit of size or, in certain cases, its MMC virtual condition.

Referencing a FOS datum at MMC has two effects on the part gaging:

1. The gage is fixed in size.
2. The part may be loose (shift) in the gage.

Figure 6-8 shows an example of a FOS datum feature referenced at MMC. Note that the datum feature simulator (gage element) that represents the true geometric counterpart is a fixed size. It is equal to the MMC of the datum feature.

Design Tip
Using the MMC modifier reduces manufacturing and gaging expense. When the function of part features is assembly, the MMC modifier is often used in the datum references of a geometric control.

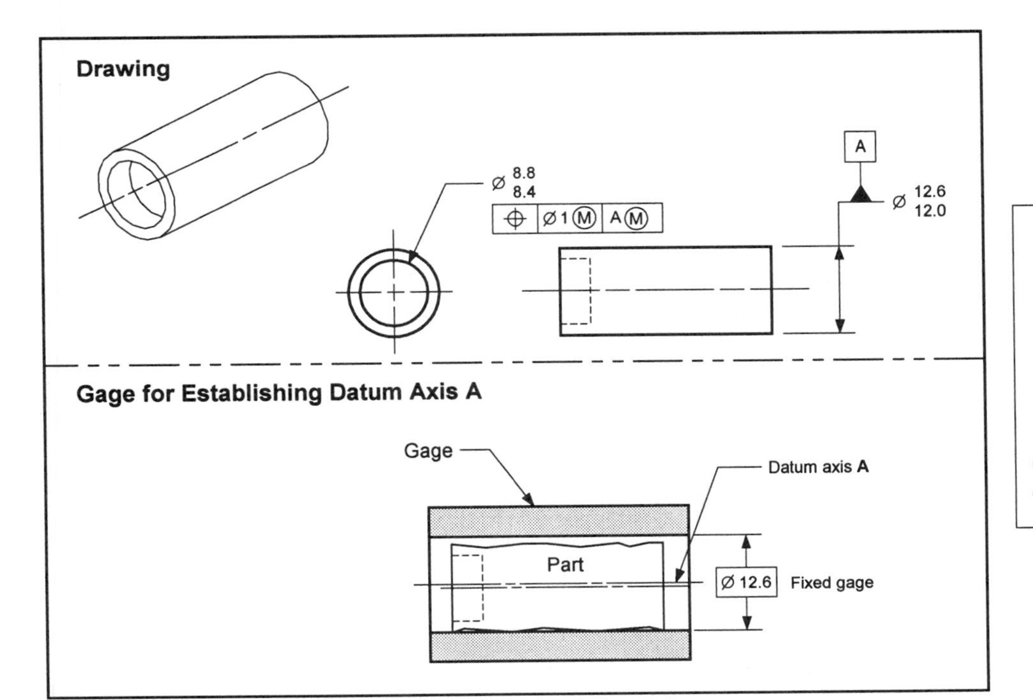

FIGURE 6-8 Effects of Using the MMC Modifier

For more info. . .
See Paragraph 4.5.4 of Y14.5.

A ***special-case FOS datum*** is when a FOS datum feature is referenced at MMC, but simulated in the gage at a boundary other than MMC. There are two cases where special-case FOS datums apply:

1. Where a straightness control is applied to a FOS datum feature
2. Where secondary or tertiary datum features of size in the same datum reference frame are controlled by a location or orientation control with respect to this higher ranking datum

Figure 6-9 shows examples of the two cases for special-case FOS datums.

For more info. . .
See Paragraph 4.5.4.1 of Y14.5.

For more info. . .
See Paragraph 4.5.4.2 of Y14.5.

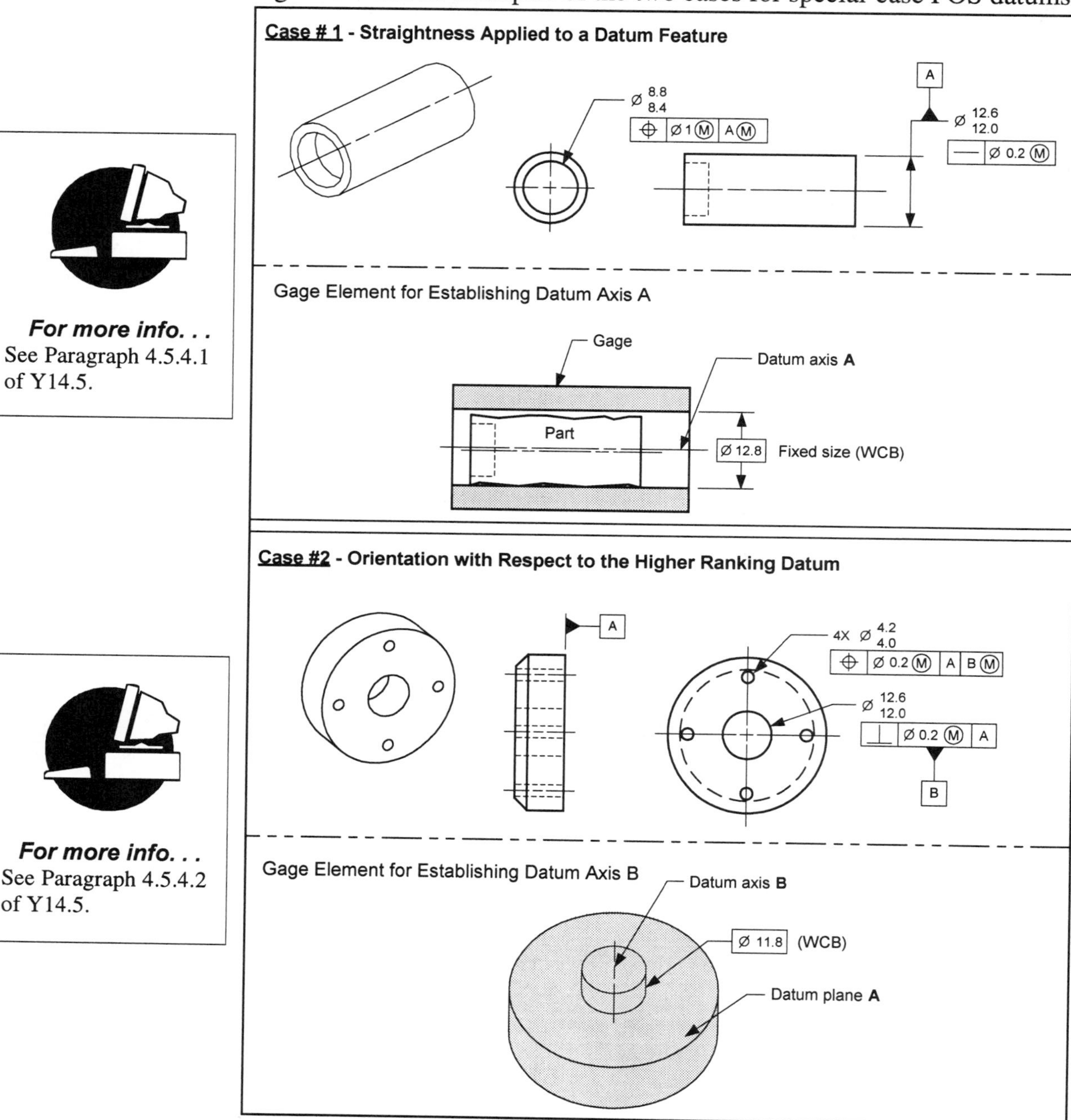

FIGURE 6-9 Special-Case FOS Datum Examples

Datum Shift

Whenever a FOS datum feature is referenced at MMC, the gage element (datum feature simulator) that simulates the perfect feature counterpart is fixed in size. Since the gage is fixed in size, but the part datum feature may vary within its size limits, there may be some looseness between the part and the gage. ***Datum shift*** is the allowable movement, or looseness, between the part datum feature and the gage. Datum shift may result in additional tolerance for the part.

For more info. . .
See Paragraph 4.5.6.3 of Y14.5.

Figure 6-10 shows an example of datum shift. When a part datum feature is at MMC, there is no datum shift allowed. As the actual mating envelope of the part datum feature departs from MMC towards LMC, a datum shift becomes available. The amount of datum shift is equal to the amount the datum feature departs from MMC. The maximum amount of datum shift possible is equal to the difference between the gage size and the LMC of the datum feature. The chart in Figure 6-10 shows the amount of datum shift for various datum feature sizes.

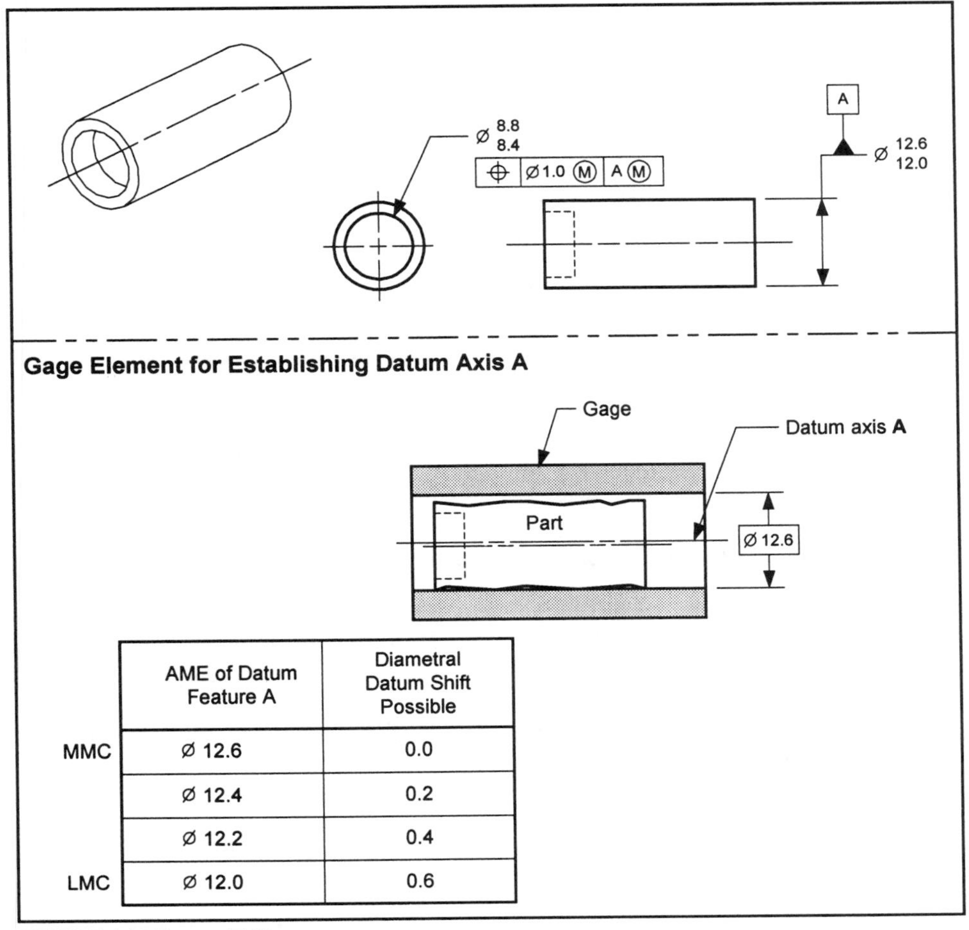

	AME of Datum Feature A	Diametral Datum Shift Possible
MMC	Ø 12.6	0.0
	Ø 12.4	0.2
	Ø 12.2	0.4
LMC	Ø 12.0	0.6

FIGURE 6-10 Datum Shift

Datum shift is similar to bonus. Like bonus, datum shift is an additional tolerance that is available under certain conditions. The actual amount of datum shift for a part can be additive to the stated tolerance in the feature control frame. The chart in Figure 6-11 shows how datum shift can result in additional tolerance for a part feature.

Author's Comment
Datum shift is not available on features of size that are gaged simultaneously. For example, there is no datum shift between the holes of a pattern of holes.

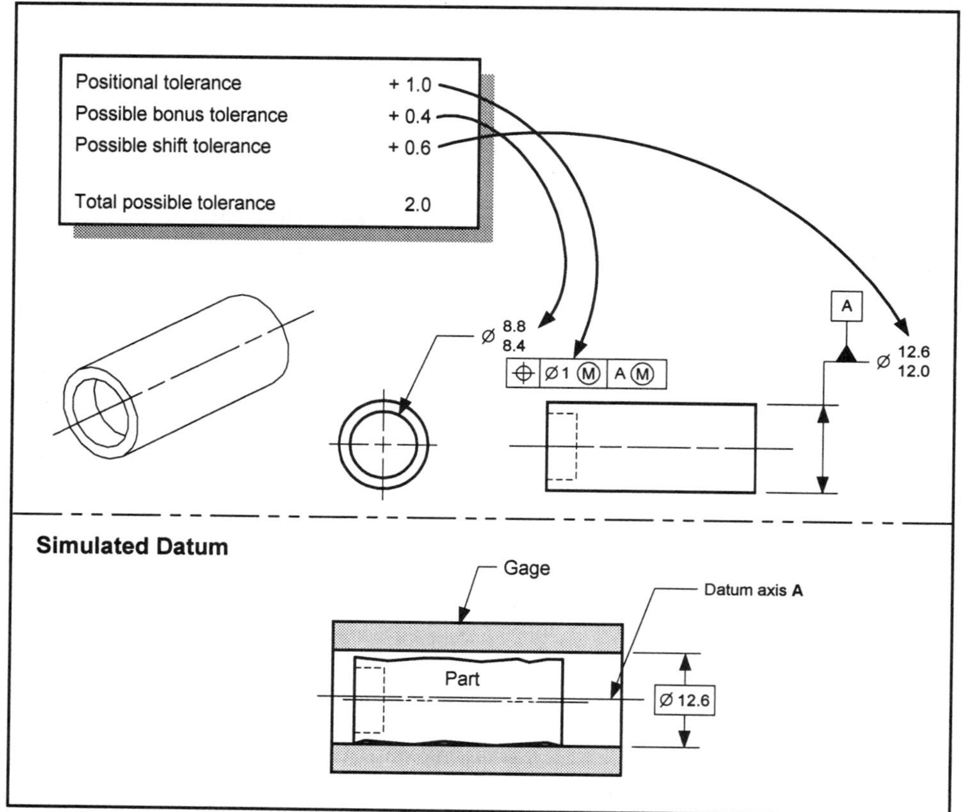

FIGURE 6-11 Datum Shift Example

Author's Comment
The datum shift concept also applies when an LMC modifier is used.

TECHNOTE 6-3 Datum Shift

- Datum shift can result in an additional tolerance for a geometric tolerance.
- Datum shift is only permissible when an MMC modifier is shown in the datum portion of the feature control frame.
- Datum shift results when the AME of the datum feature departs from MMC.
- The maximum allowable datum shift is the difference between the gage size (for the datum) and the LMC size of the datum feature.

The datum shift concept also applies with special-case FOS datums. When a special-case FOS datum is involved with datum shift, it can increase the amount of additional tolerance permissible. The chart in Figure 6-12 shows the amounts of datum shift available for various actual mating sizes of the datum feature. Note that when the datum feature is at MMC, there may be datum shift available. The amount of datum shift available is equal to the difference between special-case FOS datum and MMC when the datum feature is at MMC. In the case of a special-case FOS datum, datum shift only exists if the datum feature is straighter than its allowable straightness error.

Author's Comment
The datum shift concept for a special-case FOS datum also applies when the LMC modifier is used.

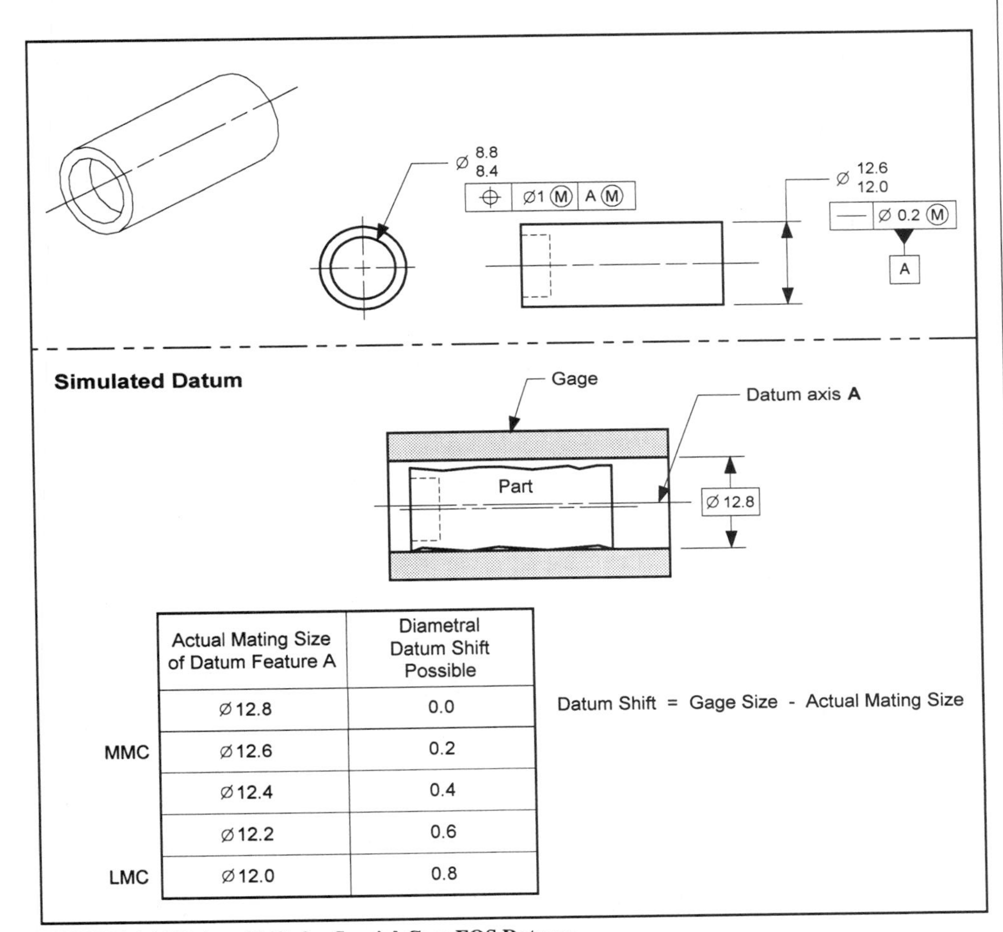

	Actual Mating Size of Datum Feature A	Diametral Datum Shift Possible
	Ø12.8	0.0
MMC	Ø12.6	0.2
	Ø12.4	0.4
	Ø12.2	0.6
LMC	Ø12.0	0.8

FIGURE 6-12 Datum Shift for Special-Case FOS Datums

TECHNOTE 6-4 Datum Shift for Special-Case FOS Datum

When a special-case FOS datum is referenced at MMC, datum shift may be possible when the datum feature is at MMC.

FOS DATUM FEATURE APPLICATIONS (MMC)

Datum Axis MMC Primary

When a diameter is designated as a datum feature and referenced in a feature control frame as primary at MMC, a fixed-gage element may be used as the datum feature simulator. The size of the fixed-gage element is equal to the MMC (or in certain cases, worst-case boundary) of the datum feature. The datum axis is the axis of the datum feature simulator. Depending upon the actual mating size of the datum feature, a datum shift may be available. An example of an external diameter as a datum feature, referenced as primary at MMC, is shown in Figure 6-13*A*. An example of an internal diameter as a datum feature, referenced as primary at MMC, is shown in Figure 6-13*B*.

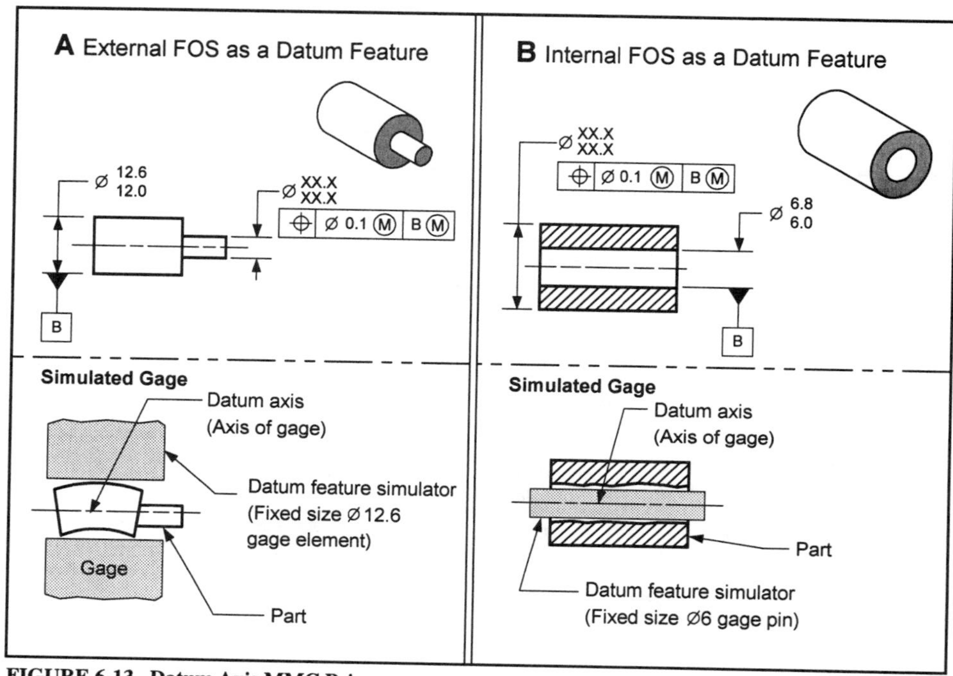

FIGURE 6-13 Datum Axis MMC Primary

Datum Centerplane MMC Primary

When a FOS that consists of parallel planes is designated as a datum feature and referenced in a feature control frame as primary at MMC, a fixed-gage element may be used as the datum feature simulator. The size of the fixed-gage element is equal to the MMC (or in certain cases, worst-case boundary) of the datum feature. The datum centerplane is the centerplane of the datum feature simulator. Depending upon the actual mating size of the datum feature, a datum shift may be available. An example of an internal FOS as a datum feature, referenced as primary at MMC, is shown in Figure 6-14*A*. An example of an external FOS as a datum feature, referenced as primary at MMC, is shown in Figure 6-14*B*.

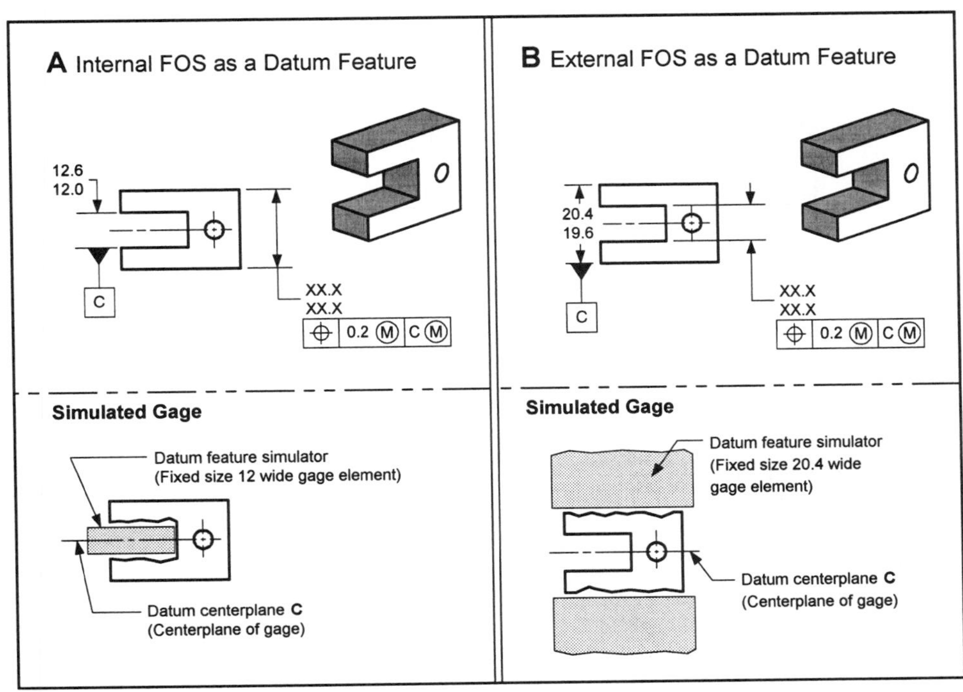

FIGURE 6-14 Datum Centerplane MMC Primary

Datum Axis MMC Secondary

When a part is oriented by a surface and located by a diameter, it is common to have the surface and diameter designated as datum features. Figure 6-15 shows an example. When referencing the datums with the face primary and the diameter secondary (MMC), the following conditions apply:

- The part will have a minimum of three-point contact with the primary datum plane.
- The datum feature simulator will be a fixed size, in this case, the worst-case boundary of datum feature *B*.
- The datum axis is the axis of the datum feature simulator.
- The datum axis is perpendicular to the primary datum plane.
- Depending upon the datum feature actual mating size, a datum shift may be available.

If angular relationship is important, a tertiary datum is necessary.

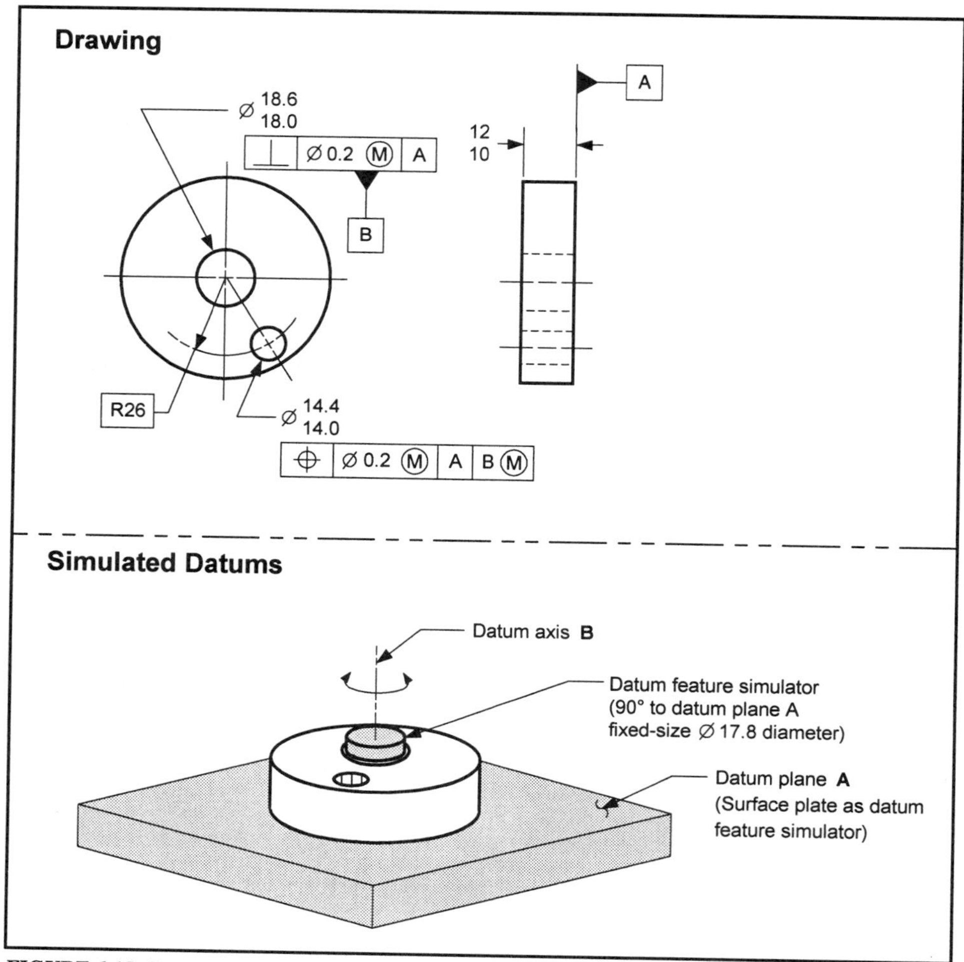

FIGURE 6-15 Datum Axis MMC Secondary

Datum Axis Secondary, Datum Centerplane Tertiary (MMC)

When a part is oriented by a surface, located by a diameter, and has an angular relationship relative to a FOS, it is common to have the surface, diameter, and FOS designated as datum features. Figure 6-16 shows an example. When referencing the datums with the face primary, diameter secondary (MMC), and slot tertiary (MMC), the following conditions apply:

- The part will have a minimum of three points of contact with the primary datum plane.
- The datum feature simulators will be fixed-size gage elements.
- The datum axis is the axis of the datum feature simulator.
- The datum axis is perpendicular to the primary datum plane.
- Depending upon the datum feature's actual mating size, a datum shift may be available.
- Second and third datum planes are to be associated with the datum axis.
- The tertiary datum centerplane is the centerplane of the tertiary datum feature simulator.

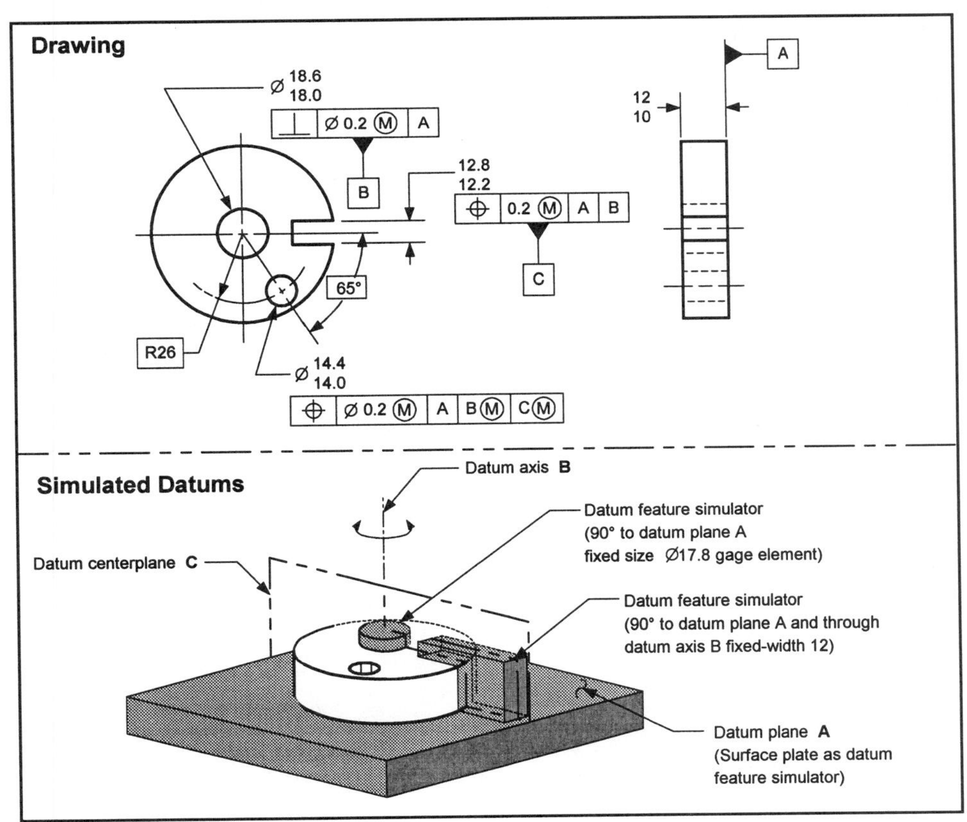

FIGURE 6-16 Datum Axis MMC Secondary, Datum Centerplane Tertiary

Author's Comment
Using a pattern of holes as a datum feature is only introduced in this text; however, it is a common occurrence in industry.

Datum Axis from a Pattern of Holes, MMC Secondary

When a part is oriented by a surface and located by a hole pattern, it is common to have the surface and the hole pattern designated as datum features. Figure 6-17 shows an example. When referencing the datums with the face primary and a hole pattern secondary (MMC), the following conditions apply:

- The part will have a minimum of three points of contact with the primary datum plane.
- The datum feature simulators will be fixed-size gage elements equal to the virtual condition of each hole diameter. In this case, individual datum axes are established at the basic location of each hole. Datum axis *B* is at the theoretical center of the pattern. A second and third datum plane exist from the datum axis.
- When the part is mounted on the primary datum surface, the pattern of holes establishes the second and third datum planes of the datum reference frame. A tertiary datum reference is not necessary since all six degrees of freedom are controlled.

Design Tip
When using a pattern of features as datum features, they must be referenced at MMC in feature control frames that reference them.

Depending upon the actual mating size and actual location and orientation of the datum features, datum shift may be available.

When using a pattern of holes as a datum feature, the spacing between the holes and the squareness of the holes needs to be specified. This is often done by using a tolerance of position control with a single datum reference.

For more info. . .
See Paragraph 4.5.8 of Y14.5.

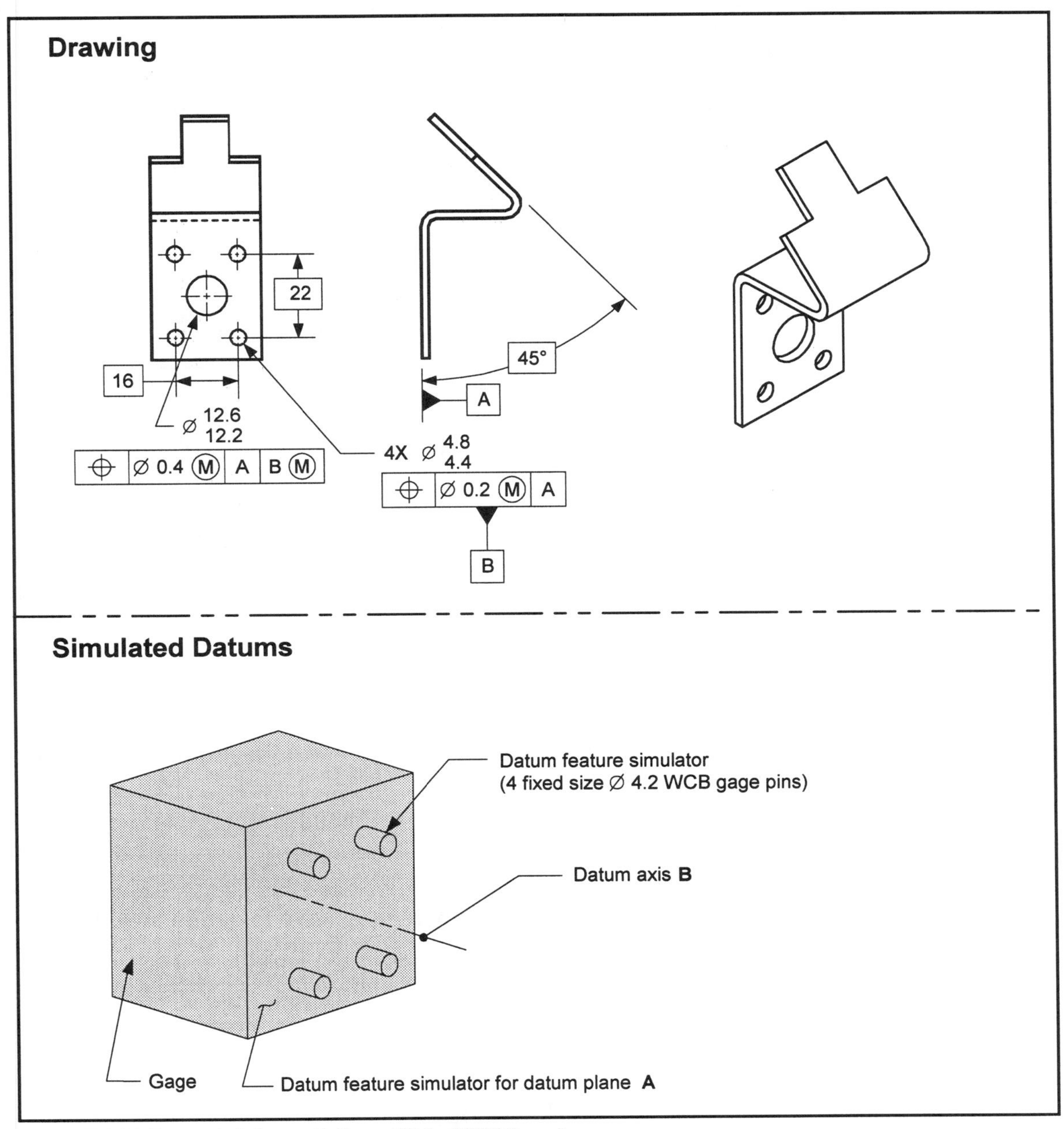

FIGURE 6-17 Datum Axis from a Pattern of Holes MMC Secondary

Datum Sequence

When interpreting datum reference frames that involve both feature of size datum features and planar datum features, the datum sequence plays a major role in the final part tolerances. In the top section of Figure 6-18, the datum portion of the feature control frame is left blank. In the bottom portion of Figure 6-18, the datum reference frame is completed in three different possible sequences.

For more info. . .
See Paragraph 4.5.6 of Y14.5.

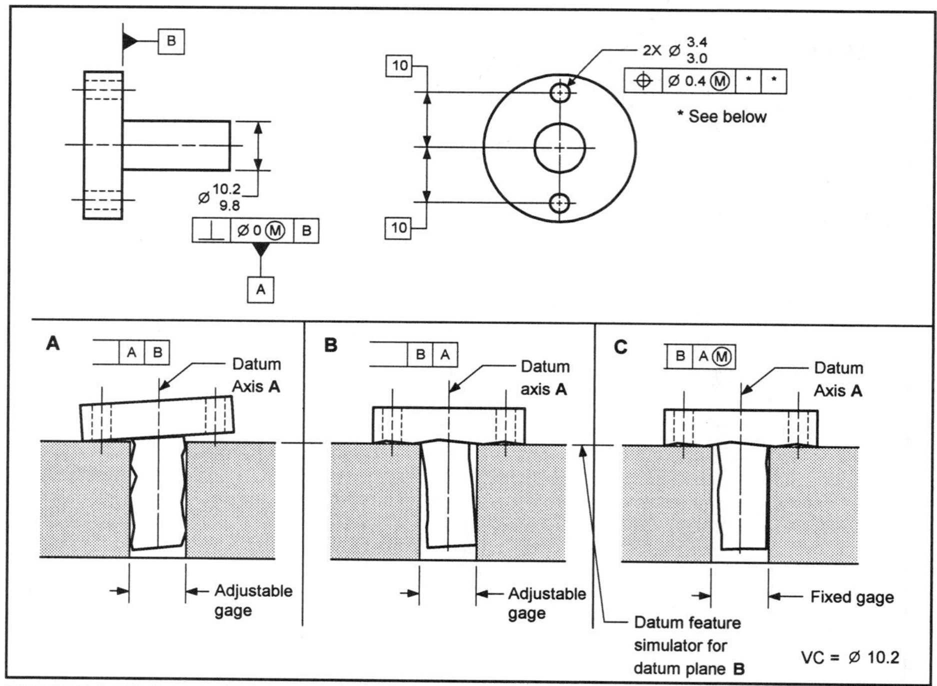

FIGURE 6-18 Datum Sequence Application

In panel *A*, the datum sequence is *A* primary at RFS and *B* secondary.

- An adjustable gage is required, and no datum shift is permissible on datum feature *A*.
- The part is oriented in the gage by datum feature *A*.
- Datum feature *B* will have a minimum of one point of contact with its datum feature simulator.
- The orientation of the holes will be relative to datum axis *A*.

In panel *B*, the datum sequence is *B* primary and *A* secondary at RFS.

- Datum feature *B* will have a three-point contact with its datum plane.
- The part is oriented in the gage by datum feature *B*.
- The orientation of the holes will be relative to datum plane *B*.
- An adjustable gage is required, and no datum shift is permissible on datum feature *A*.

In panel *C*, the datum sequence is *B* primary and *A* secondary at MMC.

- When simulating datum *A*, a fixed gage is allowed, and the gage allows datum shift.
- Datum feature *B* will have a three-point contact with its datum plane.
- The part is oriented in the gage by datum feature *B*.
- The orientation of the holes will be relative to datum plane *B*.

Summary

A summarization of datum information is shown in Figure 6-19.

Concept	Can be applied to a		Is applicable when datum is referenced at		
	Surface	FOS	MMC	LMC	RFS*
Datum Shift	No	Yes	Yes	Yes	No
Datum Targets	Yes	Yes	No	No	Yes
Ⓜ Modifier	No	Yes	Yes		
* Per Rule #2					

FIGURE 6-19 Summarization of Datums

VOCABULARY LIST

New Terms Introduced in this Chapter

Coaxial datum features
Coaxial diameters
Datum shift
Special-case FOS datum

Study Tip
Read each term. If you don't recall its meaning, look it up in the chapter.

ADDITIONAL RELATED TOPICS

Topic	ASME Y14.5M-1994 Reference
• Specifying datum features at LMC	Paragraph 4.5.5
• Screw threads, gears, and splines as datum features	Paragraph 4.5.9
• Simultaneous requirement	Paragraph 4.5.12

Author's Comment
These topics, plus advanced coverage of many of the topics introduced in this text, will be covered in my new book on advanced GD&T concepts.

QUESTIONS AND PROBLEMS

Use the word list on the right to fill in the blanks. (There are more words than blanks.)

1. Two FOS types that can be used as a datum features are ____________ and ____________.

2. When a diameter is used as a datum feature, a datum ____________ is established. When a planar FOS is used as a datum feature, a datum ____________ is established.

3. Two ways a feature control frame can communicate the size condition for a datum reference are ____________ and the use of a material condition ____________.

4. Three ways a diameter can be specified as a datum feature are:

   - The datum identification symbol can be touching the ____________ of a diameter.
   - The datum identification symbol can be touching the beginning of a ____________ line.
   - The datum identification symbol can be touching a ____________.

5. Two ways a planar FOS can be specified as a datum feature are:

   - The datum identification symbol can replace one side of a ____________ line and arrowhead.
   - The datum identification symbol can be touching the ____________ line on the opposite side of the dimension line arrowhead of a FOS.

Word List

Axis
Centerplane
Diameter
Dimension
Extension
Feature control frame
Hidden
Leader
Modifier
Planar
Point
Rule #1
Rule #2
Surface

6. For the feature control frame shown below, draw the datum feature simulator for establishing datum axis *A*.

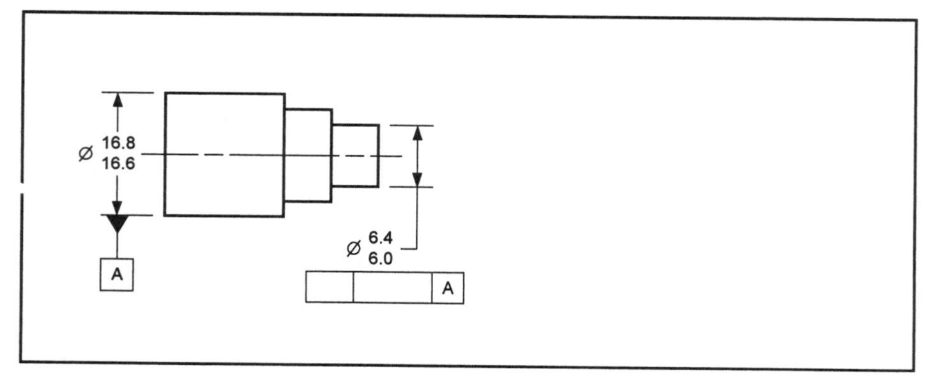

7. Use the drawing below. For the feature control frame shown, draw the datum feature simulator for establishing datum axis *A*.

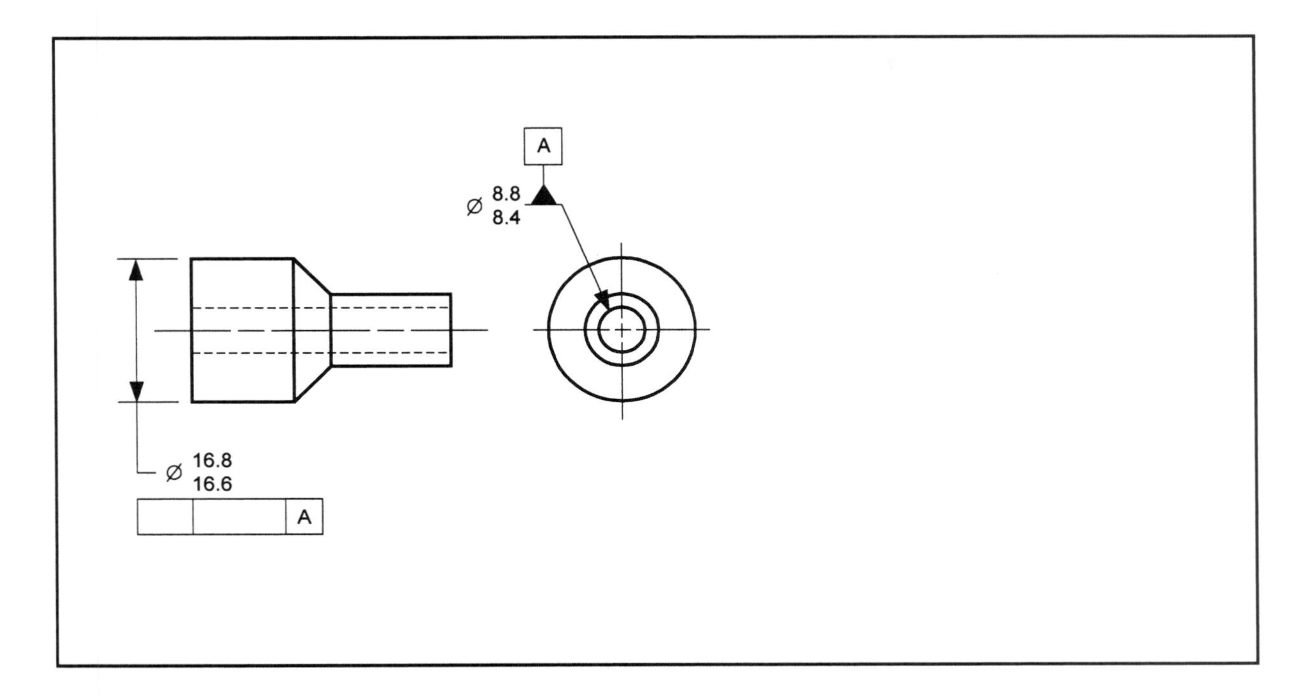

8. Use the drawing below. For the feature control frame shown, draw the datum feature simulator for establishing datum plane *B* and datum axis *A*.

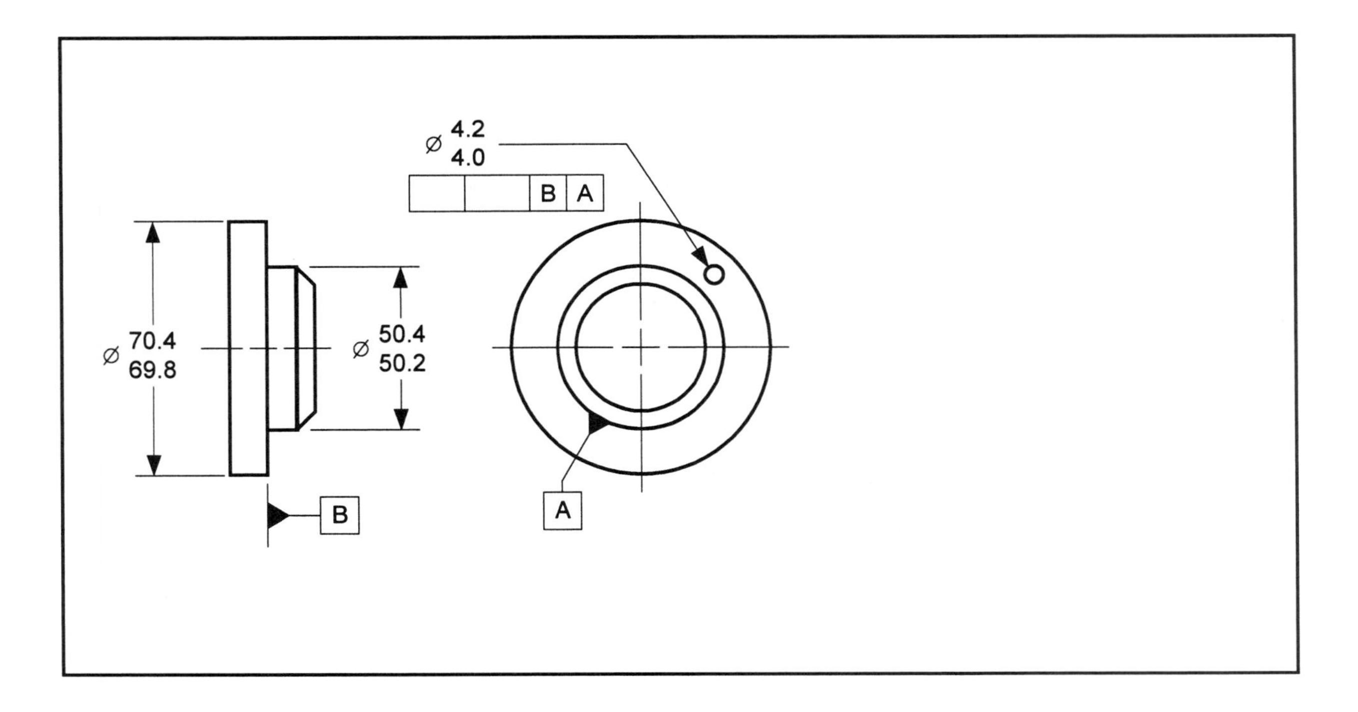

9. Use the drawing below. For the feature control frame shown, draw the datum feature simulator for establishing datum plane *B*, datum axis *A*, and datum centerplane *C*.

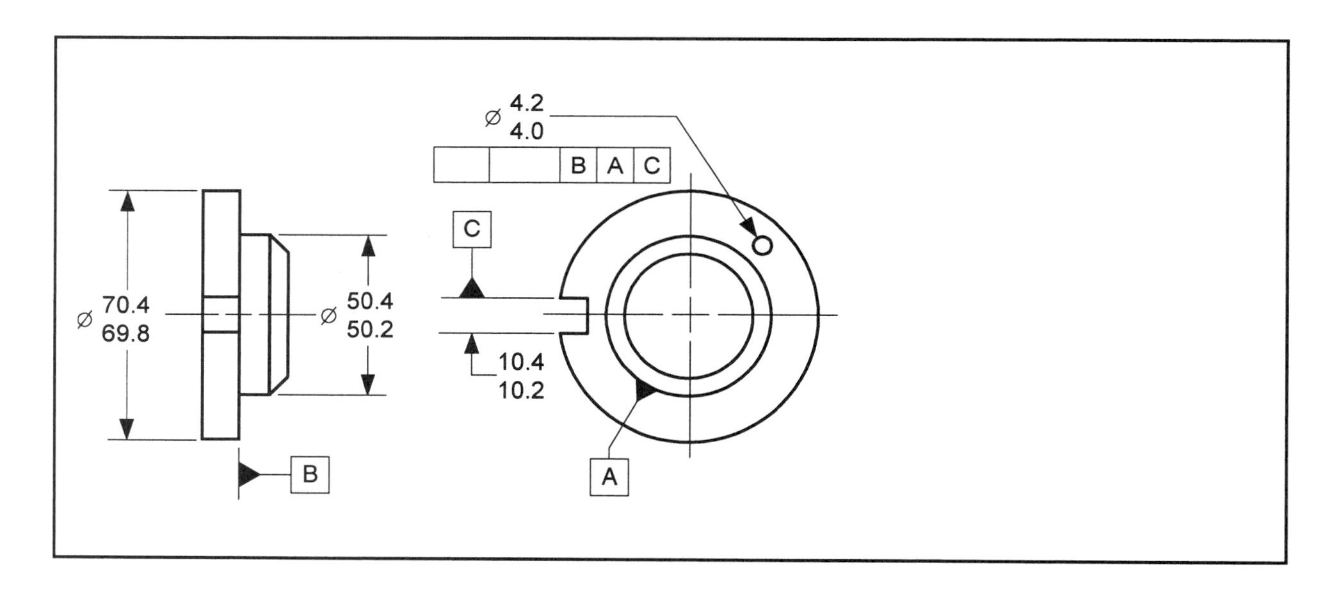

10. Describe the term, "coaxial datum features."

____

____

11. Explain the effect of using an MMC modifier in the datum portion of a feature control frame.

____

____

12. Describe the concept of "datum shift."

____

____

13. List two cases where a datum feature referenced at MMC is simulated at special-case FOS datum.

____

____

____

____

14. In a feature control frame, what denotes that datum shift is permissible?

____

15. Use the drawing below. For the feature control frame shown, draw and dimension the datum feature simulator for establishing datum axis *A*.

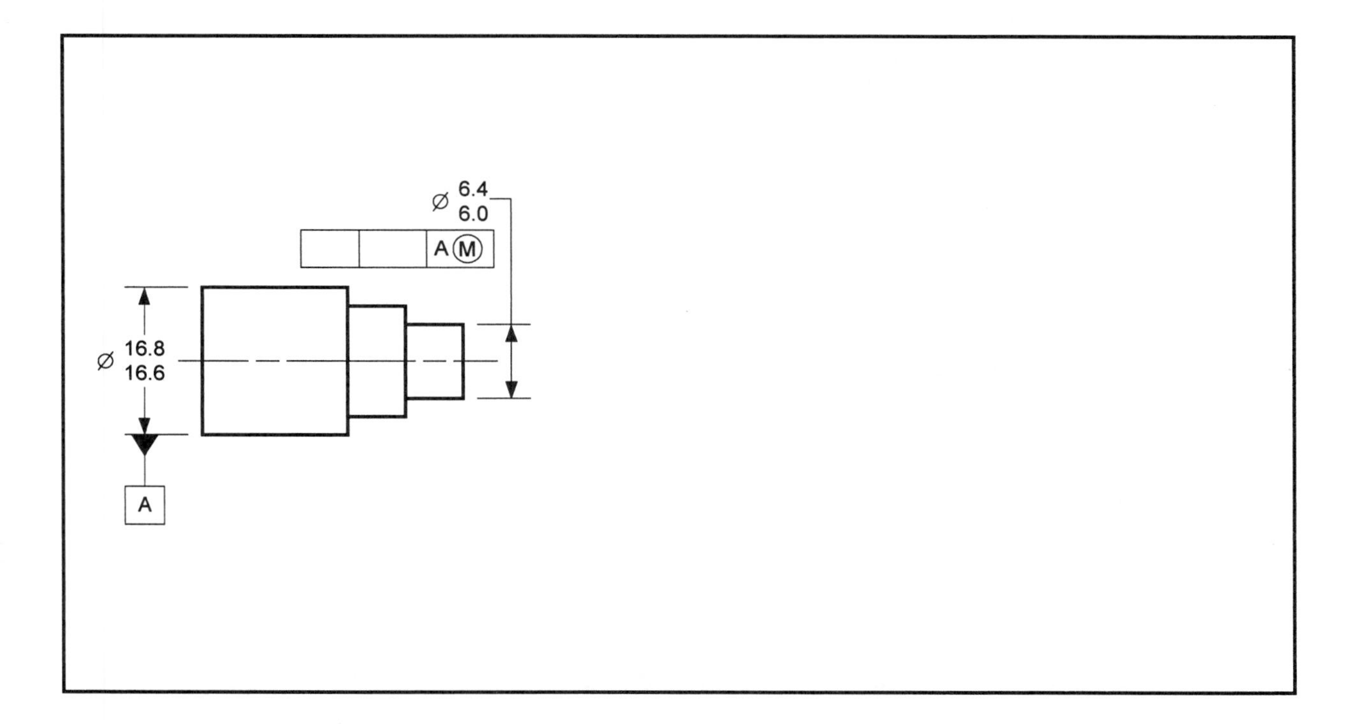

16. Using the drawing below, draw and dimension the datum feature simulator for establishing datum axis *A*.

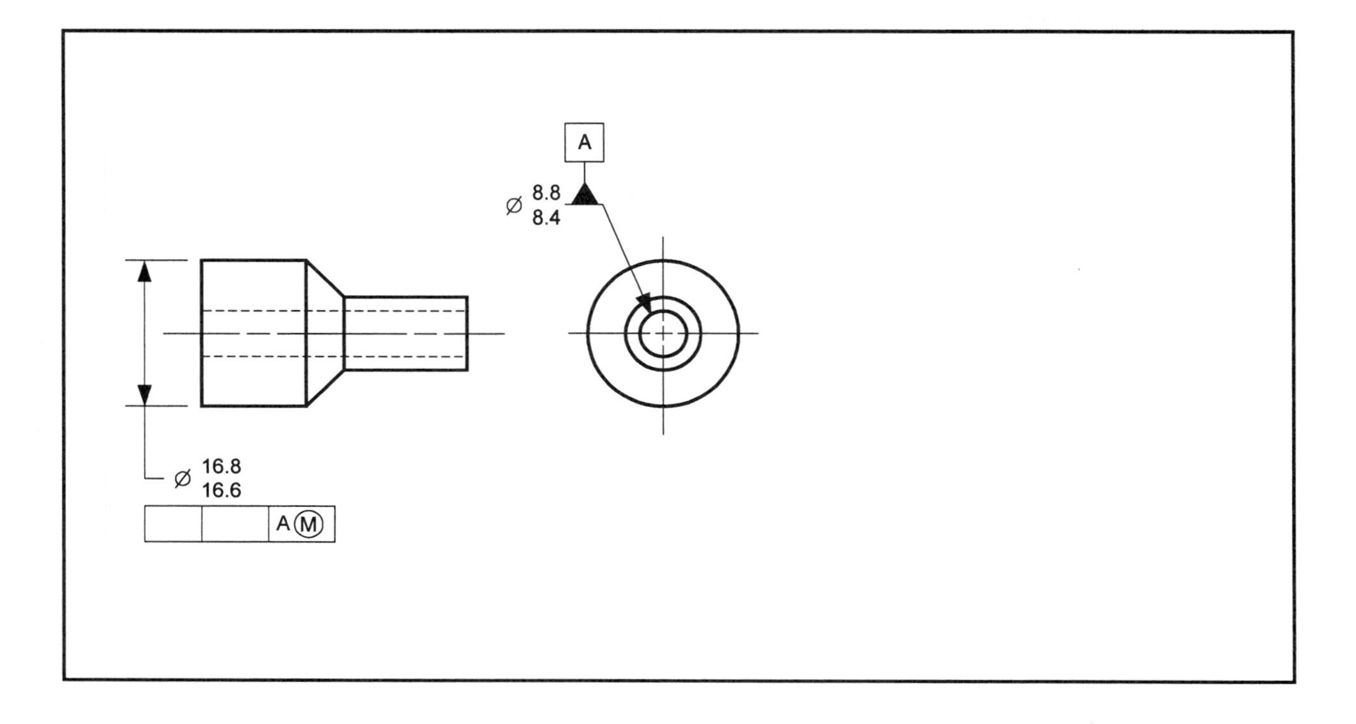

17. Using the drawing below, draw and dimension the datum feature simulator for establishing datum axis *A*.

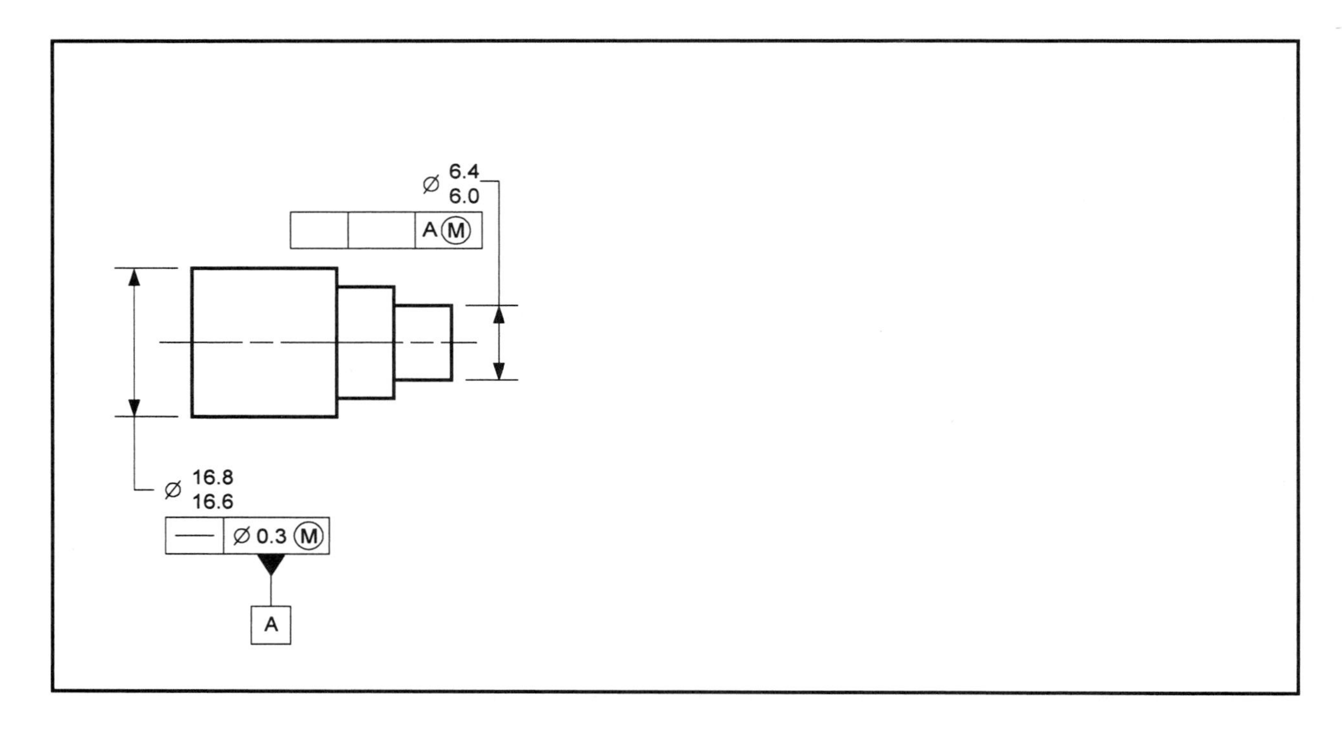

18. Use the drawing below. For the feature control frame shown, draw and dimension the datum feature simulator for establishing datum axis *A*.

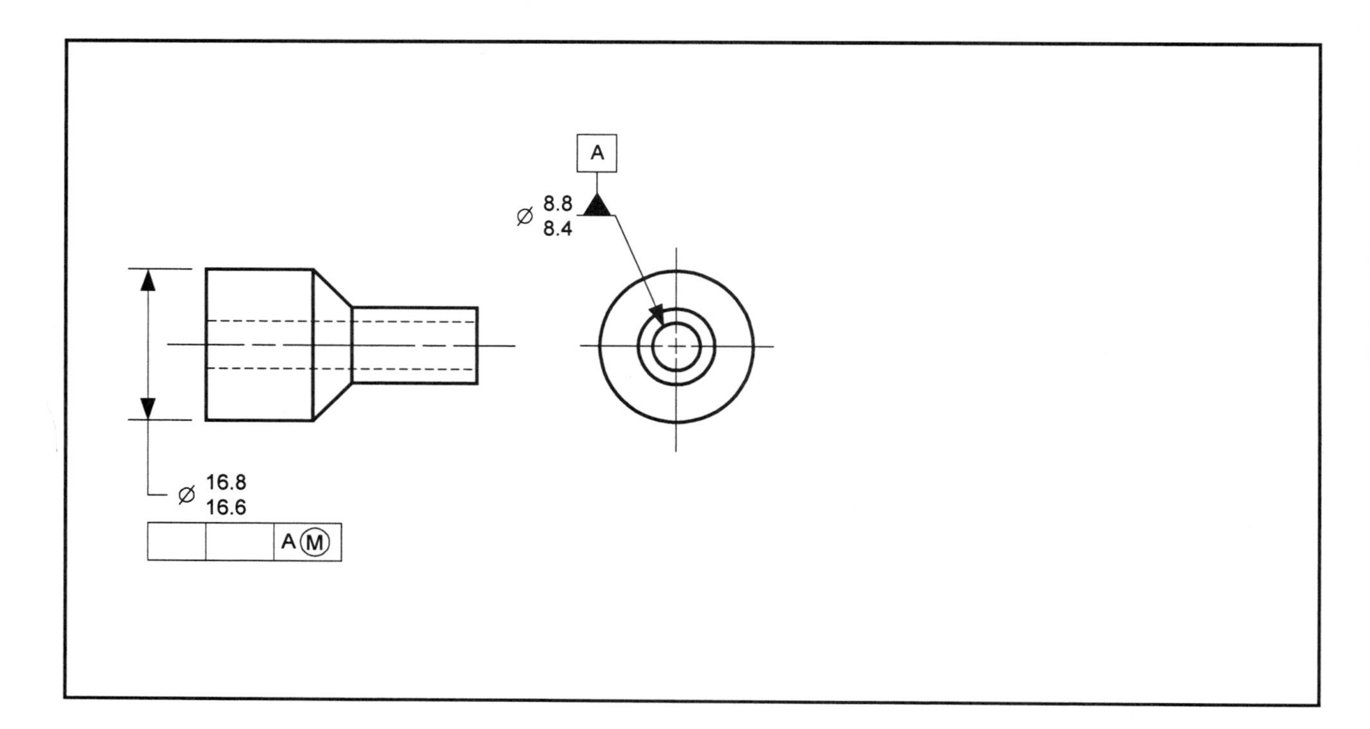

19. Use the drawing below. For the feature control frame shown, draw and dimension the datum feature simulator for establishing datum axis *B*.

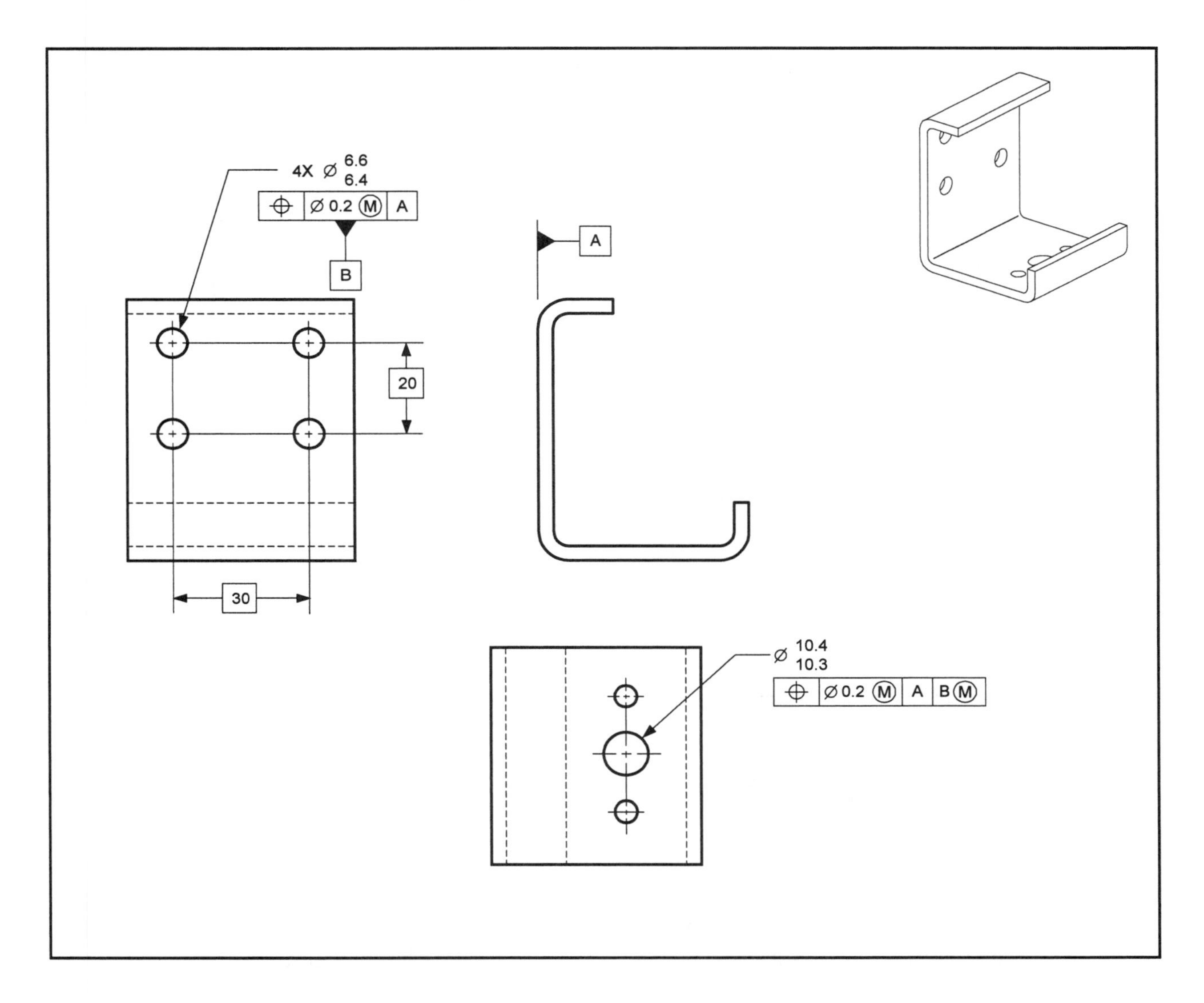

20. Fill in the chart below.

Using the drawing from question. . .	The maximum permissible datum shift is. . .
15	
16	
17	
18	
19	

21. Using the drawing below, draw the part on the datum feature simulator to represent the specified datum sequence for the positional tolerance labeled "1" and for the positional tolerance labeled "2."

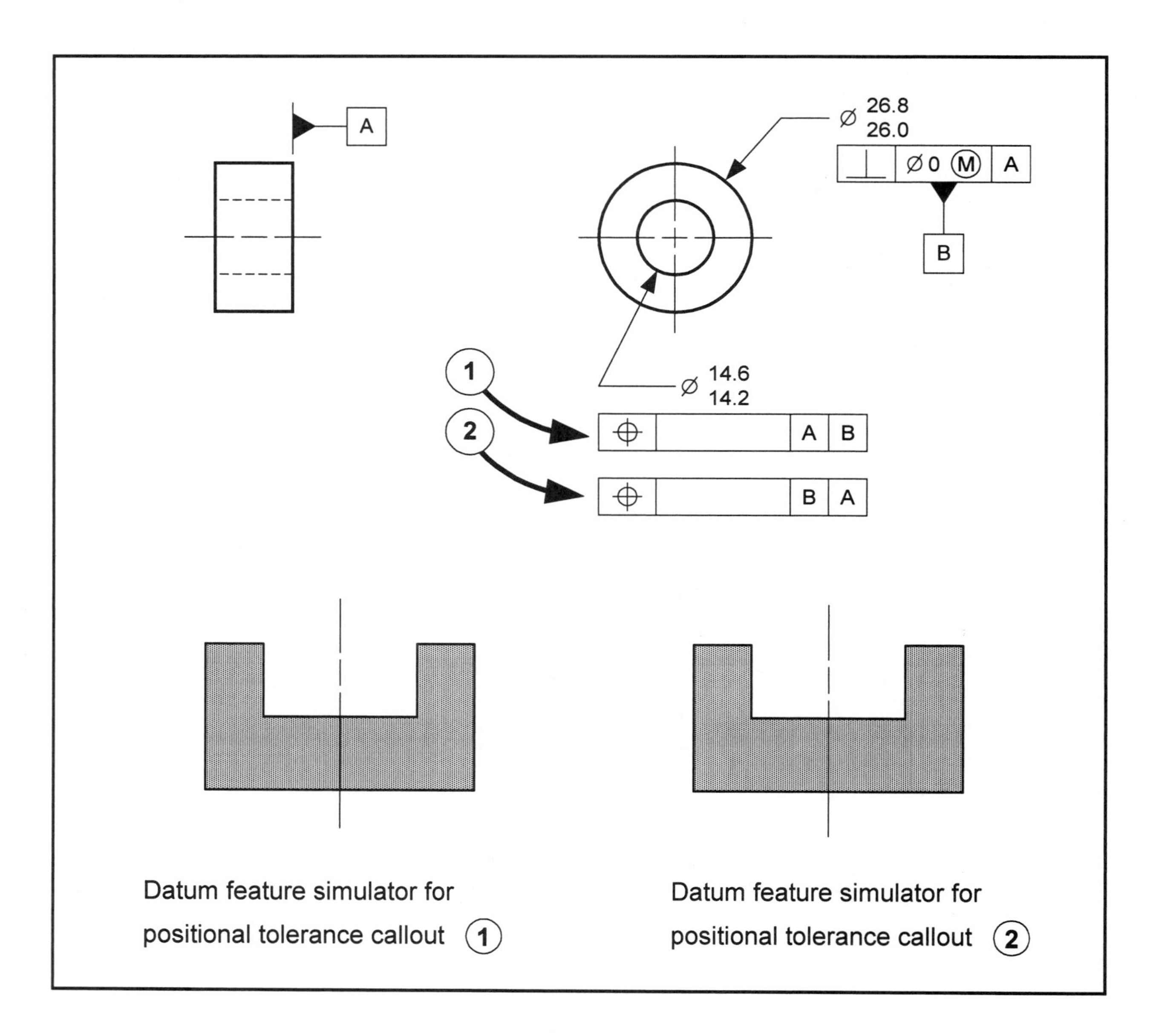

22. Using the drawing below, specify the datums according to the following instructions:

- Specify the axis of the 12.2 dia. as datum *B*.
- Specify the centerplane of the 12.0 slot as datum *C*.
- Specify the pattern of two 6.4-6.8 dia. holes as datum *D*.
- Specify the centerplane of the 38.0-38.6 width as datum *E*.
- Specify the axis of the 4 dia. hole as datum *F*.

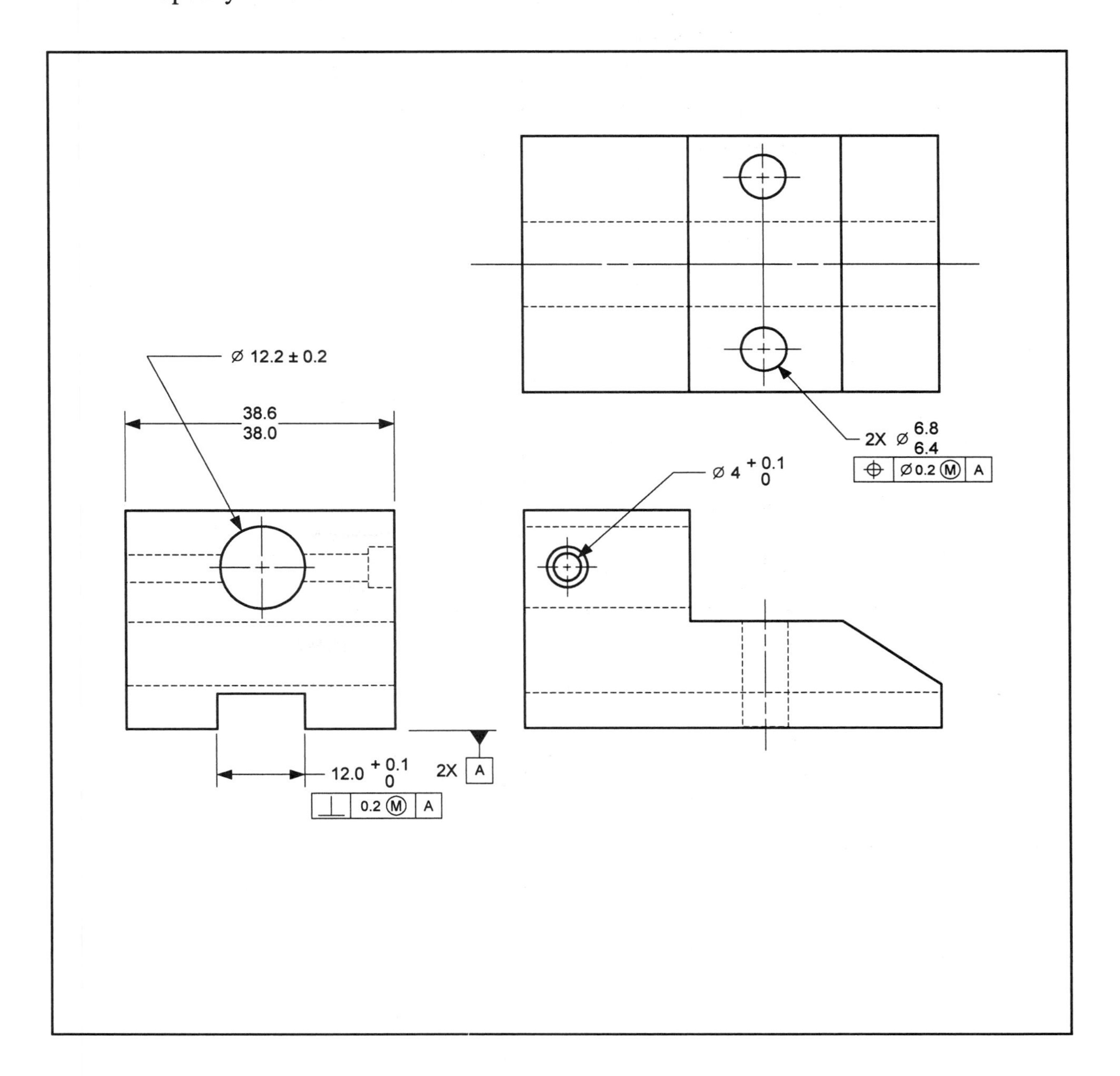

Chapter 7

Orientation Controls

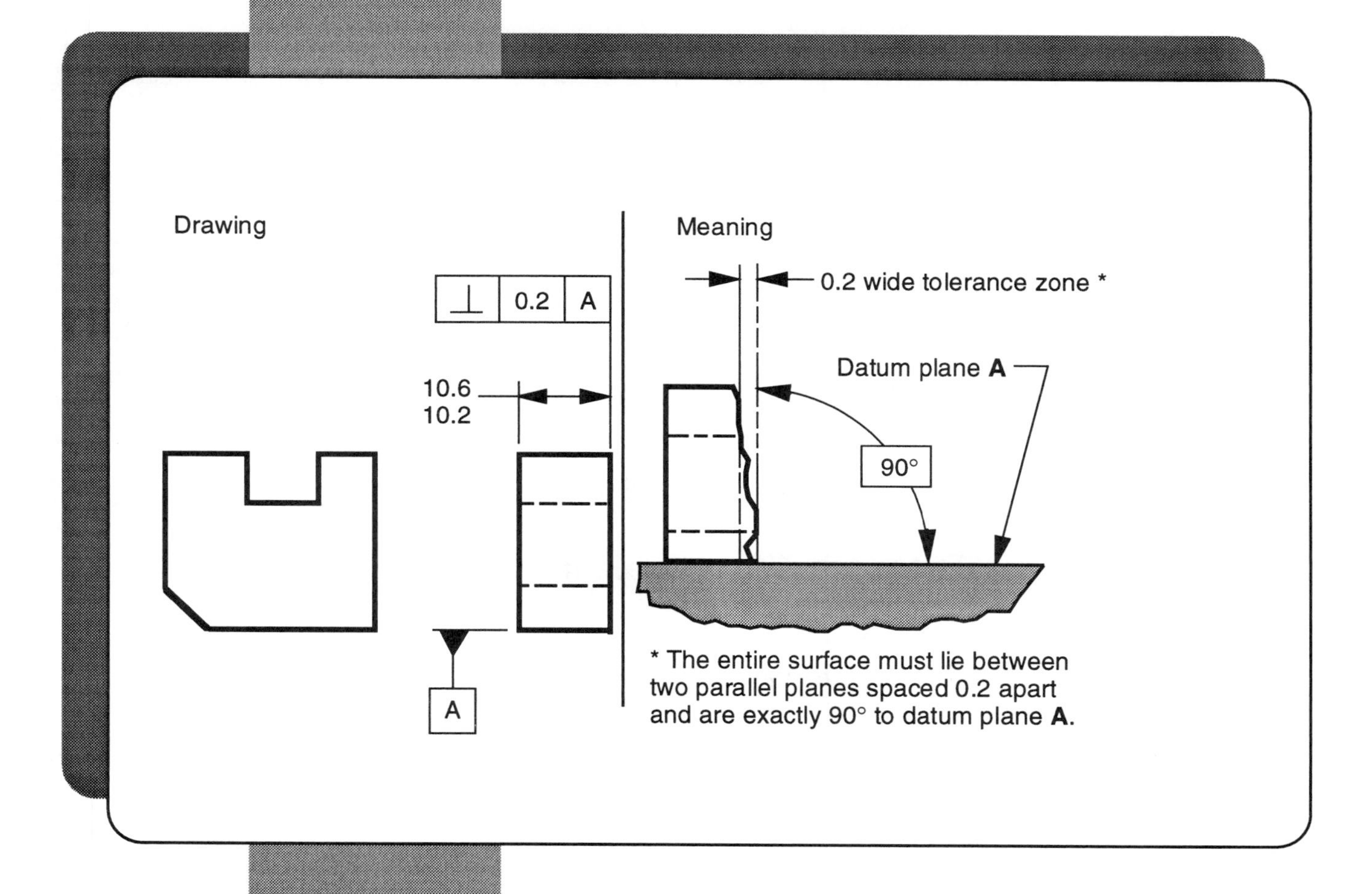

INTRODUCTION

Design Tip
Remember—
orientation controls do not <u>locate</u> features.

This chapter explains the concepts involved in defining the orientation of part features. An orientation control defines the perpendicularity (squareness), angularity, and parallelism of part features. Orientation controls are used when the limits prescribed by other dimensions and geometric controls do not provide sufficient control for the function or interchangability of a product. Orientation controls must contain a datum reference. The three orientation controls and their symbols are shown in Figure 7-1.

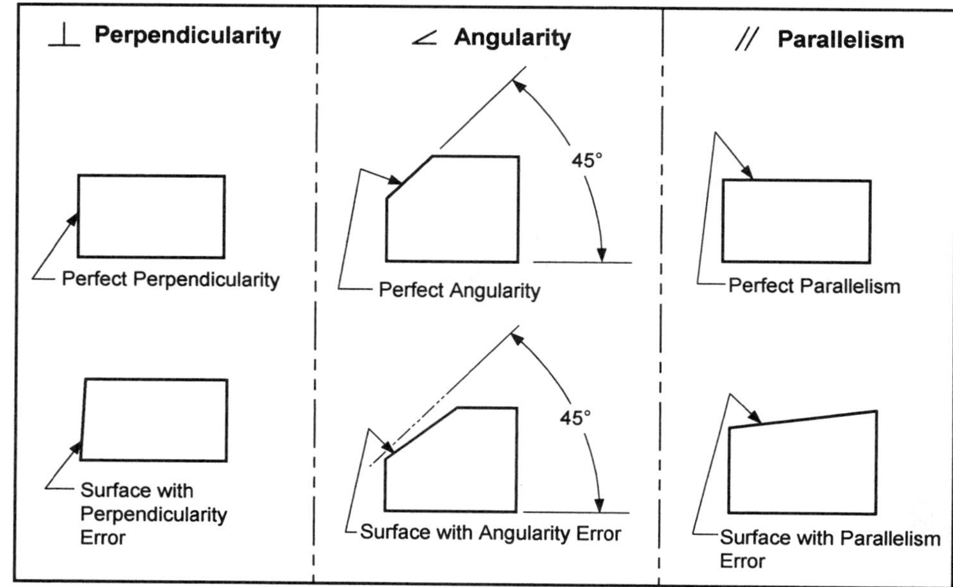

FIGURE 7-1 Orientation Controls

CHAPTER GOALS AND OBJECTIVES

There are Three Goals in this Chapter:

7-1. Interpret the perpendicularity control.
7-2. Interpret the angularity control.
7-3. Interpret the parallelism control.

Performance Objectives that Demonstrate Mastery of These Goals

Upon completion of this chapter, each student should be able to:

Goal 7-1 (pp. 178-186)

- Describe what controls the tolerance on implied right angles.
- Describe two common tolerance zone shapes for a perpendicularity control.
- Describe the tolerance zone for a perpendicularity control applied to a planar surface.
- Explain how perpendicularity of a surface affects the flatness of the surface.
- Explain the effects of using multiple datum references with a perpendicularity control.
- Explain how to control the perpendicularity of the axis/centerplane of a FOS.
- Determine when a perpendicularity control affects the WCB of a FOS.
- Draw a gage for verifying perpendicularity at MMC.
- List two indirect perpendicularity controls.
- Determine if a perpendicularity specification is legal.
- Describe how a perpendicularity control could be inspected.

Study Tip

Take a few minutes to fully understand these objectives. When reading this chapter, look for information to help you master these objectives.

Goal 7-2 (pp. 187-191)

- Describe the tolerance zone for an angularity control applied to a surface.
- Describe two common tolerance zone shapes for an angularity control.
- Explain how angularity of a surface affects the flatness of the surface.
- Determine when an angularity control affects the WCB of a FOS.
- Explain how to control the angularity of the axis/centerplane of a FOS.
- List two indirect angularity controls.
- Determine if an angularity specification is legal.

Goal 7-3 (pp. 192-198)

- Describe how parallelism is controlled when no symbol is shown.
- Describe the tolerance zone for a parallelism control applied to a surface.
- Describe two common tolerance zone shapes for a parallelism control.
- Describe how to locate the tolerance zone for a parallelism applied to a surface.
- Explain how parallelism of a surface affects the flatness of the surface.
- Explain how to control the parallelism of the axis/centerplane of a FOS.
- Determine when a parallelism control affects the WCB of a FOS.
- Interpret the effects of using the tangent plane modifier with a parallelism control.
- List two indirect parallelism controls.
- Determine if a parallelism specification is legal.

PERPENDICULARITY CONTROL

Author's Comment
My experience indicates that in many cases implied 90° angles are not verified during inspection. This leads me to believe that it is better to specify the tolerance on implied 90° angles whenever possible.

Implied Right (90°) Angles

Wherever two lines on a drawing are shown at 90°, they are an implied 90° angle. The tolerance for an implied 90° angle comes from the titleblock tolerance (on some drawings it is contained in the general notes). See Figure 7-2*A*. This method works satisfactorily for some drawings, but it contains two shortcomings. First, the tolerance zone is fan-shaped; it increases the farther it gets from the origin of the angle. The second shortcoming is that implied 90° angles lack a datum reference; the part could use either side of the 90° angle to begin the measurement, which may affect functional relationships.

In Figure 7-2*B*, the part could be inspected in two different ways: using the long side as the datum feature to check the angular relationship of the short side, or using the short side as the datum reference to check the angular relationship of the long side.

Design Tip
Analyze all implied 90° angles to determine how much tolerance each angle could have based on the function of the part. Remember, tolerances should be as large as possible to keep manufacturing costs low.

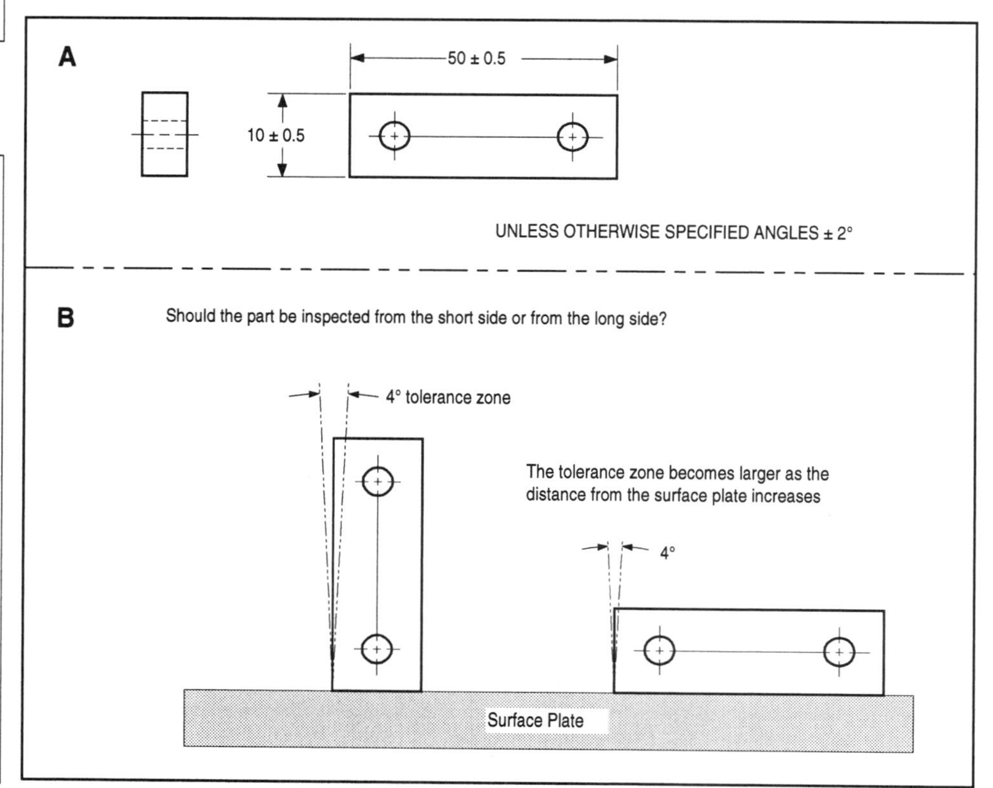

FIGURE 7-2 Implied Right (90°) Angles

TECHNOTE 7-1 Implied 90° Angles

Unless otherwise specified, the tolerances on implied 90° angles are controlled by the titleblock tolerance (or in a general note).

Definition of Perpendicularity

Perpendicularity is the condition that results when a surface, axis, or centerplane is exactly 90° to a datum. A ***perpendicularity control*** is a geometric tolerance that limits the amount a surface, axis, or centerplane is permitted to vary from being perpendicular to the datum.

For more info. . .
See Fundamental Dimensioning Rule #5 on page 12.

Perpendicularity Tolerance Zones

The two common tolerance zones for a perpendicularity control are:

1. Two parallel planes
2. A cylinder

The following applications show these tolerance zones and discuss their use.

For more info. . .
See Paragraph 6.6.4.1 of Y14.5.

Perpendicularity Applications

Most perpendicularity applications fall into one of three general cases:

1. Perpendicularity applied to a surface
2. Perpendicularity applied to a planar FOS
3. Perpendicularity applied to a cylindrical FOS

In Figure 7-3, a perpendicularity control is applied to a surface. This is the most common application of perpendicularity. When perpendicularity is applied to a surface, the following four conditions apply:

1. The shape of the tolerance zone is two parallel planes that are perpendicular to the datum plane.
2. The tolerance value of the perpendicularity control defines the distance between the tolerance zone planes.
3. All the elements of the surface must be within the tolerance zone.
4. The perpendicularity tolerance zone limits the flatness of the toleranced feature.

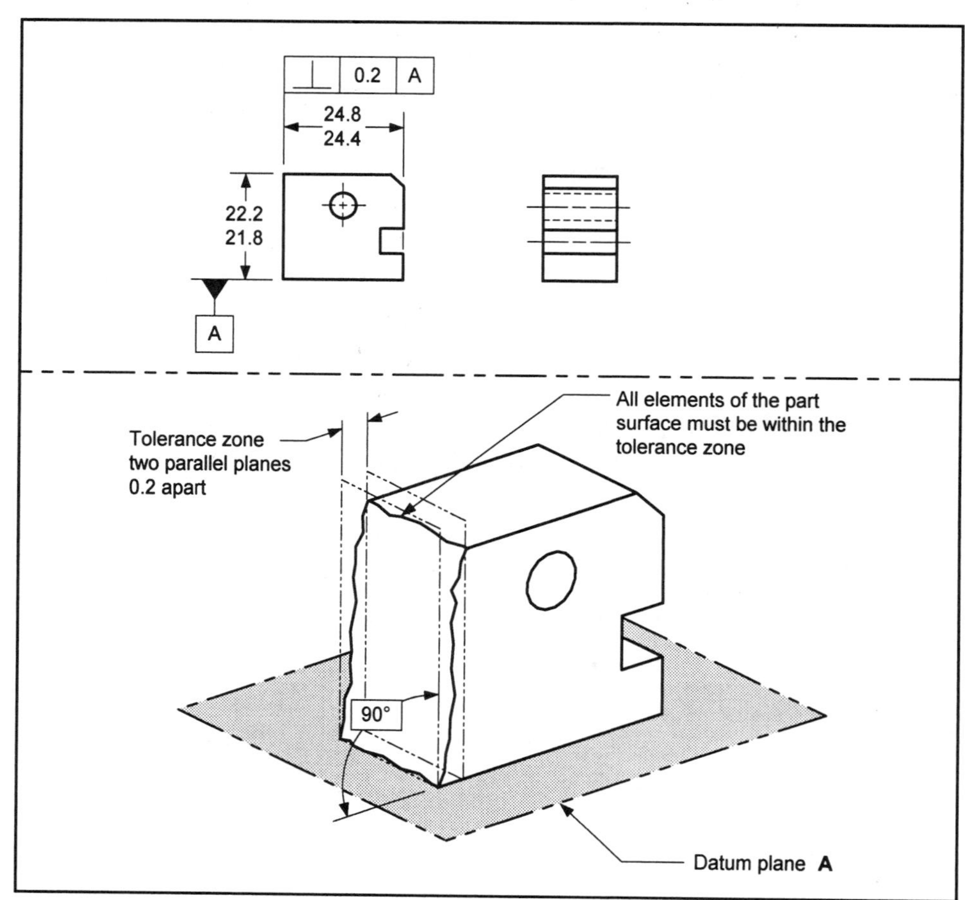

FIGURE 7-3 Perpendicularity Applied to a Surface

In Figure 7-4, a perpendicularity control is applied to a surface. In this application, the perpendicularity control contains two datum references. When two datum references are used in a perpendicularity control, the tolerance zone is perpendicular to two datum planes, and all the conditions from Figure 7-3 apply.

TECHNOTE 7-2 Perpendicularity Applied to a Surface

- The shape of the tolerance zone is two parallel planes that are perpendicular to the datum plane(s).
- The distance between the planes is equal to the perpendicularity tolerance zone.
- All elements of the toleranced surface must be within the tolerance zone.
- The perpendicularity tolerance zone limits the flatness of the toleranced feature.

Author's Comment

A third tolerance zone shape is possible for perpendicularity; it is two parallel lines. It is not very common, and it is beyond the scope of this text. For more information on this concept, see the additional related topics on page 199.

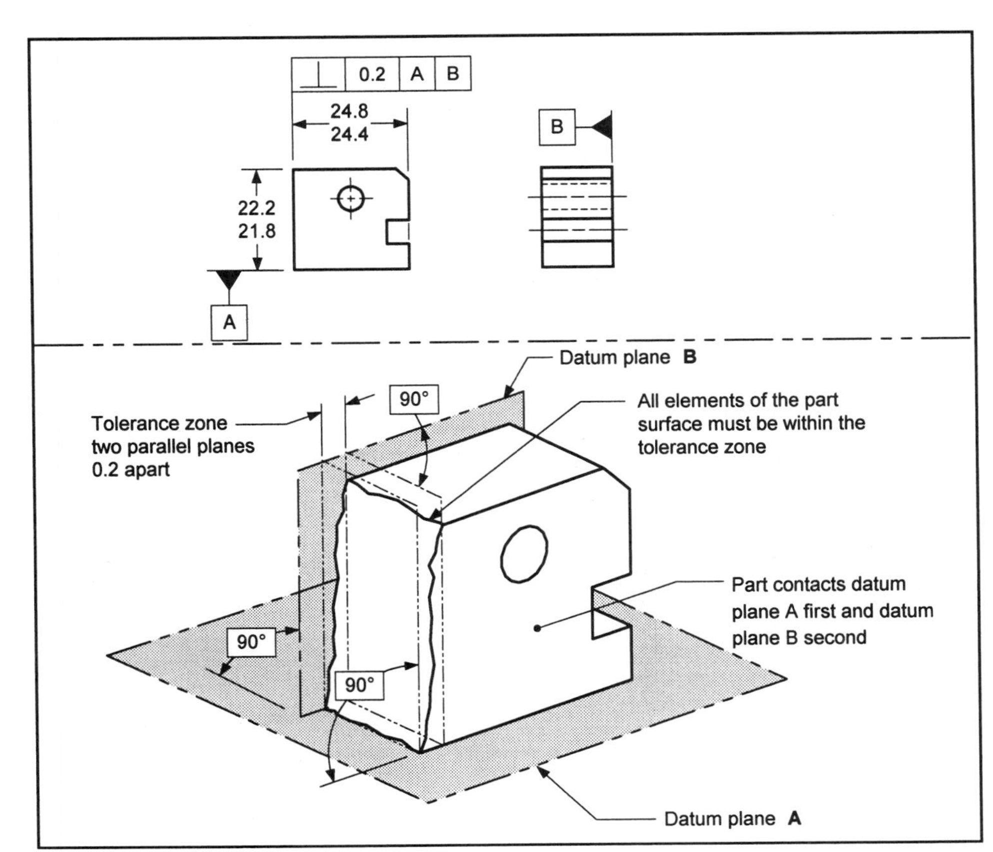

FIGURE 7-4 Perpendicularity with Two Datum References

In Figure 7-5, a perpendicularity control that contains the MMC modifier is applied to a planar FOS. This type of geometric control is often used to ensure the function of assembly. When perpendicularity is applied to a planar FOS and contains the MMC modifier, the following conditions apply:

- The shape of the tolerance zone is two parallel planes that are perpendicular to the datum plane.
- The tolerance value of the perpendicularity control defines the distance between the tolerance zone planes.
- The centerplane of the AME of the FOS must be within the tolerance zone.
- A bonus tolerance is permissible.
- A fixed gage may be used to verify the perpendicularity control.

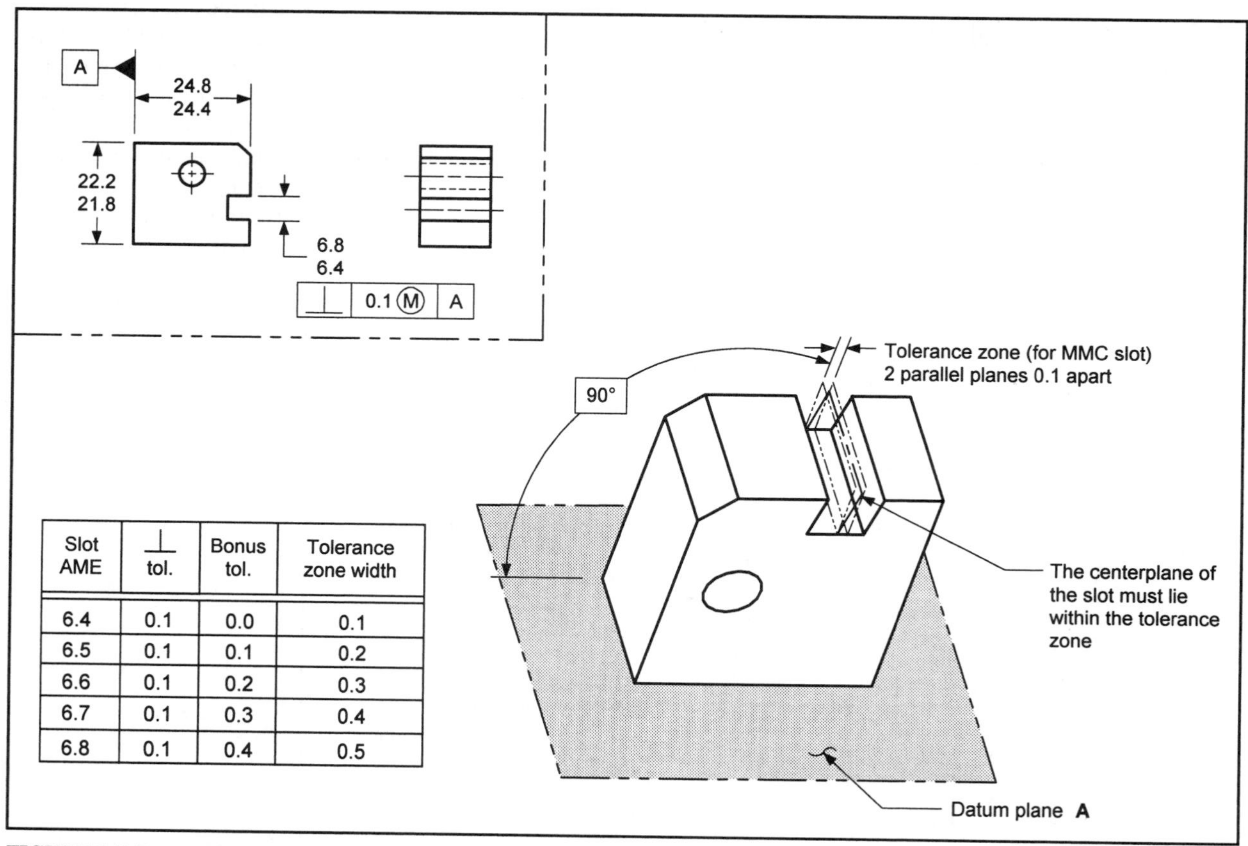

Slot AME	⊥ tol.	Bonus tol.	Tolerance zone width
6.4	0.1	0.0	0.1
6.5	0.1	0.1	0.2
6.6	0.1	0.2	0.3
6.7	0.1	0.3	0.4
6.8	0.1	0.4	0.5

FIGURE 7-5 Perpendicularity Applied to a Planar FOS

In Figure 7-6, a perpendicularity control that contains the MMC modifier is applied to a cylindrical FOS. This type of geometric control is often used to ensure the function of assembly. When a perpendicularity control is applied to a cylindrical FOS, it controls the axis of the FOS. In Figure 7-6, the following conditions apply:

- The tolerance zone is a cylinder that is perpendicular to the datum plane.
- The tolerance value of the perpendicularity control defines the diameter of the tolerance zone cylinder.
- The axis of the diameter must be within the tolerance zone (when the FOS is at MMC).
- A bonus tolerance is permissible.
- The WCB of the diameter is affected.
- A fixed gage may be used to verify the perpendicularity control.

When calculating bonus tolerance for a FOS that is toleranced with an orientation control, the actual mating envelope is oriented relative to the primary datum.

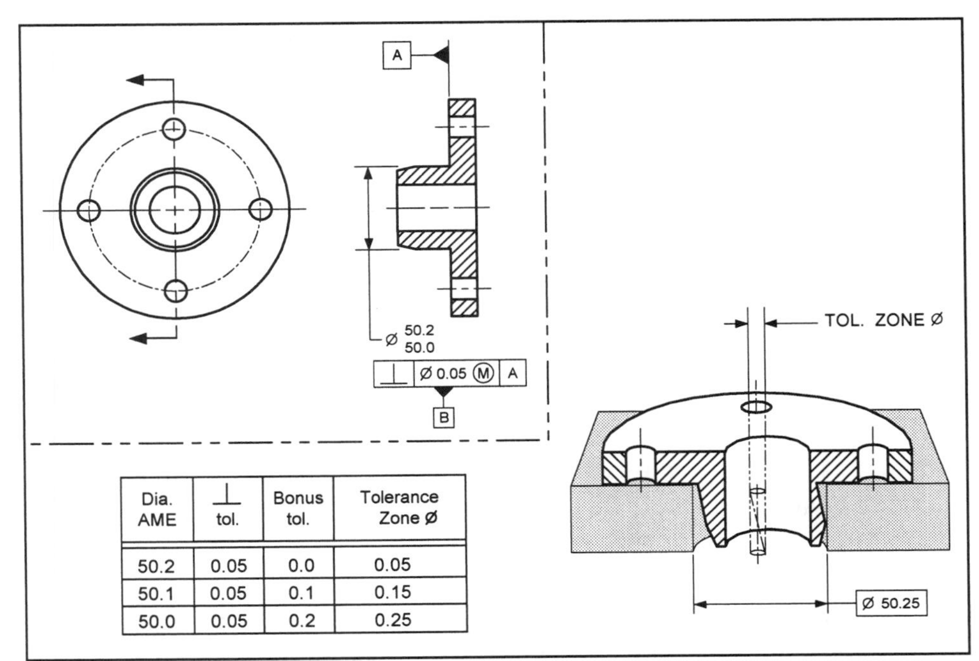

Dia. AME	⊥ tol.	Bonus tol.	Tolerance Zone Ø
50.2	0.05	0.0	0.05
50.1	0.05	0.1	0.15
50.0	0.05	0.2	0.25

FIGURE 7-6 Perpendicularity Applied to a Diametrical FOS

Design Tip

A common application of perpendicularity applied to a diameter is to control a secondary datum feature relative to the primary datum.

A gage used to verify perpendicularity (at MMC) is shown in Figure 7-6. It has a surface that serves as a simulated datum feature for datum plane *A*. The gage contains a hole that verifies the perpendicularity of the hub diameter. The gage hole diameter is equal to the virtual condition of the hub.

TECHNOTE 7-3 Perpendicularity (at MMC) Applied to a FOS

- The tolerance zone is a cylinder or two parallel planes.
- The axis or centerplane must be within the tolerance zone.
- A bonus tolerance is permissible.
- A fixed gage may be used to verify the perpendicularity control.

When a perpendicularity control is applied to a surface, the WCB of the toleranced surface is not affected. When a perpendicularity control is applied to a FOS, the WCB of the FOS is affected. The WCB of a FOS that is toleranced with an orientation control is oriented relative to the datums specified.

Indirect Perpendicularity Controls

There are several geometric controls that can indirectly affect the perpendicularity of a part feature. Tolerance of position, runout, and profile can limit perpendicularity; however, indirect perpendicularity controls are not inspected. Their effect on perpendicularity is a result of the part surface, axis, or centerplane being within the zone for the specified geometric control. If it is desired to have the perpendicularity of a part feature inspected, a perpendicularity control should be specified. If a perpendicularity control is used, its tolerance value should be less than the tolerance value of any indirect perpendicularity controls that apply.

Legal Specification Test for a Perpendicularity Control

For a perpendicularity control to be a legal specification, it must satisfy the following conditions:

- A datum must be referenced in the feature control frame.
- If it is applied to a surface, the projected tolerance zone, diameter, MMC, and LMC modifiers may not be used in the tolerance portion of the feature control frame. If it is applied to a FOS, modifiers may be used.
- The tolerance value specified must be less than any other geometric tolerances that control the perpendicularity of the feature (for example, tolerance of position, total runout, concentricity, and profile).

Figure 7-7 shows a legal specification flowchart for a perpendicularity specification. This chart applies to RFS datum references only.

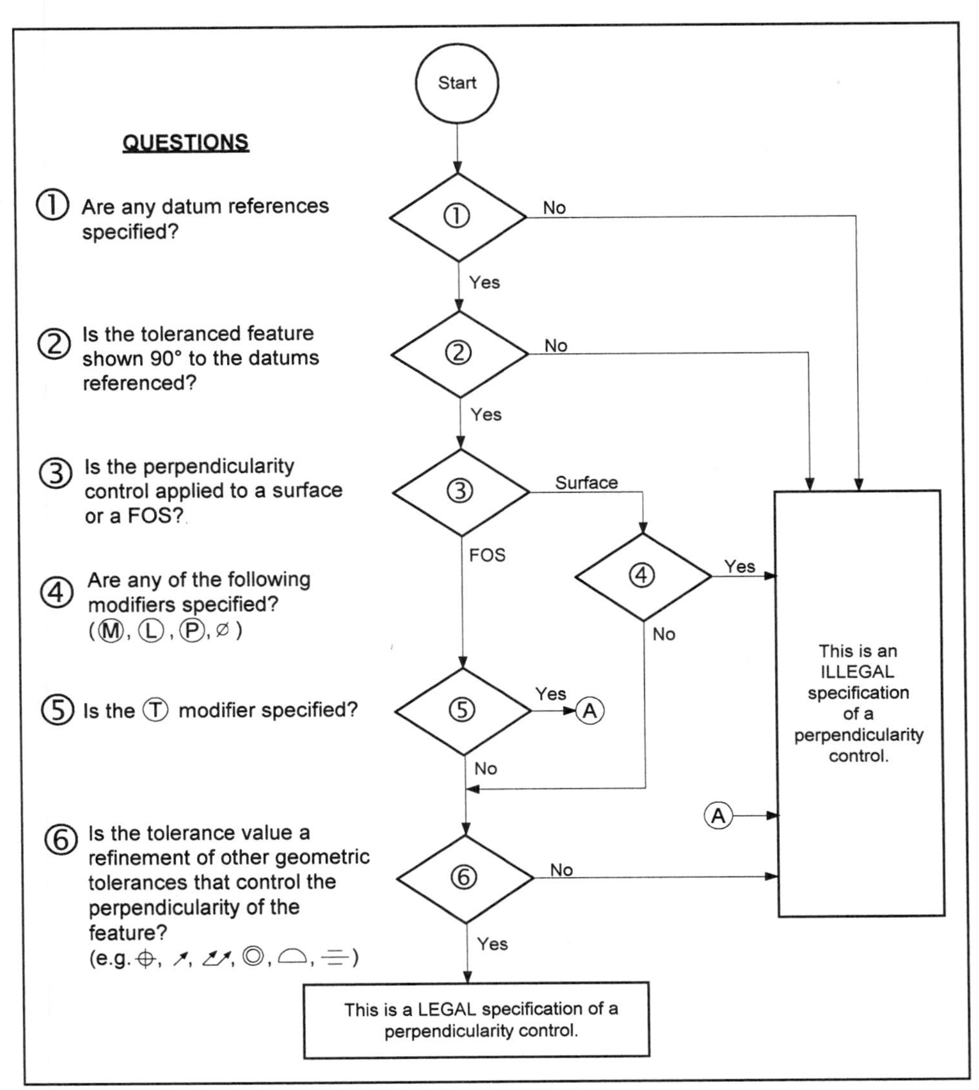

FIGURE 7-7 Legal Specification Flowchart for Perpendicularity

Author's Comment

When a perpendicularity control is applied to a cylindrical FOS, it usually contains a diameter symbol modifier.

For more info. . .
An example of how to gage perpendicularity applied to a diameter is shown in Figure 7-6.

Inspecting Perpendicularity

Figure 7-3 shows a part with a perpendicularity specification. When inspecting this part, three separate checks are required: the size of the FOS, the Rule #1 boundary, and the perpendicularity requirement. Chapter 2 discussed how to check the size and Rule #1 boundary; now we will look at how to inspect the perpendicularity requirement. One way to inspect the perpendicularity control on the part from Figure 7-3 is shown in Figure 7-8:

First, rest the part on the datum(s) specified, in this case datum *A*. Next, a precision square is placed on the datum plane and is brought into contact with the part surface being inspected. The gap between the precision square and the part surface is the perpendicularity error of the part. A gage wire with a diameter equal to the perpendicularity tolerance value is slipped into the gap between the precision square and the part surface. If the gage wire cannot fit into the gap, the surface is within its perpendicularity tolerance value.

Author's Comment
Additional measurements could be obtained by placing a measuring device through the holes on the angle plate to measure the gap between the plate and the part surface.

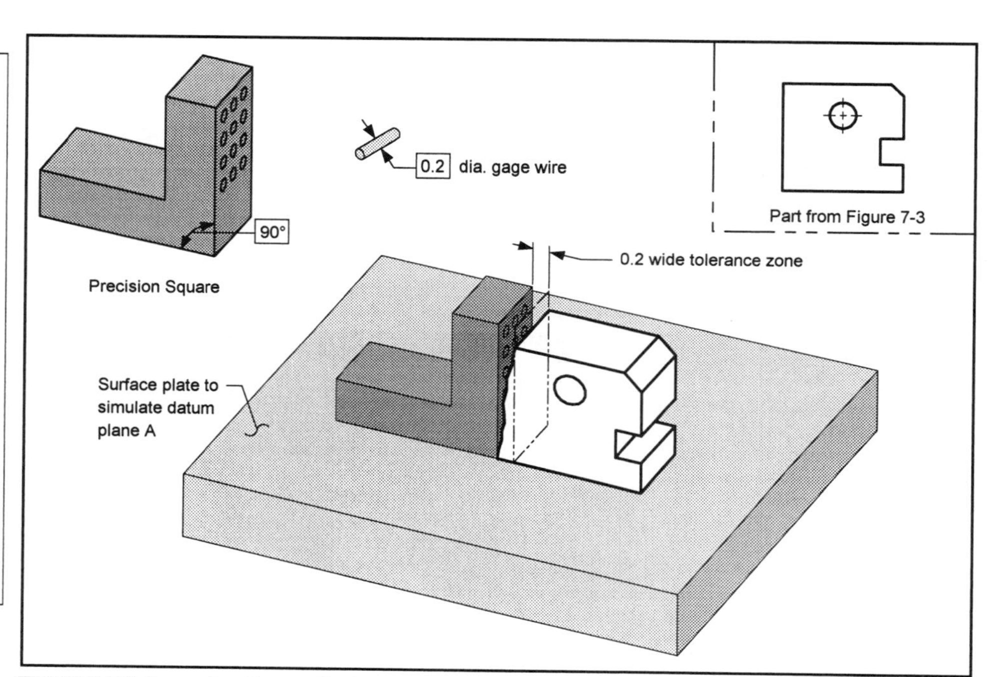

FIGURE 7-8 Inspecting Perpendicularity

ANGULARITY CONTROL

Definition

Angularity is the condition of a surface, centerplane, or axis being exactly at a specified angle. An ***angularity control*** is a geometric tolerance that limits the amount a surface, axis, or centerplane is permitted to vary from its specified angle.

For more info. . .
See Paragraph 6.6.2 of Y14.5.

Angularity Tolerance Zones

The two common tolerance zone shapes for an angularity control are:

1. Two parallel planes
2. A cylinder

Author's Comment
One aspect of angularity that is of concern to inspectors is that the specified tolerance is not in degrees, but in a parallel plane tolerance zone.

Angularity Applications

Most angularity applications fall into one of two general cases:

1. Angularity applied to a surface, or
2. Angularity applied to a cylindrical FOS

TECHNOTE 7-4 Angularity Applied To A Surface

- The shape of the tolerance zone is two parallel planes.
- The tolerance zone is oriented relative to the datum plane with a basic angle.
- All of the elements of the surface must be within the tolerance zone.
- The angularity tolerance zone also limits the flatness of the toleranced surface.

When an angularity control is applied to a surface, the WCB of the toleranced surface is not affected. When an angularity control is applied to a FOS, the WCB of the FOS is affected. The WCB of a FOS that is toleranced with an orientation control is oriented relative to the datums specified.

In Figure 7-9, an angularity control is applied to a surface; this is the most common application of angularity. In an angularity application, the part feature being controlled must be dimensioned with a basic angle relative to the datums specified. When angularity is applied to a surface, the following conditions apply:

- The shape of the tolerance zone is two parallel planes.
- The angularity control tolerance value defines the distance between the tolerance zone planes.
- All the elements of the surface must be within the tolerance zone.
- The tolerance zone is oriented relative to the datum plane by a basic angle.
- The angularity tolerance zone also limits the flatness of the toleranced surface.

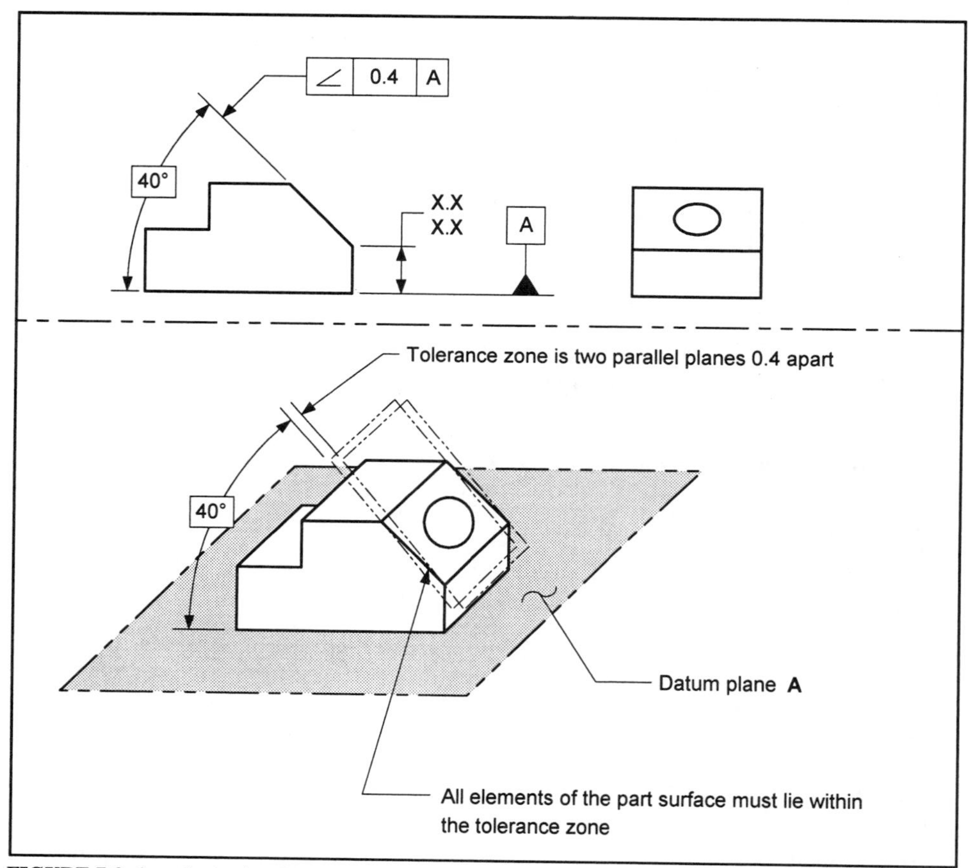

FIGURE 7-9 Angularity Applied to a Surface

In Figure 7-10, an angularity control is applied to a diametrical FOS. Note the use of the diameter modifier in the tolerance portion of the feature control frame. When angularity is applied to a diameter, it controls the orientation of the axis of the diameter. In Figure 7-10, the following conditions apply:

- The tolerance zone is a cylinder.
- The angularity control tolerance value defines the diameter of the tolerance cylinder.
- The axis of the toleranced feature must be within the tolerance zone.
- The tolerance zone is oriented relative to the datum plane by a basic angle.
- An implied 90° basic angle exists in the other direction.

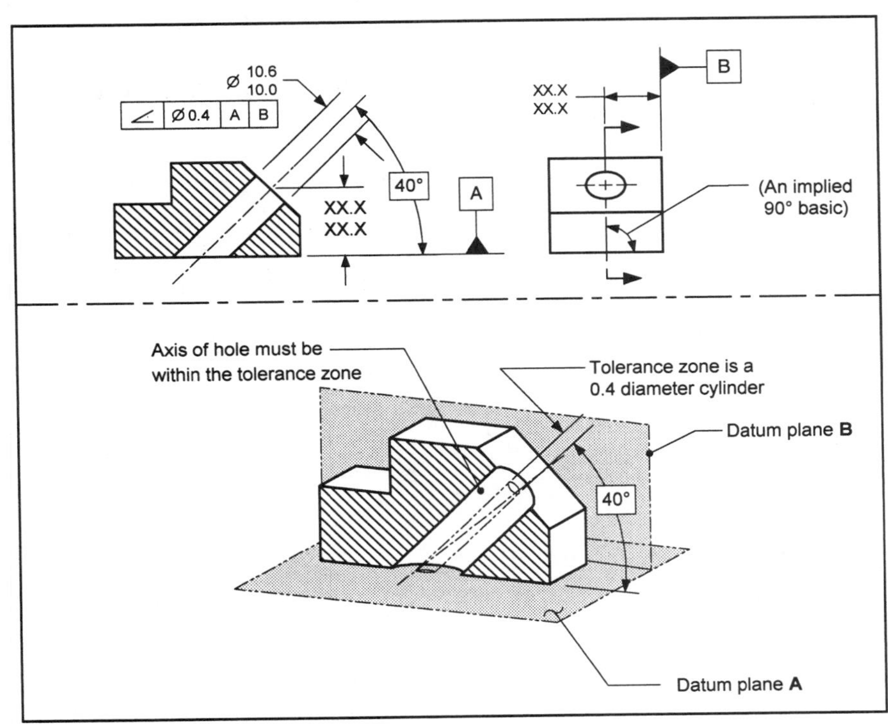

FIGURE 7-10 Angularity Applied to a FOS

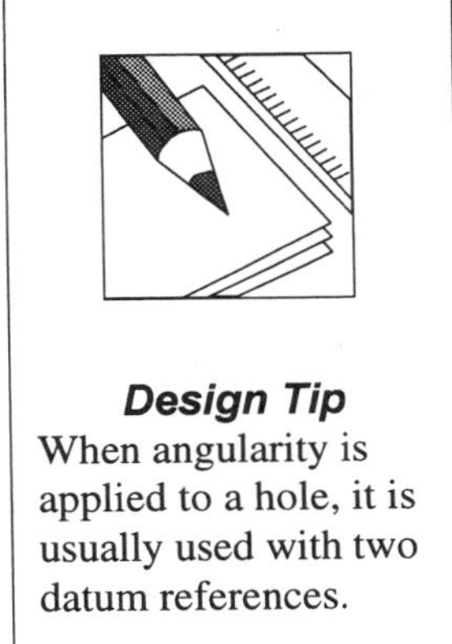

Design Tip
When angularity is applied to a hole, it is usually used with two datum references.

Indirect Angularity Controls

There are several geometric controls that can indirectly affect the angularity of a part feature: tolerance of position, total runout, and profile can limit angularity. However, indirect angularity controls are not inspected; their effect on angularity is a result of the part surface, axis, or centerplane being within the tolerance zone for the geometric control. If it is desired to have the angularity or a part feature inspected, an angularity control should be specified. If an angularity control is used, its tolerance value should be less than the tolerance value of any indirect angularity controls that apply.

Legal Specification Test for an Angularity Control

For an angularity control to be a legal specification, it must satisfy the following conditions:

- One or more datum planes, a datum axis, or centerplane must be referenced in the feature control frame.
- If it is applied to a surface, the projected toleranced zone, diameter, MMC, and LMC modifiers may not be used in the tolerance portion of the feature control frame. (If it is applied to a FOS, modifiers may be used.)
- A basic angle must be specified relative to the datums referenced.
- The tolerance value specified must be a refinement of any other geometric tolerances that control the angularity of the feature (for example, tolerance of position, runout, and profile).

Figure 7-11 shows a legal specification flowchart for an angularity control. The chart applies to RFS datum references only.

Author's Comment

When an angularity control is applied to a cylindrical FOS, it usually contains a diameter symbol modifier.

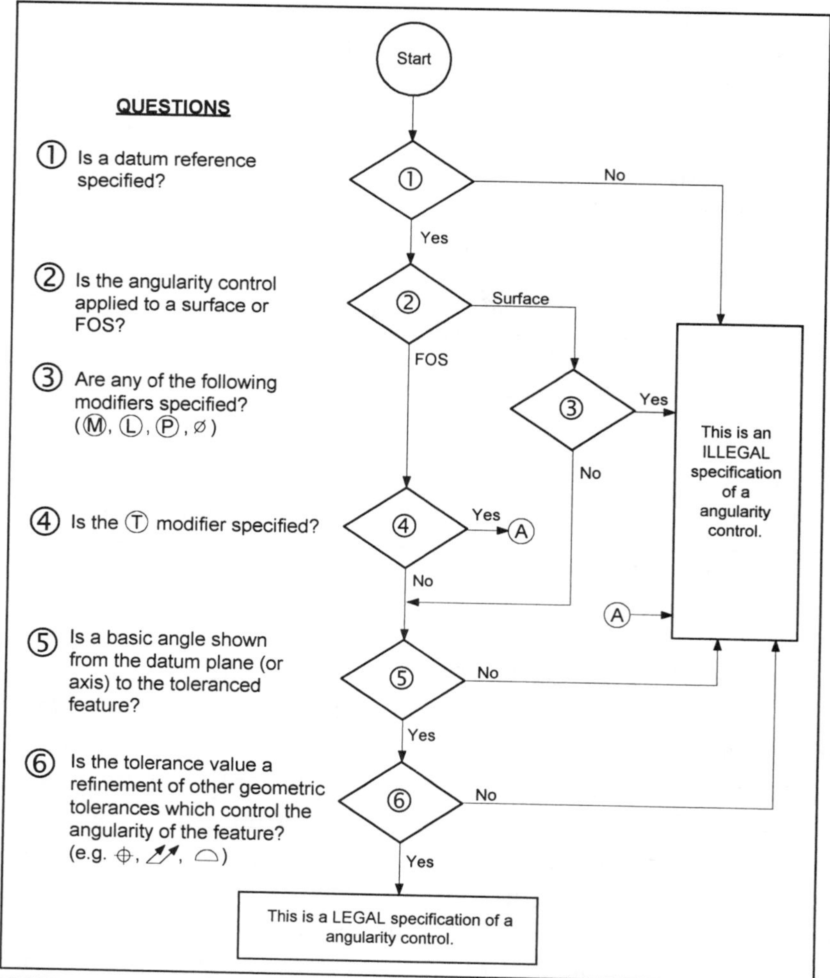

FIGURE 7-11 Legal Specification Flowchart for Angularity

Verifying Angularity

Figure 7-9 shows a part with an angularity specification. Let's look at how to verify the angularity requirement.

Figure 7-12 shows one way the angularity control could be verified:

The part is mounted on the gage equipment using a sine plate set at the basic angle to bring the toleranced surface parallel to the surface plate. A dial indicator is used to verify that the surface elements are within the angularity tolerance zone.

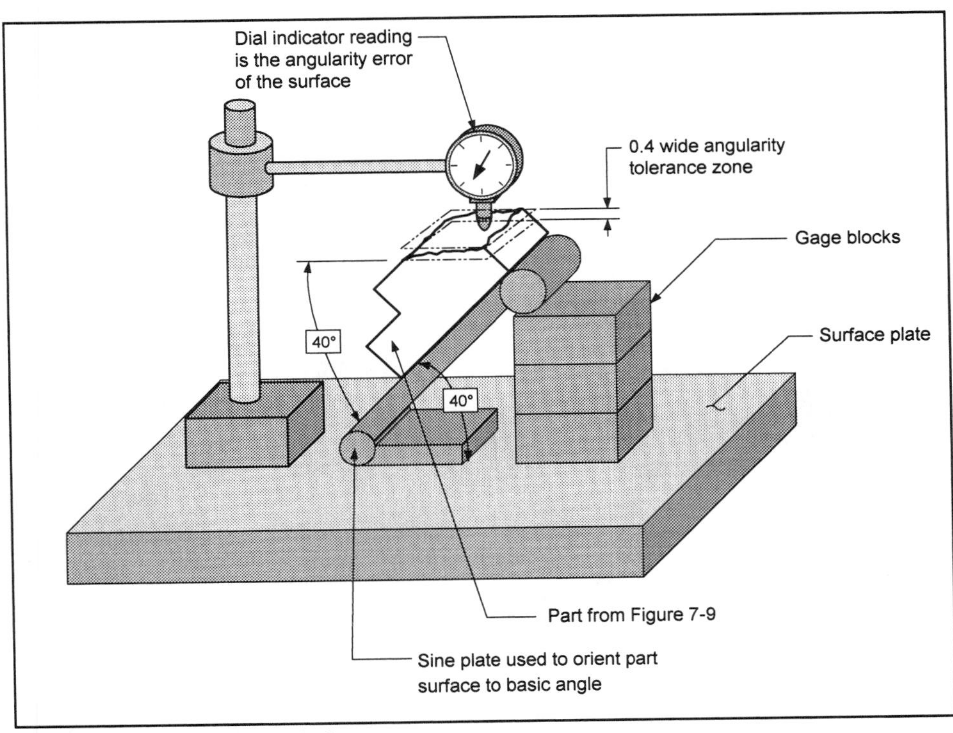

FIGURE 7-12 Verifying Angularity

PARALLELISM CONTROL

Implied Parallelism

Wherever two surfaces are shown to be parallel on a drawing, the size dimension of the surfaces controls the parallelism between the surfaces (see Figure 7-13). This method is satisfactory for some drawings, but it contains two shortcomings. The first is that the parallelism requirement is the same value as the size requirement. The second shortcoming is the lack of a datum reference. In Figure 7-13, the part could be inspected from either side. This would produce different measurements by different inspectors.

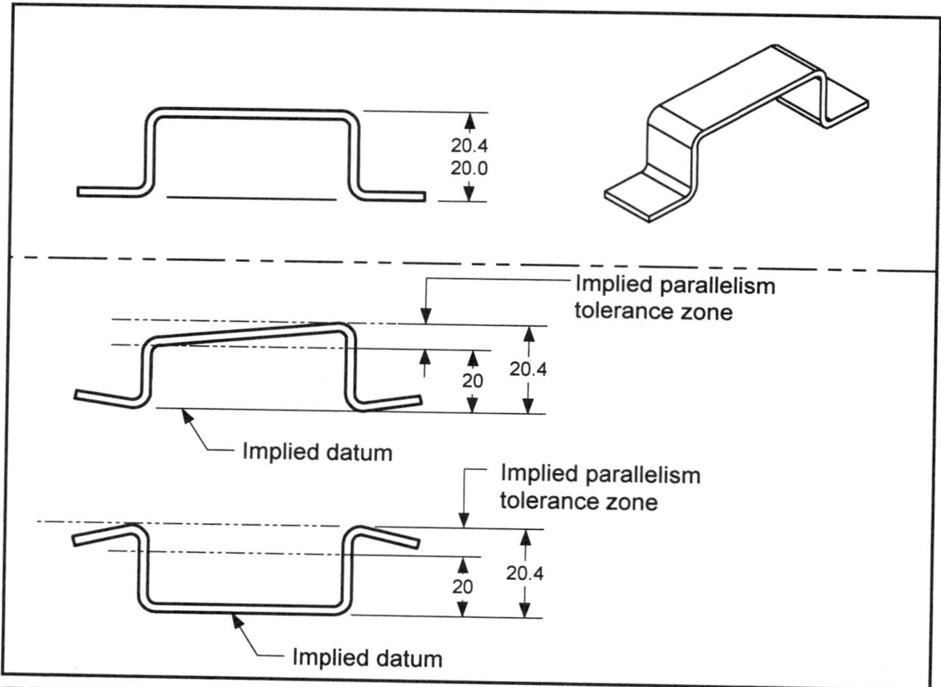

FIGURE 7-13 Implied Parallelism

TECHNOTE 7-5 Implied Parallelism

Unless otherwise specified, the parallelism of two surfaces shown as parallel on a drawing is controlled by the limits of the dimension between the surfaces.

For more info. . .
See Paragraph 6.6.3 of Y14.5.

Definition of a Parallelism Control

Parallelism is the condition that results when a surface, axis or centerplane is exactly parallel to a datum. A ***parallelism control*** is a geometric tolerance that limits the amount a surface, axis, or centerplane is permitted to vary from being parallel to the datum.

Parallelism Tolerance Zones

The two common tolerance zones for a parallelism control are:

1. Two parallel planes
2. A cylinder

The following applications show these tolerance zones and discuss their use.

Author's Comment
The third tolerance zone shape possible for parallelism is two parallel lines. It is not very common, and it is beyond the scope of this text.

Parallelism Applications

Most parallelism applications fall into one of two general cases:

1. Parallelism applied to a surface, or
2. Parallelism applied to a diameter (MMC)

In Figure 7-14, a parallelism control is applied to a surface. This is the most common application of parallelism. When parallelism is applied to a surface, the following conditions apply:

- The tolerance zone is two parallel planes that are parallel to the datum plane.
- The tolerance zone is located within the limits of the size dimension.
- The tolerance value of the parallelism control defines the distance between the tolerance zone planes.
- All the elements of the surface must be within the tolerance zone.
- The parallelism tolerance zone limits the flatness of the toleranced feature.

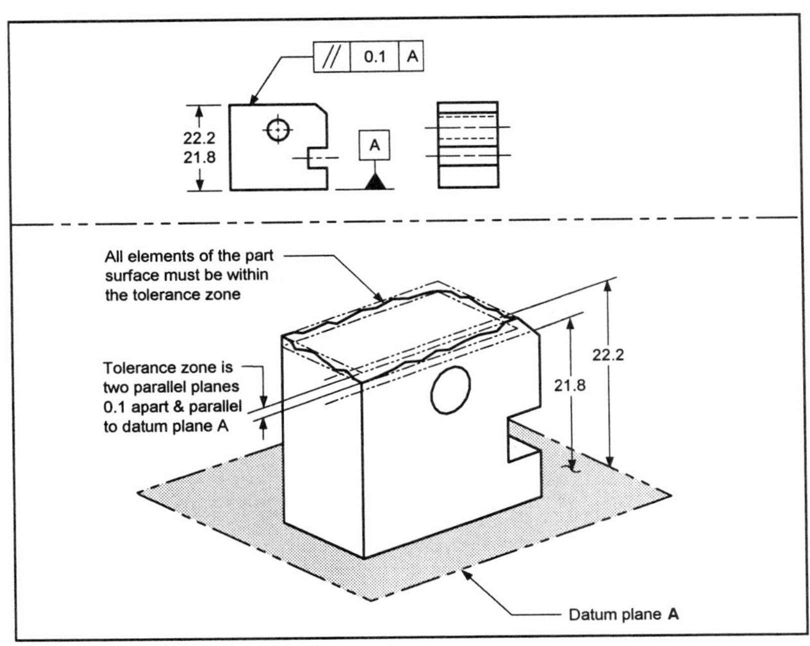

FIGURE 7-14 Parallelism Applied to a Surface

In Figure 7-15, a parallelism control that contains the MMC modifier is applied to a cylindrical FOS. This type of geometric control is often used to ensure the function of assembly. The parallelism control is applied to a diameter and contains the MMC modifier. The following conditions apply:

- The tolerance zone is a cylinder that is parallel to the datum plane.
- The tolerance value of the parallelism control defines the diameter of the tolerance zone cylinder.
- The axis of the diameter must be within the tolerance zone (when the FOS is at MMC).
- A bonus tolerance is permissible.
- A fixed gage may be used to verify the parallelism control.
- The WCB (or virtual condition) of the hole is affected.

When calculating bonus tolerance for a FOS that is toleranced with an orientation control, the actual mating envelope is oriented relative to the datums specified.

The gage for verifying parallelism (at MMC) is shown in Figure 7-15. The gage contains a surface that serves as a simulated datum feature for datum plane *A*. The gage also contains a pin that verifies the parallelism of the hole. The gage pin diameter is equal to the virtual condition of the hole (10.2 - 0.1 = 10.1). The gage pin must slide to accommodate the tolerance for the location of the hole.

When a parallelism control is applied to a surface, the WCB of the toleranced surface is not affected. When a parallelism control is applied to a FOS, the WCB of the FOS is affected. The WCB of a FOS that is toleranced with an orientation control is oriented relative to the datums specified.

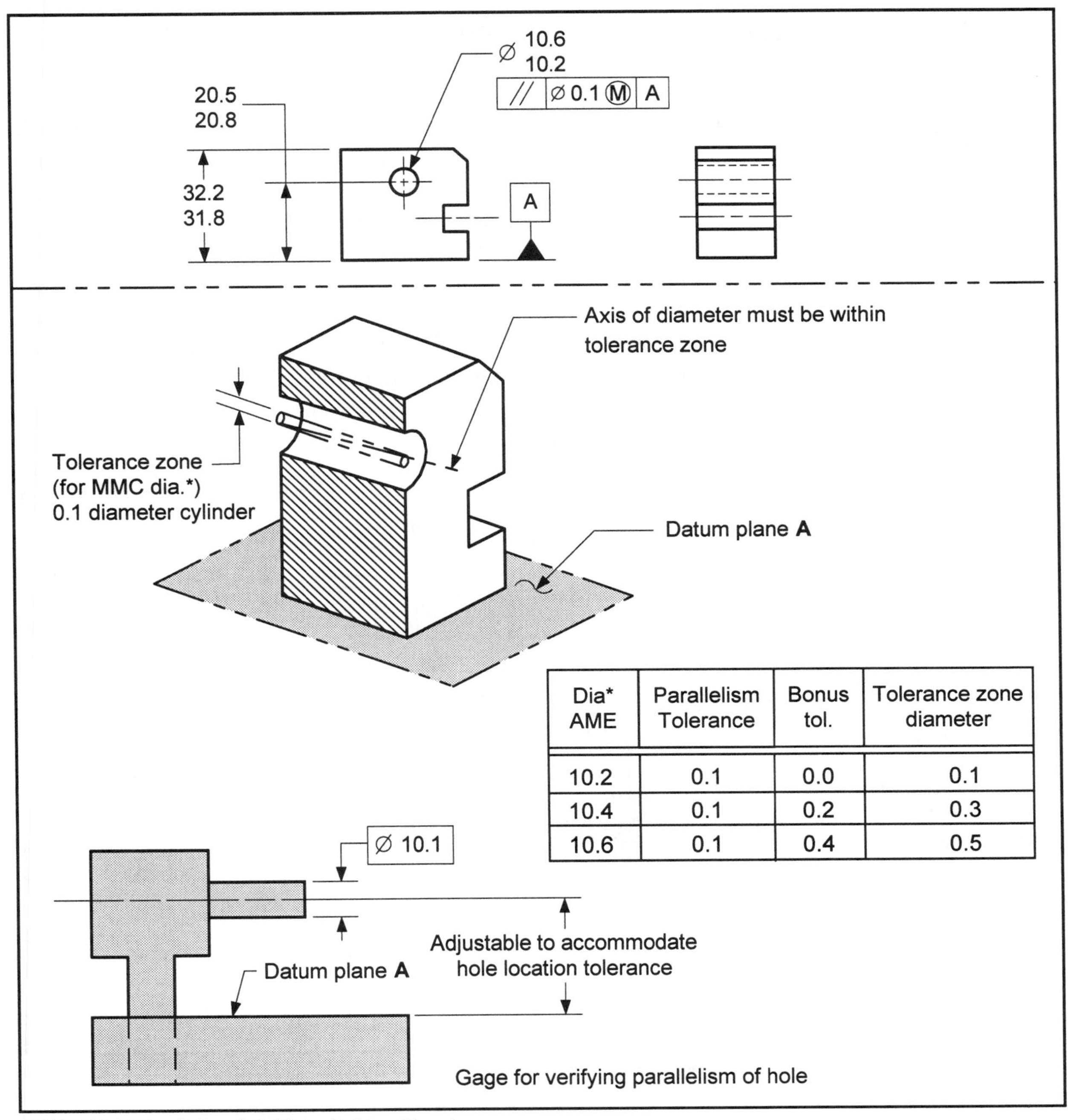

Dia* AME	Parallelism Tolerance	Bonus tol.	Tolerance zone diameter
10.2	0.1	0.0	0.1
10.4	0.1	0.2	0.3
10.6	0.1	0.4	0.5

FIGURE 7-15 Parallelism (MMC) Applied to a Diameter

TECHNOTE 7-6 Parallelism (at MMC) Applied to a FOS

- The tolerance zone is a cylinder (or two parallel planes).
- The axis (or centerplane) must be within the tolerance zone.
- A bonus tolerance is permissible.
- A fixed gage may be used to verify the parallelism control.
- The WCB of the FOS is affected.

Parallelism with the Tangent Plane Modifier

Another use for a parallelism application is with the tangent plane modifier. The tangent plane modifier denotes that only the tangent plane established by the high points of the controlled surfaces must be within the parallelism tolerance zone. When the tangent plane modifier is used in parallelism callouts, the flatness of the toleranced surfaces is not controlled. Figure 7-16 shows a parallelism application that uses the tangent plane modifier. The following conditions apply:

- The tolerance zone is two parallel planes.
- The tangent plane established by the high points of the surface(s) must be within the 0.1 parallelism tolerance zone.
- The flatness of the toleranced surface is not controlled by the parallelism control.

Author's Comment
The tangent plane modifier can also be used with the perpendicularity and angularity controls.

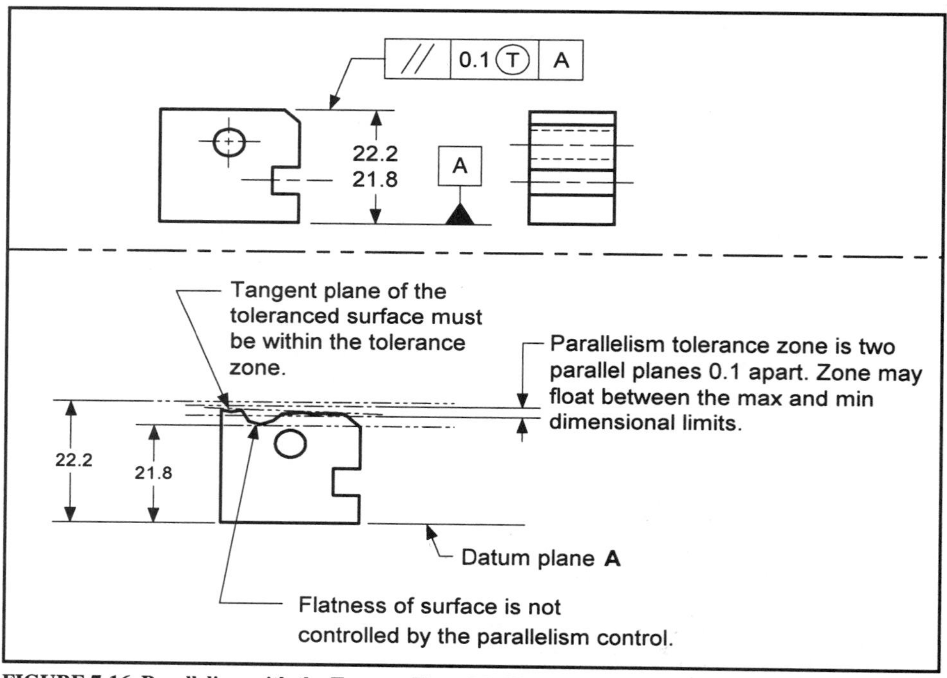

FIGURE 7-16 Parallelism with the Tangent Plane Modifier

Design Tip
In many cases, using the tangent plane modifier can reduce manufacturing costs. When using an orientation control, evaluate the requirements for the form of the surface. If the flatness tolerance could be greater than the orientation tolerance, consider using the tangent plane modifier.

TECHNOTE 7-7 Parallelism with the Tangent Plane Modifier

When the tangent plane modifier is used in a parallelism control:

- Only the tangent plane of the toleranced surface must be within the parallelism tolerance zone.
- The flatness of the surface is not controlled by the parallelism callout.

For more info. . .
See Paragraph 6.6.1.3 in Y14.5.

Indirect Parallelism Controls

There are several geometric controls that can indirectly affect the parallelism of a part feature: tolerance of position, total runout, and profile can limit parallelism in certain cases. However, indirect parallelism controls are not inspected; their effect on parallelism is a result of the part surface, axis, or centerplane being within the zone for the specified geometric control. If a parallelism control is used, its tolerance value should be less than the tolerance value of any indirect parallelism controls that apply.

Legal Specification Test for a Parallelism Control

For a parallelism control to be a legal specification, it must satisfy the following conditions:

- One or more datum planes, a datum axis, or centerplane must be referenced in the feature control frame.
- If it is applied to a surface, the LMC or MMC modifiers may not be used in the tolerance portion of the feature control frame.
- The tolerance value specified must be less than any other geometric tolerances that control the parallelism of the feature (for example, tolerance of position, runout, and profile).

Figure 7-17 shows a legal specification flowchart for a parallelism control.

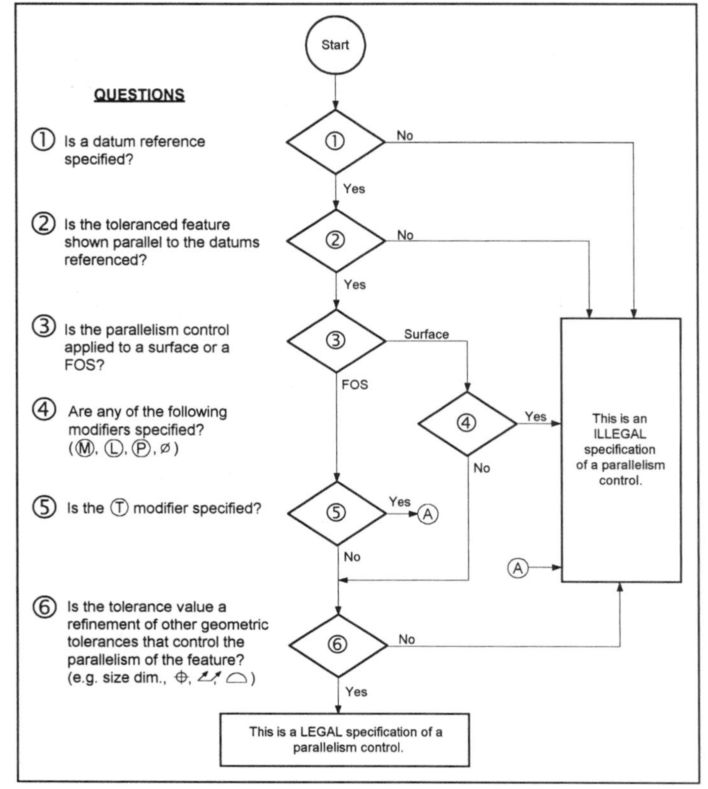

FIGURE 7-17 Legal Specification Flowchart for Parallelism

Author's Comment

When a parallelism control is applied to a cylindrical FOS, it usually contains a diameter symbol modifier.

Verifying Parallelism

Figure 7-14 shows a part with a parallelism specification. Let's look at how to inspect the parallelism requirement.

One way the parallelism controls could be inspected is shown in Figure 7-18:

The part is placed on a surface plate. A dial indicator is used to find the maximum variation between the high and low points of the toleranced surface. The difference between the max. and min. dial indicator reading is the parallelism error of the part surface.

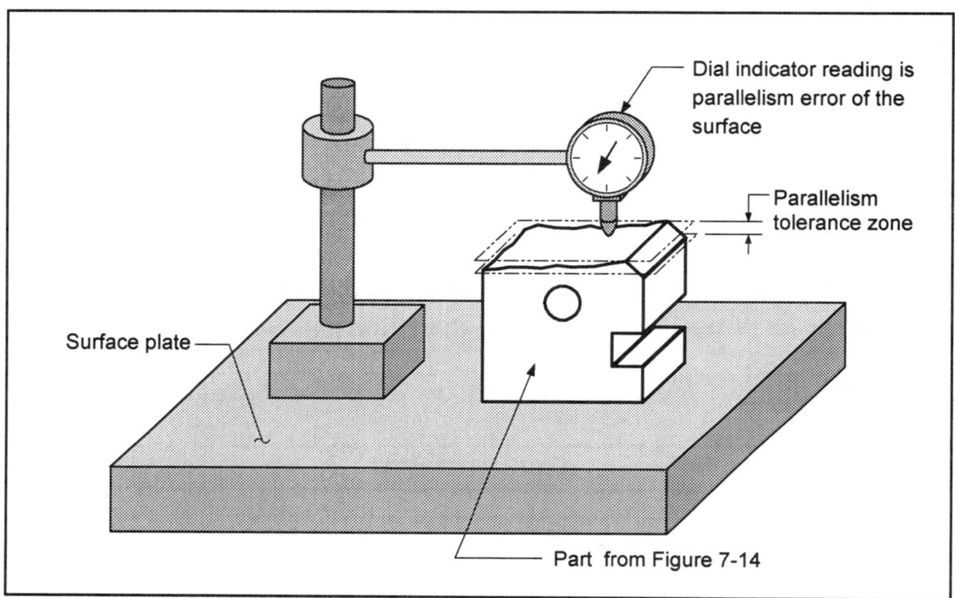

FIGURE 7-18 Verifying Parallelism

Summary

A summarization of orientation control information is shown in Figure 7-19.

Symbol	Datum reference required	Can be applied to a Surface	Can be applied to a FOS	Can affect WCB	Can use Ⓜ or Ⓛ modifier	Can be applied at RFS	Can use Ⓣ modifier
⊥	Yes	Yes	Yes	Yes*	Yes*	Yes**	Yes•
∠	Yes	Yes	Yes	Yes*	Yes*	Yes**	Yes•
//	Yes	Yes	Yes	Yes*	Yes*	Yes**	Yes•
* When applied to a FOS	** Automatic per Rule #2	• When applied to a surface					

FIGURE 7-19 Summarization of Orientation Controls

VOCABULARY LIST

New Terms Introduced in this Chapter

Angularity
Angularity control
Parallelism
Parallelism control
Perpendicularity
Perpendicularity control

Study Tip
Read each term. If you don't recall its meaning, look it up in the chapter.

ADDITIONAL RELATED TOPICS

Topic	ASME Y14.5M-1994 Reference
• Use of notes to modify tolerance zones	Paragraph 6.6.1

Author's Comment
These topics, plus advanced coverage of many of the topics introduced in this text, will be covered in my new book on advanced GD&T concepts.

QUESTIONS AND PROBLEMS

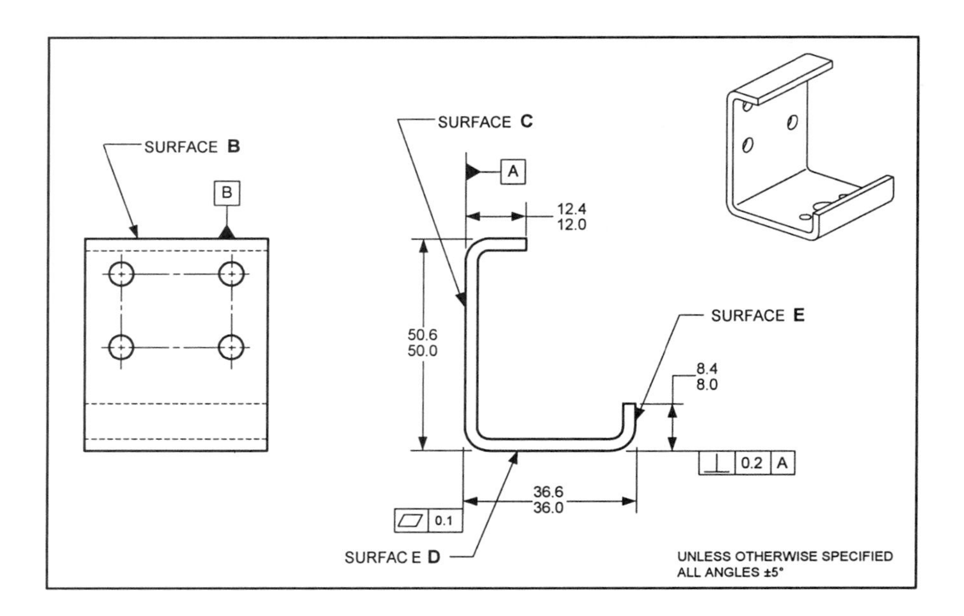

Questions 1-5 refer to the drawing above.

1. What are the shape and size of the tolerance zone for the perpendicularity control?

___

___

2. Fill in the chart below.

The flatness of surface. . .	Is limited to. . .
B	
C	
D	
E	

3. What controls the squareness between surfaces *D* and *E*?

___

4. Describe how the tolerance zone for the perpendicularity callout is oriented. ___

___

5. If the perpendicularity callout was revised to | ⊥ | 0.2 | A | B |, what effect would this have on the tolerance zone?

___

___

6. List two common tolerance zone shapes for a perpendicularity control.

____

____

7. List three conditions that exist when a perpendicularity control is applied to a surface.

____

____

____

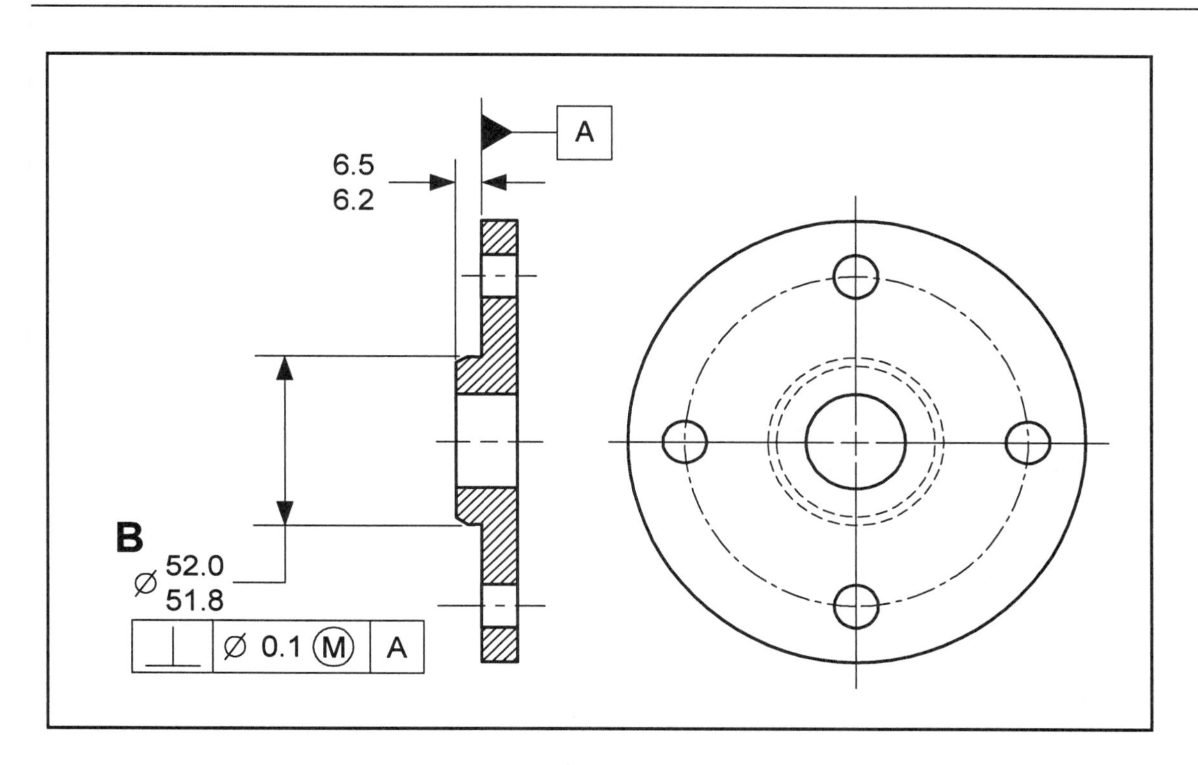

Questions 8-10 refer to the drawing above.

8. What is the size and shape of the perpendicularity control tolerance zone?_______________

____

9. Fill in the chart below.

If the actual size of dia. *B* is...	The bonus tolerance possible is...	The perpendicularity tolerance zone diameter would be...
52.0		
51.9		
51.8		

10. In the space below, draw and dimension the gage for verifying the perpendicularity callout.

11. List three conditions that exist when a perpendicularity control (at MMC) is applied to a FOS.____

____

____

12. For each perpendicularity control shown below, indicate if it is a legal specification. If a control is illegal, explain why.

A. | ⊥ | 0.1 | A | ____________________

B. | ⊥ | ⌀ 0.1 | A | ____________________

C. | ⊥ | ⌀ 0.1 Ⓜ | Z | ____________________

D. | ⊥ | ⌀ 0.1 Ⓢ | Z | ____________________

E. | ⊥ | 0.1 Ⓜ | B | ____________________

F. | ⊥ | ⌀ 0.1 Ⓜ | ____________________

13. List two common tolerance zone shapes for an angularity control.

____

____

14. List three conditions that exist when an angularity control is applied to a surface.

____

____

____

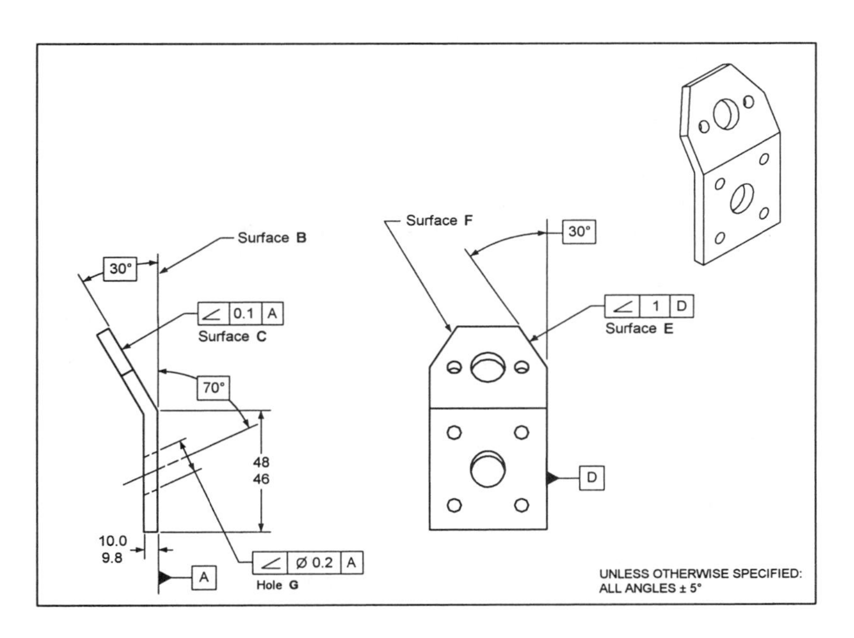

Questions 15-20 refer to the drawing above.

15. What is the shape and size of the tolerance zone for the angularity control applied to surface *C*?____

16. What controls the orientation of the angularity tolerance zone for surface *C*?

____

17. Fill in the chart below.

The angularity of surface. . .	Is limited to. . .
B	
C	

18. Is the angularity control on surface *C* a legal specification?____________ If no, explain.

____

19. Is the angularity control on surface *E* a legal specification? __________ If no, explain.

____

20. Is the angularity control on hole *G* a legal specification? __________ If no, explain.

____

21. List three conditions that exist when angularity is applied to a diameter.

____

____

____

22. List two common tolerance zone shapes for a parallelism control.

____

____

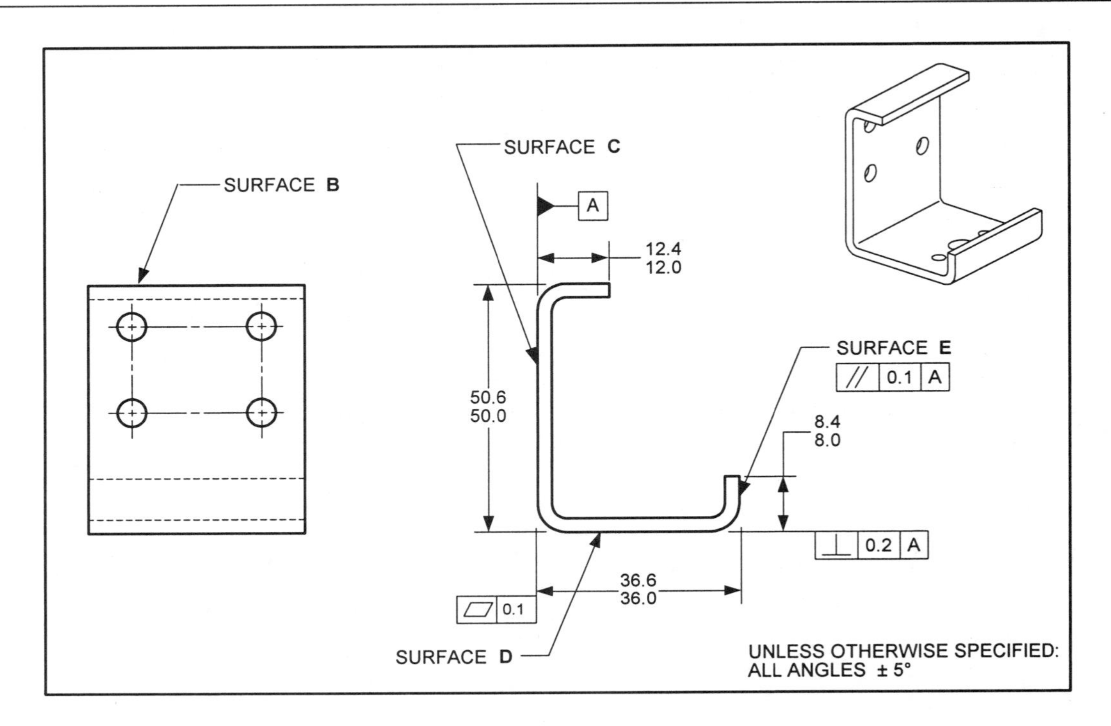

Questions 23-26 refer to the drawing above.

23. What is the shape and size of the tolerance zone for the parallelism control?_______________

____

24. Fill in the chart below.

The parallelism of surface. . .	Is limited to. . .
B	
C	
D	
E	

25. What controls the parallelism between surfaces *B* and *D*? _______________

____

26. Describe how the tolerance zone for the parallelism control is oriented and located. _______
____

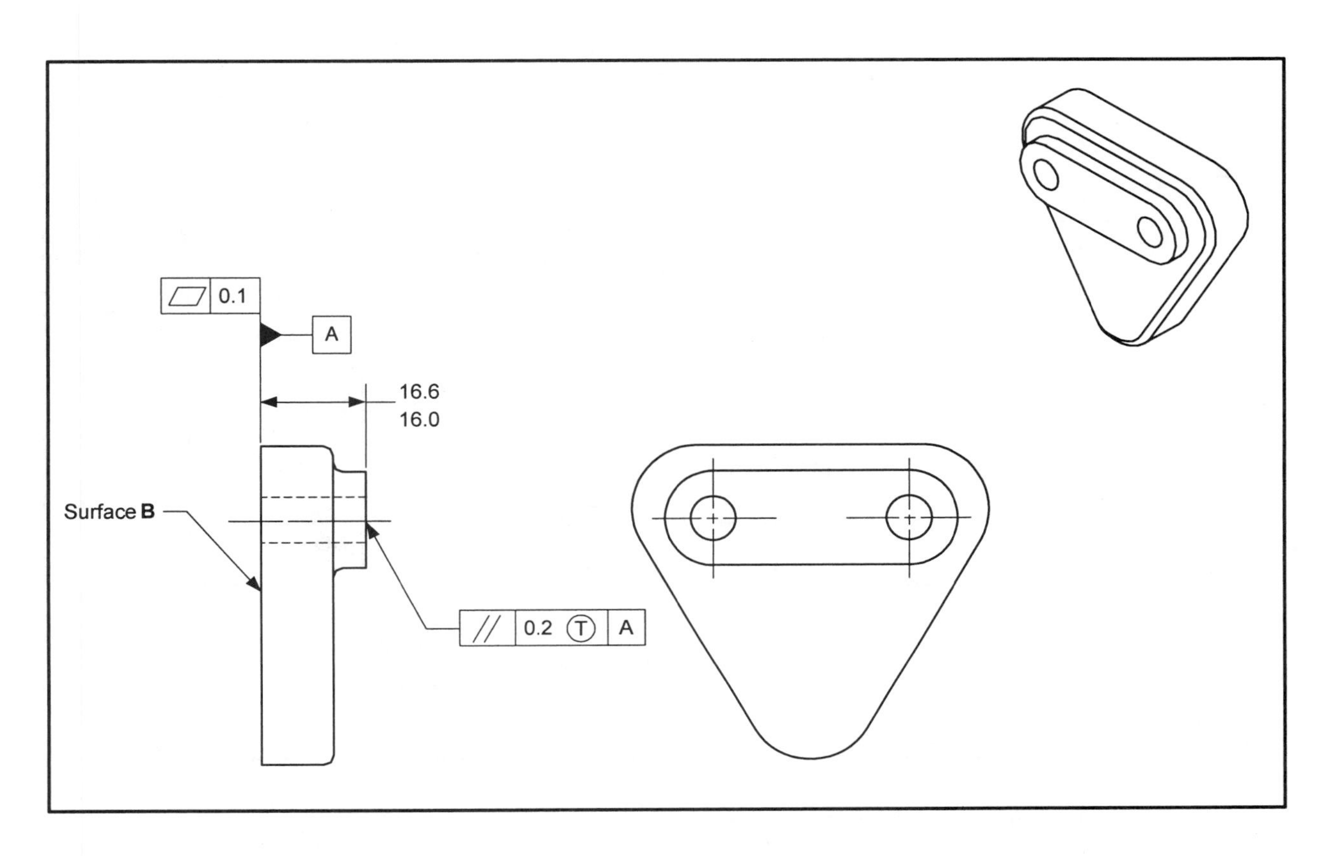

Questions 27-30 refer to the drawing above.

27. The flatness of surface *B* is limited to ____________________________.

28. The flatness of the raised pad is limited to ____________________________.

29. Describe the shape and size of the tolerance zone for the parallelism control.
____
____

30. What effect does the Ⓣ modifier inside the parallelism control have on the part?
____
____

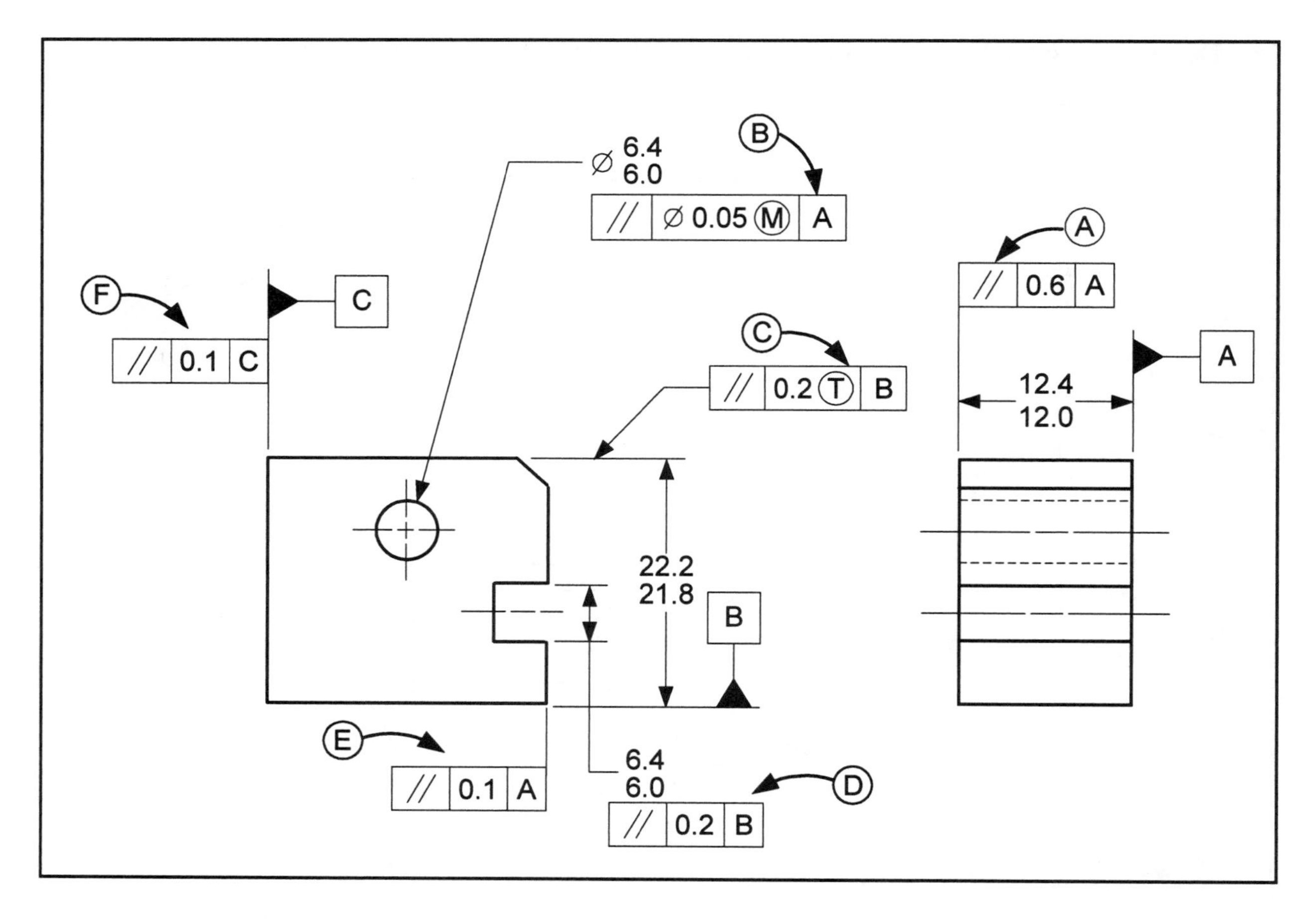

Question 31 refers to the drawing above.

31. For the geometric control listed at each location shown above, indicate if it is a legal specification. If a control is illegal, explain why.

Ⓐ ____

Ⓑ ____

Ⓒ ____

Ⓓ ____

Ⓔ ____

Ⓕ ____

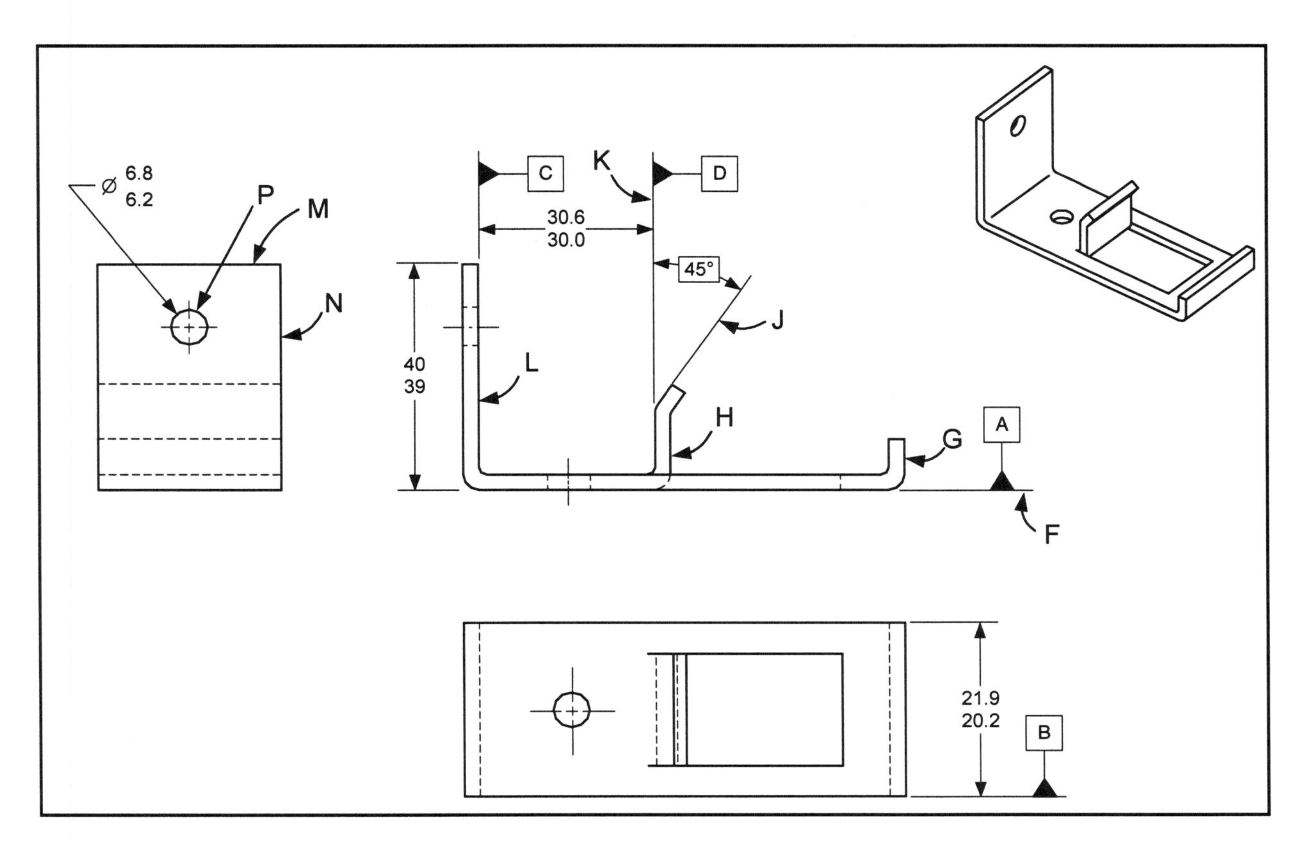

32. Use the instructions below to complete the drawing above.

   a. Add a control to surface *L* that limits its perpendicularity relative to datums *A* & *B* to within 0.2.

   b. Add a control to surface *K* that limits its parallelism relative to datum *C* within 0.2.

   c. Add a control to surface *J* that limits its angularity relative to datum *D* within 0.2.

   d. Add a control to surface *F* that limits its flatness within 0.1.

   e. Add a control to surface *G* that limits its perpendicularity relative to datum *A* within 0.6.

   f. Add a control to diameter *P* that limits its perpendicularity relative to datum *C* within a 0.2 cylindrical tolerance zone when the diameter is at MMC.

   g. Add a control to surface *M* that limits its parallelism relative to datum *A* within 0.2. Use a tangent plane modifier.

   h. Add a control to surface *N* that limits its parallelism relative to datum *B* within 0.4.

Chapter 8

Tolerance of Position, Part 1

Ø 5.6 Theoretical boundary (virtual condition) tolerance zone

Datum plane **C**

12.4
12.0

16.4
16.0

Datum plane **A**

Datum plane **B**

INTRODUCTION

This chapter will help you to read and understand drawings that use tolerance of position (TOP) controls. TOP controls are used to define the location of a FOS. The symbol for TOP is shown in Figure 8-1. TOP also indirectly controls the orientation of a FOS relative to the primary datum shown in the feature control frame.

 Tolerance of Position (TOP)

FIGURE 8-1 Tolerance of Position (TOP)

CHAPTER GOALS AND OBJECTIVES

There are Three Goals in this Chapter:

8-1. Understand the fundamental concepts of tolerance of position: the definitions and conventions, the advantages, and the basic theories.
8-2. Interpret RFS and MMC tolerance of position applications.
8-3. Draw cartoon gages for tolerance of position (MMC) applications.

Performance Objectives that Demonstrate Mastery of These Goals

Upon completion of this chapter, each student should be able to:

Study Tip
Take a few minutes to fully understand these objectives. When reading this chapter, look for information to help you master these objectives.

Goal 8-1 (pp. 211-221)

- Define the term, "true position."
- Describe a TOP control.
- List two types of implied basic relationships common with TOP.
- List six advantages of TOP.
- List four types of relationships that can be controlled with TOP.
- Describe when the MMC modifier should be specified in a TOP control.
- Explain the virtual condition boundary theory for TOP.
- Explain the axis theory for TOP.

Goal 8-2 (pp. 221-234)

- Describe two common tolerance zone shapes for a TOP (RFS).
- List three conditions that apply when a TOP control is applied at RFS.
- Describe the tolerance zone in TOP (RFS) applications.
- Calculate the WCB of a FOS controlled with TOP at RFS.
- List three conditions that exist when an MMC modifier is used in a TOP application.

- Describe the tolerance zone in TOP (MMC) applications.
- Calculate the amount of bonus tolerance available for a TOP application.
- Calculate the amount of datum shift available in a coaxial diameter TOP application.
- Determine if a TOP specification is legal.

Goal 8-3 (pp. 235-239)
- Define the term, "functional gage."
- List five benefits of a functional gage.
- Define the term, "cartoon gage."
- Draw and dimension a cartoon gage for a TOP application.

TOP GENERAL INFORMATION

Definitions and Conventions

True position is the theoretically exact location of a FOS as defined by basic dimensions. A ***tolerance of position (TOP) control*** is a geometric tolerance that defines the location tolerance of a FOS from its true position. When specified on an RFS basis, a TOP control defines a tolerance zone that the center, axis, or centerplane of the AME of a FOS must be within. When specified on an MMC or LMC basis, a TOP control defines a boundary—often referred to as the virtual condition—that may not be violated by the surface or surfaces of the considered feature.

Where it is desired to specify a TOP on an RFS basis, the feature control frame does not show any modifiers. RFS is the default condition for all geometric tolerances. Where it is desired to specify a TOP on an MMC or LMC basis, the appropriate modifier is shown in the tolerance portion of the feature control frame. MMC and LMC modifiers may also be specified in the datum portion of the feature control frame where desired and appropriate. See Figure 8-2.

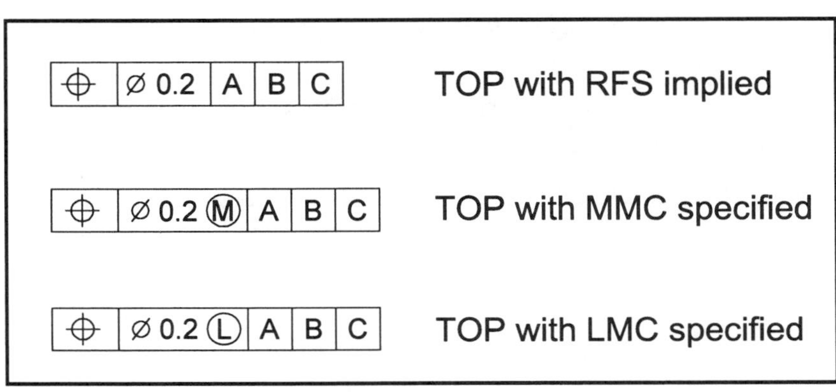

FIGURE 8-2 TOP Feature Control Frames

Whenever a TOP control is specified, the theoretically exact location of the axis or centerplane of the feature of size must be defined with basic dimensions. The theoretically exact location of a FOS as defined by basic dimensions is called the true position of the FOS. An example of a TOP tolerance zone and its true position are shown in Figure 8-3.

Author's Comment
Whenever a geometric control with datum references is used, it controls the orientation of the toleranced feature relative to the primary datum referenced.

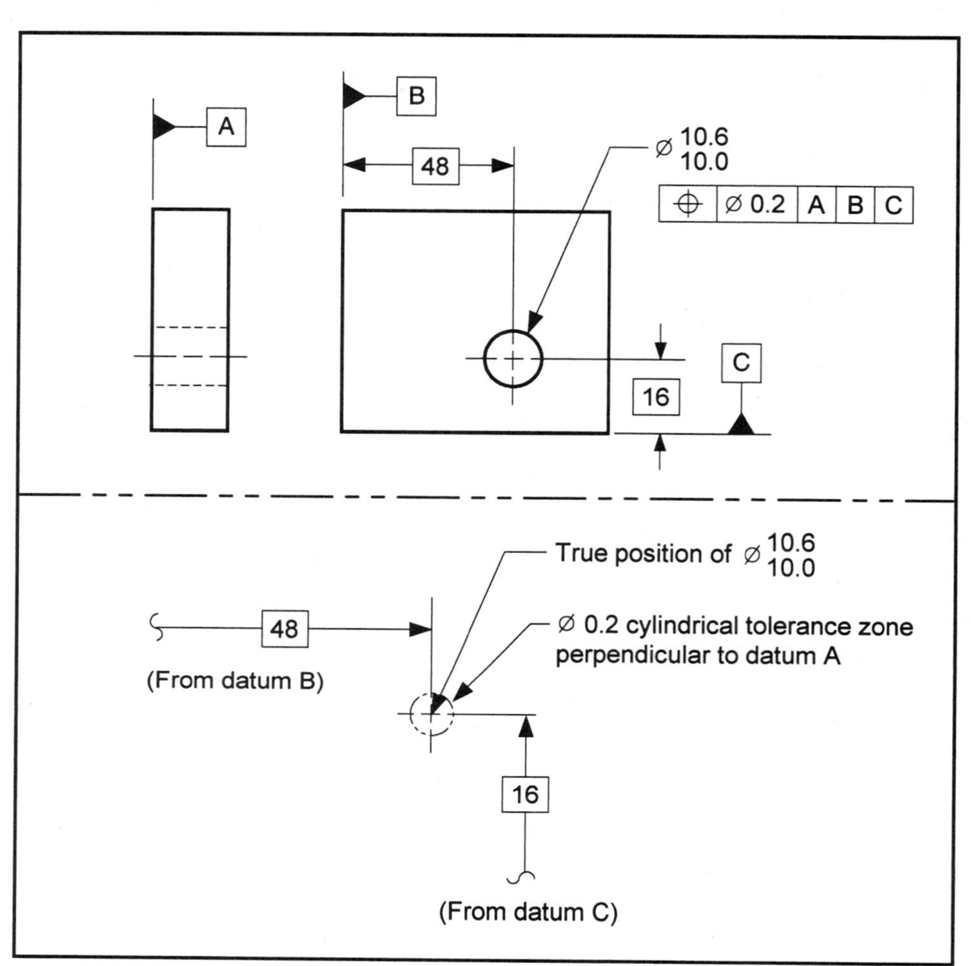

FIGURE 8-3 TOP Tolerance Zone

TECHNOTE 8-1 True Position

When a TOP callout is specified, the true position of the FOS is the theoretically exact location as defined by the basic dimensions.

Basic dimensions define the true position of the toleranced FOS relative to the datums referenced in the feature control frame. In certain cases, the basic dimensions in a TOP application are not specified; they are implied. There are two types of implied basic dimensions common in TOP applications:

Implied basic 90° angles—A 90° basic angle applies where centerlines of features in a pattern (or surfaces shown at right angles on a drawing) are located and defined by basic dimensions and no angle is specified. An example is shown in Figure 8-4.

Implied basic zero dimension—Where a centerline or centerplane of a FOS is shown in line with a datum axis or centerplane, the distance between the centerlines or centerplanes is an implied basic zero. An example is shown in Figure 8-4.

For more info. . .
Fundamental Dimensioning Rule #6 is shown on page 12.

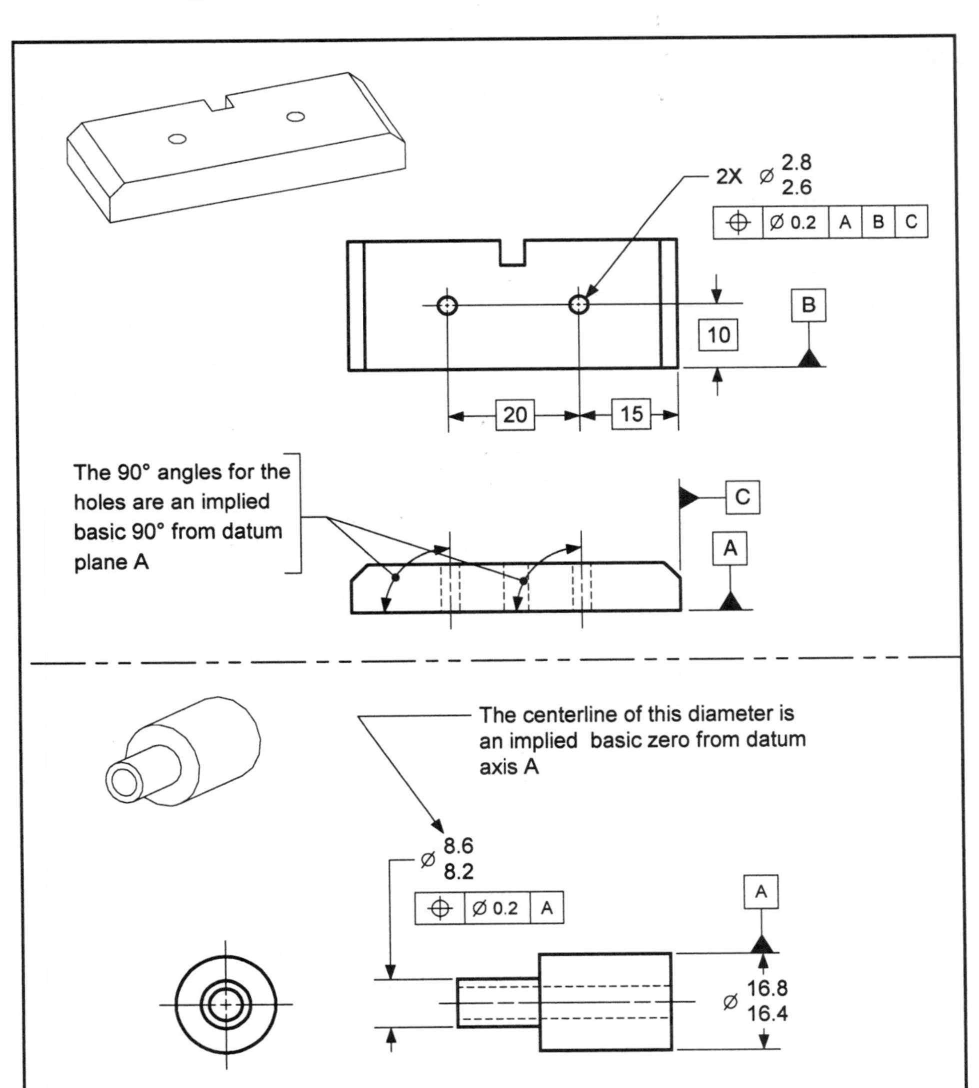

FIGURE 8-4 Implied Basic 90° Angles and Implied Basic Zero Dimension

Advantages of TOP

In comparison with coordinate tolerancing, TOP offers many advantages. Six important advantages are that TOP:

1. Provides larger tolerance zones; cylindrical tolerance zones are 57% larger than square zones
2. Permits additional tolerances—bonus and datum shift
3. Prevents tolerance accumulation
4. Permits the use of functional gages
5. Protects the part function
6. Lowers manufacturing costs

These advantages will become obvious as TOP is explained in this chapter; however, a few statements to highlight the advantages of TOP are in order.

First, the use of cylindrical tolerance zones is common with TOP controls. The advantage of cylindrical tolerance zones is shown in Figure 8-5. The 57% additional tolerance for the location of a FOS gained by using a cylindrical tolerance zone is a significant advantage for manufacturing. Additional tolerances—like bonus and datum shift—can easily add 50-100% or more additional tolerance in certain cases without affecting function. (Examples are shown later in this chapter.) Cylindrical tolerance zones, bonus tolerances, and datum shift all contribute to a clear interpretation and lower manufacturing costs.

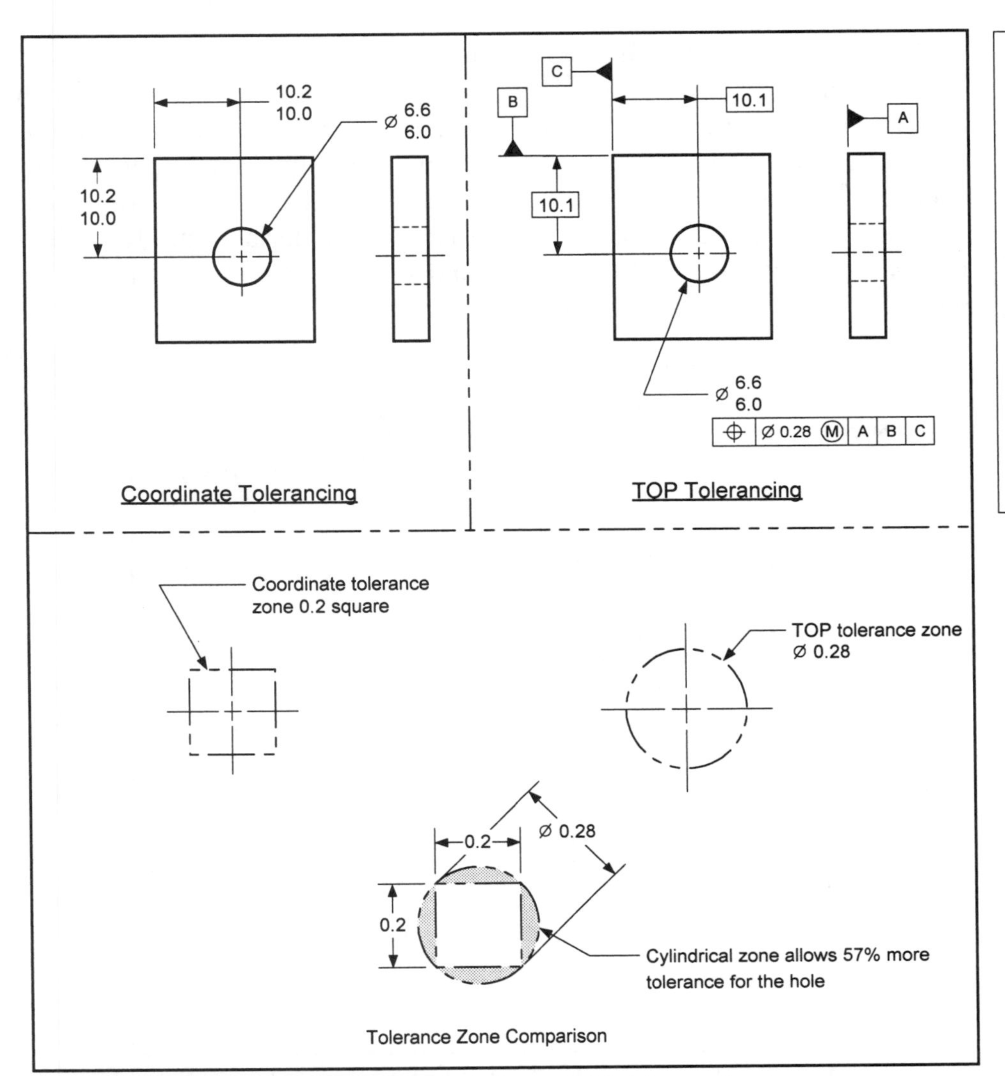

FIGURE 8-5 Comparison of Coordinate Tolerancing and TOP Tolerance Zones

Author's Comment

With coordinate tolerancing, it is not stated whether the holes are defined from surfaces or surfaces are defined from the holes.

Types of Part Relationships that Can be Controlled with TOP

TOP is commonly used to control four types of part relationships:

1. The distance between features of size, such as holes, bosses, slots, tabs, etc.
2. The location of features of size (or patterns of features of size) such as holes, bosses, slots, tabs, etc.
3. The coaxiality between features of size
4. The symmetrical relationship between features of size

Guide to TOP Modifier Usage

When specifying TOP controls, the designer must specify under which material condition the control is to apply. Figure 8-6 provides a guide for determining when the MMC or LMC condition should be specified or when the RFS condition should be invoked. Note that the function of the FOS being toleranced is the primary criteria for material condition selection. Also, the relative cost to produce and verify a FOS is most favorable when the MMC modifier is used.

Modifier	Commonly used in these functional applications	Bonus or datum shift permissible	Relative cost to produce and verify
Ⓜ	· Assembly · Location of a non-critical FOS	Yes	Lowest
Ⓛ	· Minimum wall thickness · Minimum part distance · Minimum machine stock · Alignment	Yes	Greater than MMC; less than RFS
RFS invoked by showing no modifier	· To control a symmetrical relationship · When the effects of bonus or datum shift will be detrimental to the function of the part · To control minimum machine stock. · Centering · Alignment	No	Highest

FIGURE 8-6 Guide for Selecting Modifiers in TOP Controls Based on Product Function

When considering the functions of a FOS, it is often found that assembly with other parts is required; therefore, the MMC modifier is the most commonly used modifier in TOP controls. Also, the MMC modifier is the least expensive option for producing and verifying a FOS.

TOP THEORIES

Two theories can be used to visualize the effects of a TOP control:

1. The ***virtual condition boundary theory***—A theoretical boundary limits the location of the surfaces of a FOS.
2. The ***axis theory***—The axis (or centerplane) of a FOS must be within the tolerance zone.

Both theories are useful and—in most cases—equivalent. However, the axis theory is most common in RFS TOP applications, and the boundary theory is most common in MMC tolerance of position applications.

The Virtual Condition Boundary Theory

To illustrate the virtual condition boundary theory, let's examine the conditions that result from a TOP at MMC applied to a hole. In this type of application, the specified tolerance of position applies when the hole is at MMC. The hole must be within its specified limit of size, and it must be located so no element of its surface will be inside a theoretical boundary. The theoretical boundary is centered about the true position of the hole. The diameter of the theoretical boundary is equal to the MMC of the FOS minus the TOP tolerance value. The theoretical boundary is the virtual condition (or gage pin diameter) of the hole. Figure 8-7 shows an example.

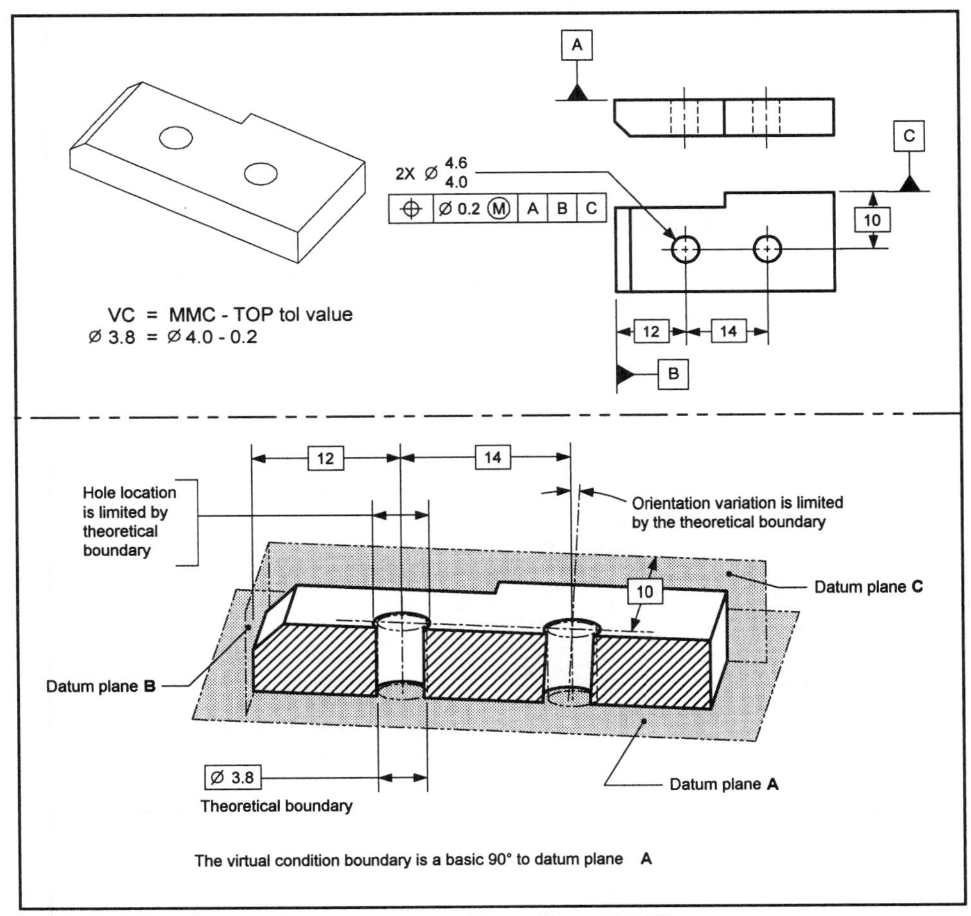

FIGURE 8-7 TOP Virtual Condition Boundary Theory (Internal FOS)

Author's Comment

As a convention in this text, a boundary tolerance zone is used in MMC applications, and an axis tolerance zone is used in RFS applications.

A TOP is also an indirect orientation control. The theoretical boundary is oriented relative to the primary datum referenced in the TOP callout. The gage pin that controls the location of the FOS also limits the orientation of the FOS. Figure 8-7 shows an example.

The virtual condition boundary theory also applies to external features of size. The theoretical boundary for a TOP (applied at MMC) of an external FOS is equal to the MMC of the FOS plus the TOP tolerance value. The location and orientation of the FOS is limited by the TOP control. Figure 8-8 shows an example.

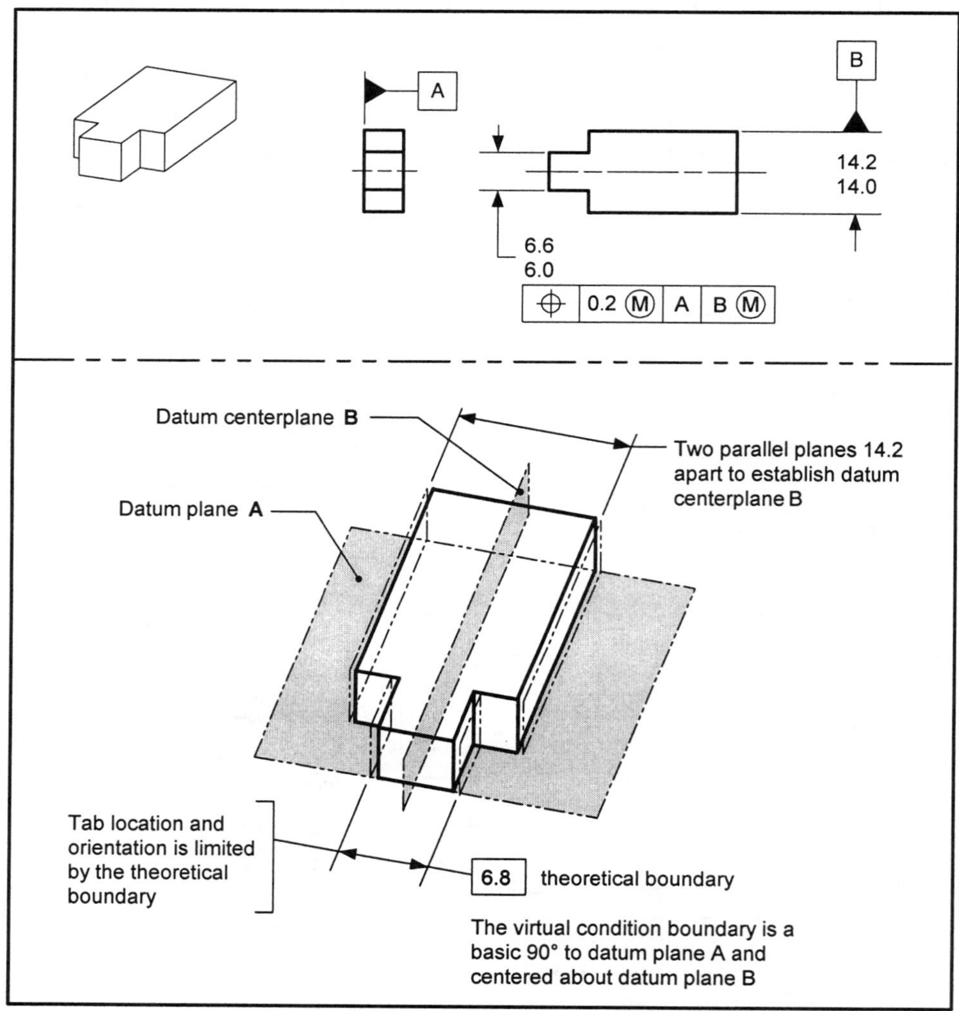

FIGURE 8-8 TOP Virtual Condition Boundary Theory (External FOS)

TECHNOTE 8-2 Virtual Condition Boundary Theory of TOP

A theoretical boundary (virtual condition) limits the location of the surfaces of a FOS.

The Axis Theory

The axis theory is often used when TOP is applied on an RFS basis. To illustrate the axis theory, let's examine the conditions that result from a TOP at RFS applied to a hole. In this type of application, the specified TOP applies at whatever size the hole is produced. The hole must be within its specified limit of size, and the axis of the hole AME must be located within the TOP tolerance zone cylinder. The TOP tolerance zone cylinder is centered around the true position of the hole. The diameter of the TOP tolerance zone cylinder is equal to the tolerance value specified in the TOP callout. Figure 8-9 shows an example.

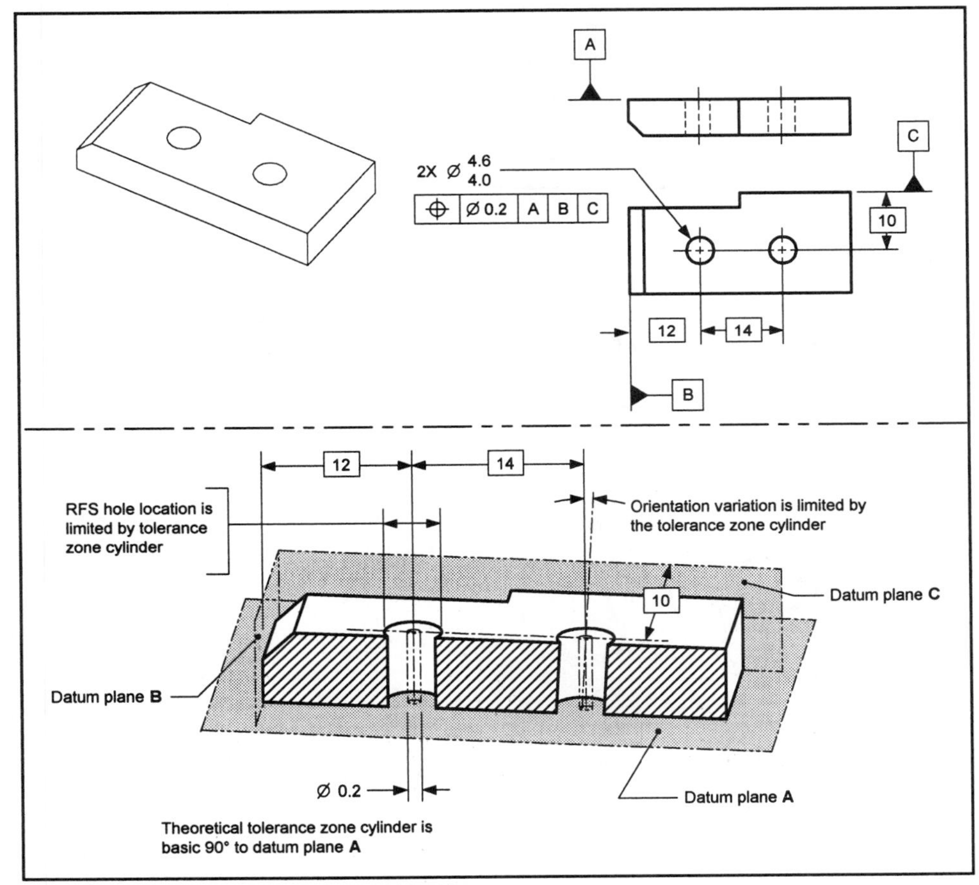

FIGURE 8-9 TOP Axis Theory (Internal FOS)

A TOP is also an indirect orientation control. The tolerance zone is oriented relative to the primary datum referenced in the TOP callout. The tolerance zone that controls the location of the FOS also limits the orientation of the FOS. Figure 8-9 shows an example.

The axis (or centerplane) theory also applies to planar features of size. The tolerance zone for a TOP (applied at RFS) of an external planar FOS is two parallel planes spaced apart a distance equal to the TOP tolerance value. The orientation and location of the centerplane of the AME of a FOS is limited by the TOP tolerance zone. Figure 8-10 shows an example.

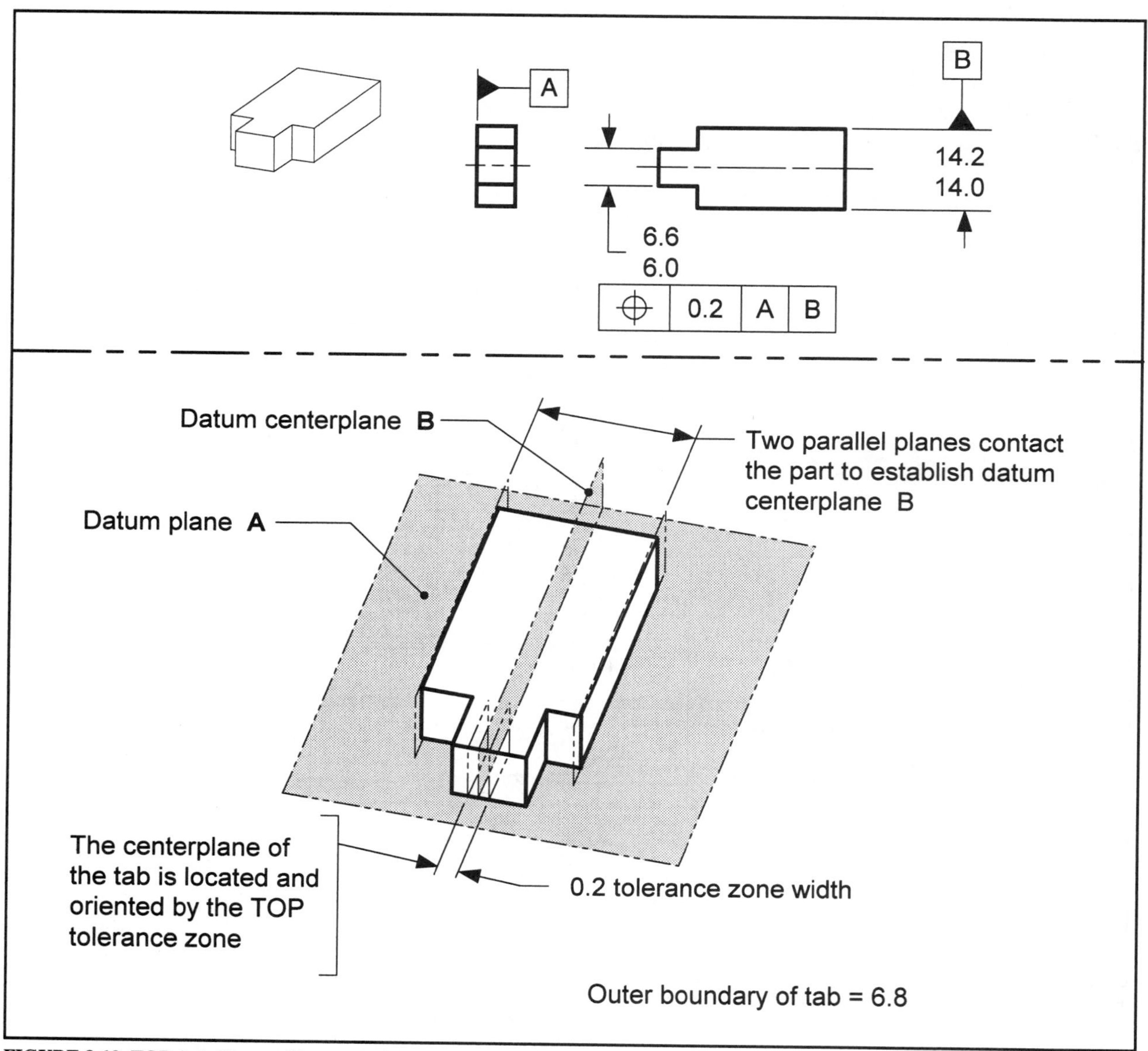

FIGURE 8-10 TOP Axis Theory (External FOS)

TECHNOTE 8-3 Axis Theory of TOP (RFS)

The axis (or centerplane) of the AME of a FOS must be within the tolerance zone. The diameter (or width) of the tolerance zone is equal to the TOP tolerance value.

COMMON TOP RFS APPLICATIONS

Where TOP is Used on an RFS Basis

In certain cases, the function of a part may require a TOP to be applied on an RFS basis. The chart in Figure 8-6 describes several applications where the RFS modifier is recommended. When a TOP is applied on a RFS basis, a closer control is imposed on the part when compared to an MMC application. Also, the inspection of the TOP requirement becomes more complex.

Whenever a TOP control is applied at RFS, three conditions are present:

1. The tolerance zone applies to the axis (or centerplane) of the FOS.
2. The tolerance value applies regardless of the size of the toleranced feature of size.
3. The requirement must be verified with a variable gage.

RFS Tolerance Zones

In tolerance of position (RFS applications), two tolerance zones are common: a fixed diameter cylinder and two parallel planes a fixed distance apart. The diameter of the tolerance zone cylinder—or the distance between the parallel planes—is equal to the tolerance value specified in the TOP callout. The location of the tolerance zone is always centered around the true position of the FOS.

Author's Comment
TOP may also use a spherical tolerance zone. This condition is rare and beyond the scope of this text.

In Figure 8-9, the TOP callout includes a diameter symbol in the tolerance portion of the feature control frame. The diameter symbol denotes a cylindrical tolerance zone. In Figure 8-10, the TOP callout does not show a diameter modifier in the feature control frame. When no diameter symbol is shown in a TOP feature control frame, the tolerance zone shape is two parallel planes.

TECHNOTE 8-4 TOP Applied at RFS

When a TOP is applied at RFS, these conditions apply:

1. The tolerance zone is an axis or centerplane zone.
2. The tolerance zone applies regardless of the feature's size.
3. Variable gaging must be used to verify the TOP requirements.

The Location of a Hole Controlled with Tolerance of Position (RFS)

In certain cases, it may be desired to control the location of a hole with a TOP at RFS. In Figure 8-11, the axis of the hole is being controlled relative to the outside surfaces of the part. In this application, the following conditions apply:

- The shape of the tolerance zone is a cylinder.
- The tolerance zone is located by the basic dimensions relative to the datum planes.
- The tolerance zone applies RFS.
- The dimension between the centerline of the hole and datum plane *A* is an implied basic 90°.
- No datum shift is permissible.
- The tolerance zone also controls the orientation of the hole relative to the primary datum reference from the TOP callout.
- Rule #1 still applies.
- The WCB of the hole is affected (6.0 - 0.2 = 5.8).

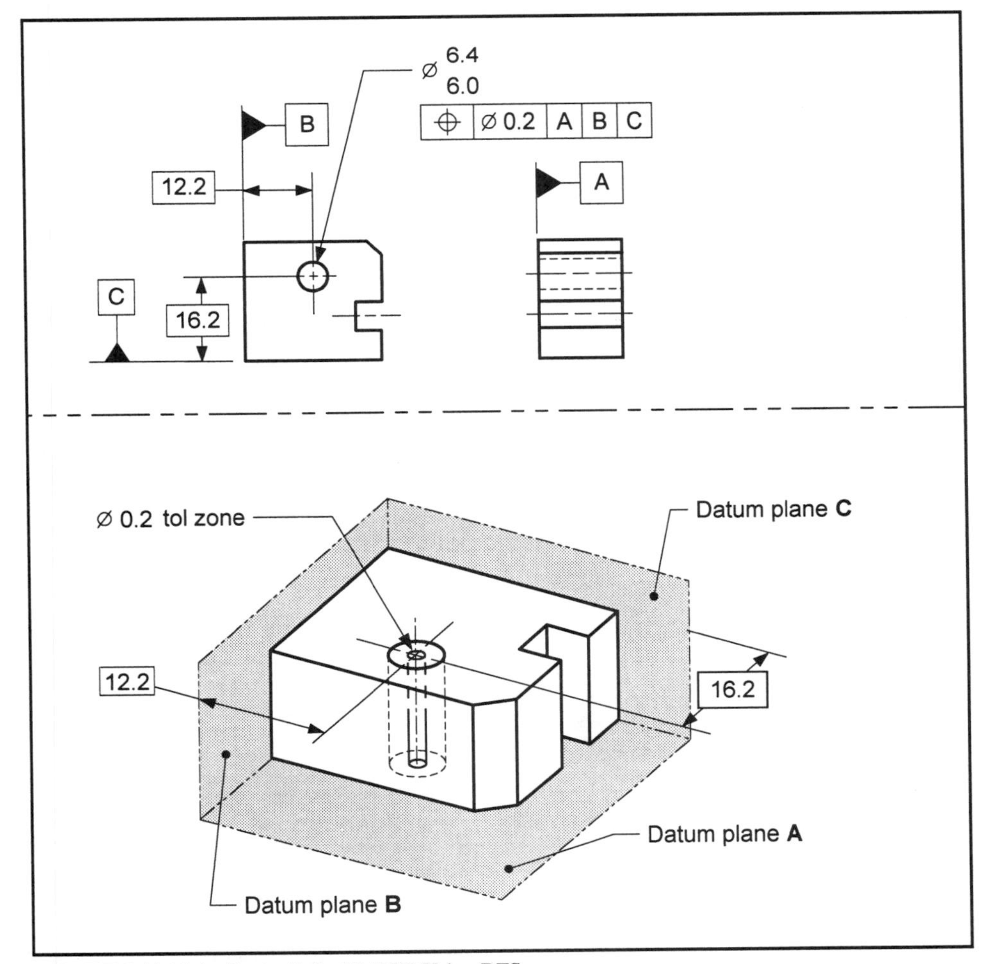

FIGURE 8-11 Hole Controlled with TOP Using RFS

The Location of a Pattern of Holes Controlled with Tolerance of Position (RFS)

In certain cases, it may be desired to control the location of a pattern of holes with TOP at RFS. In Figure 8-12, the location of the pattern of holes is being controlled relative to the edges of the part. In this application, the following conditions apply:

- The shape of each tolerance zone is cylindrical.
- The tolerance zones are located by the basic dimensions.
- The tolerance zones apply RFS.
- The tolerance zones also control the orientation of the holes relative to the primary datum reference from the TOP callout.
- The tolerance zones are at an implied basic 90° to datum *A*.
- Rule #1 applies.

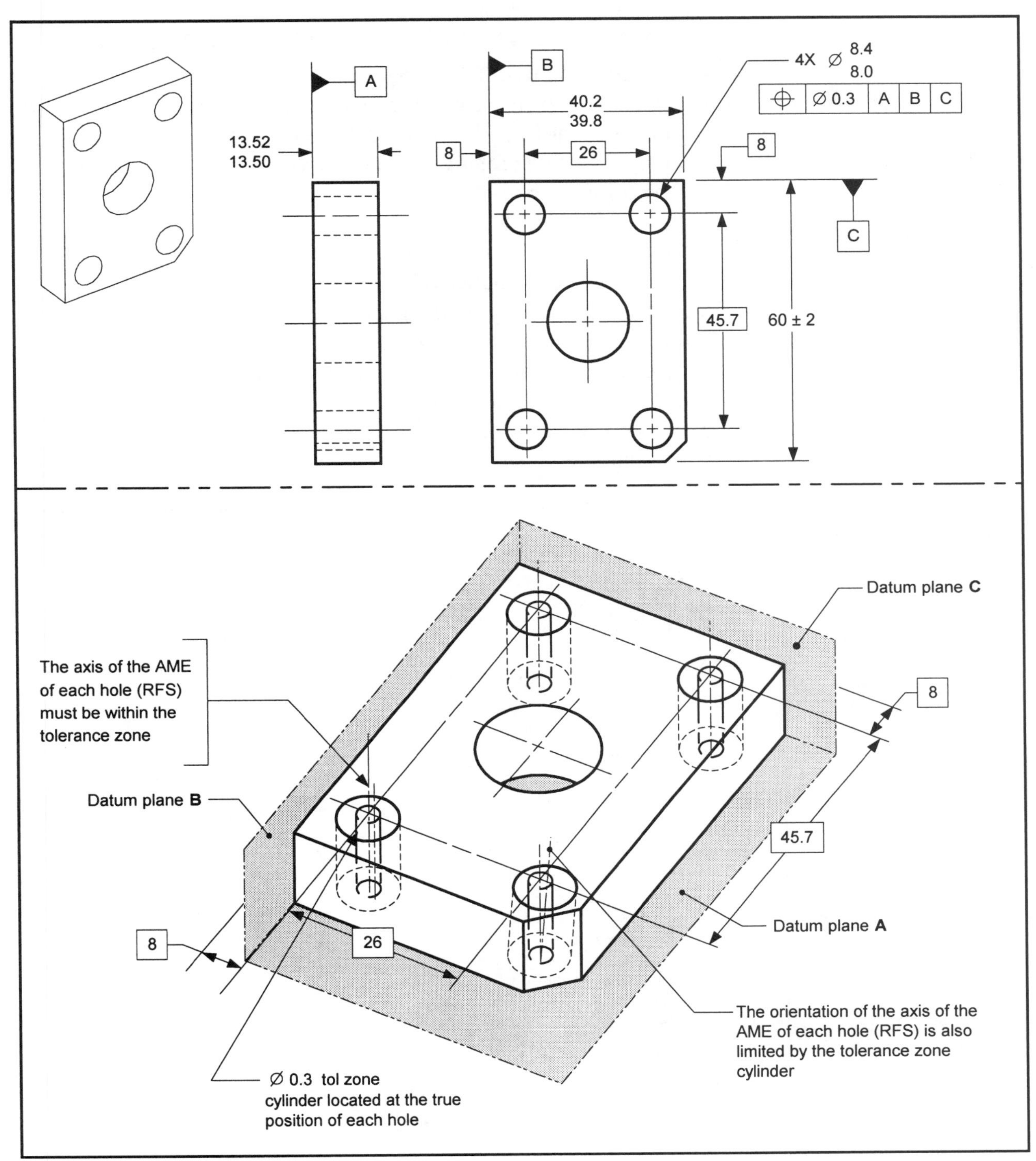

FIGURE 8-12 Pattern of Holes Controlled with TOP Using RFS

The Location of Coaxial Diameters Controlled with Tolerance of Position (RFS)

In certain cases, it may be desired to control the location of coaxial diameters with a TOP at RFS. In Figure 8-13, the location of the coaxial diameter is being controlled relative to datum axis *A*. In this application, the following conditions apply:

- The shape of the tolerance zone is cylindrical.
- The tolerance zone applies RFS.
- The dimension specifying the location of the diameter relative to the datum feature is an implied basic zero.
- The tolerance zone also limits the orientation of the toleranced diameter relative to datum axis *A*.
- There is no datum shift.
- Rule #1 applies.

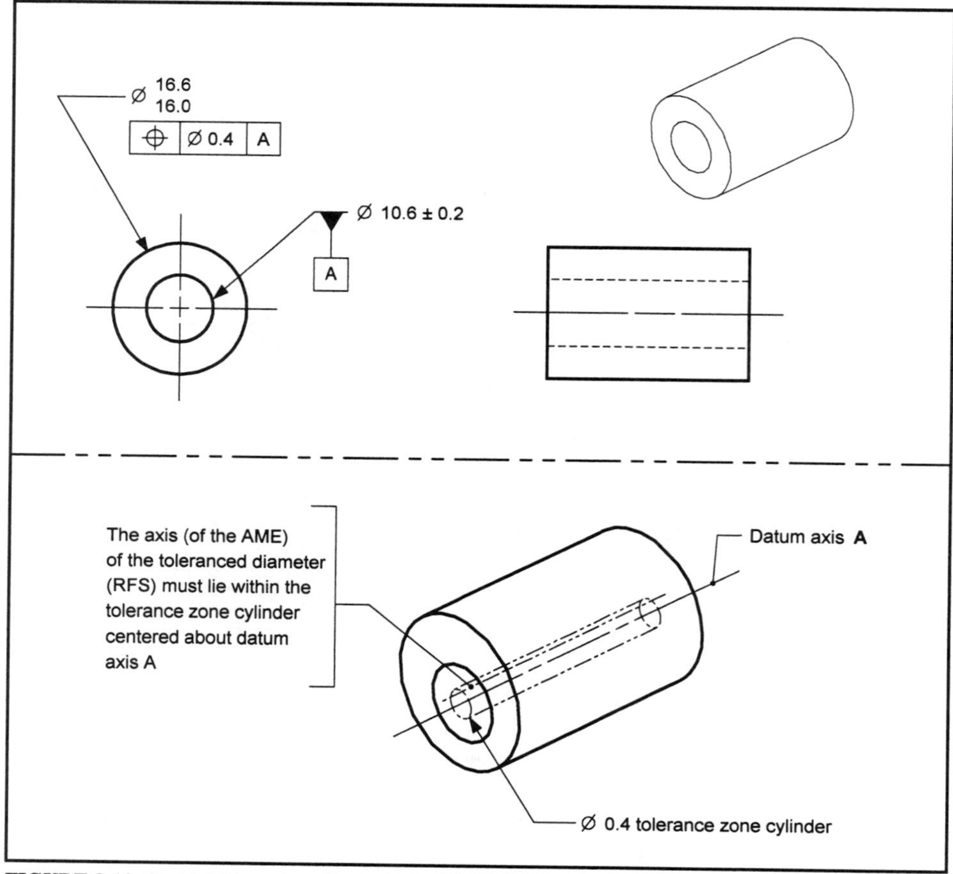

FIGURE 8-13 Coaxial Diameters Controlled with TOP (RFS)

INSPECTING TOP APPLIED AT RFS

Figure 8-11 shows a part with a TOP specification. When inspecting the hole on this part, three separate checks are required: the size of the hole, the Rule #1 boundary, and the TOP requirement. Chapter 2 explained how to check the size and Rule #1 boundary; now we will look at how to inspect the TOP requirement.

When TOP is specified on an RFS basis, it requires variable gaging to verify the requirements. A ***variable gage*** is a gage that is capable of providing a numerical reading of a part parameter. Examples of variable gages are CMMs; special dedicated variable gages; and standard measuring equipment, such as collets, height gages, expanding mandrels, and dial indicators.

One way the TOP requirement on the part from Figure 8-11 could be inspected is shown in Figure 8-14. First, the part is rested on the surface plate and the gage elements that simulate the datum reference frame. The first, second, and third part surfaces to contact the inspection equipment are defined by the datum sequence of the TOP callout. Once the part is located in its datum reference frame, the location of the hole is established. A "best fit" gage pin is placed in the toleranced hole. The gage pin represents the actual mating envelope of the hole. Next, the location of the center of the gage pin—relative to the datum reference frame—is determined. The center of the gage pin must be within the tolerance zone cylinder that is defined by the TOP callout.

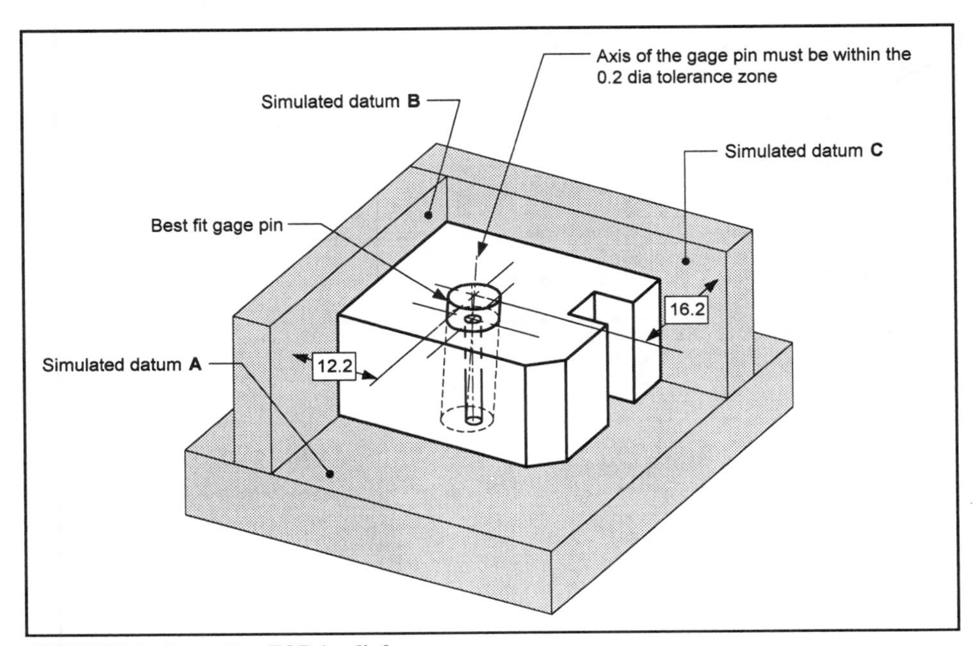

FIGURE 8-14 Inspecting TOP Applied

COMMON TOP MMC APPLICATIONS

Author's Comment
In product design, the function of assembly is very common. Therefore, the use of the MMC modifier in TOP is very common.

Where TOP is Used on an MMC Basis

In certain cases, the function of a part may indicate a TOP is to be applied on an MMC basis. This occurs when the part function is assembly or when the effects of bonus tolerance and/or datum shift would not have a detrimental effect on the function of the part. When a TOP is applied on an MMC basis, it is a more liberal control than an RFS application. Figure 8-15 shows a comparison between TOP applied at MMC and RFS.

Whenever the MMC modifier is used in a TOP control, three conditions are present in the application:

1. The tolerance zone is considered a boundary zone.
2. A bonus tolerance and/or datum shift is permissible.
3. The requirement can be verified with a functional gage.

MMC Tolerance Zones

In tolerance of position MMC applications, two tolerance zone shapes are common: a virtual condition cylindrical boundary and a virtual condition parallel plane boundary. The virtual condition boundary is often considered the gage pin (or width) size. The location of the tolerance boundary is always centered around the true position of the FOS.

	MMC	RFS
Tolerance zone	A boundary zone	An axis zone (or centerplane)
Bonus tolerance permissible	Yes	No
Gaging	Functional (fixed)	Variable

FIGURE 8-15 TOP MMC/RFS Comparison

TECHNOTE 8-5 TOP Applied at MMC

When a TOP is applied at MMC, three conditions apply:

1. The tolerance zone is considered a boundary zone.
2. A bonus tolerance and/or datum shift is permissible.
3. The requirement can be verified with a functional gage.

The Location of a Hole Controlled with TOP (MMC)

In certain cases, it may be desired to control the location of a hole with a TOP at MMC. In Figure 8-16, the hole is being controlled relative to the outside surfaces of the part. In this application, the following conditions apply:

- The shape of the tolerance zone is a virtual condition cylindrical boundary.
- The tolerance zone is located by the basic dimensions from the datum planes.
- The relationship between the centerline of the hole and datum plane *A* is an implied basic 90° angle.
- A bonus tolerance is permissible.
- The tolerance zone also controls the orientation of the hole relative to the primary datum reference of the TOP callout.
- Rule #1 applies.

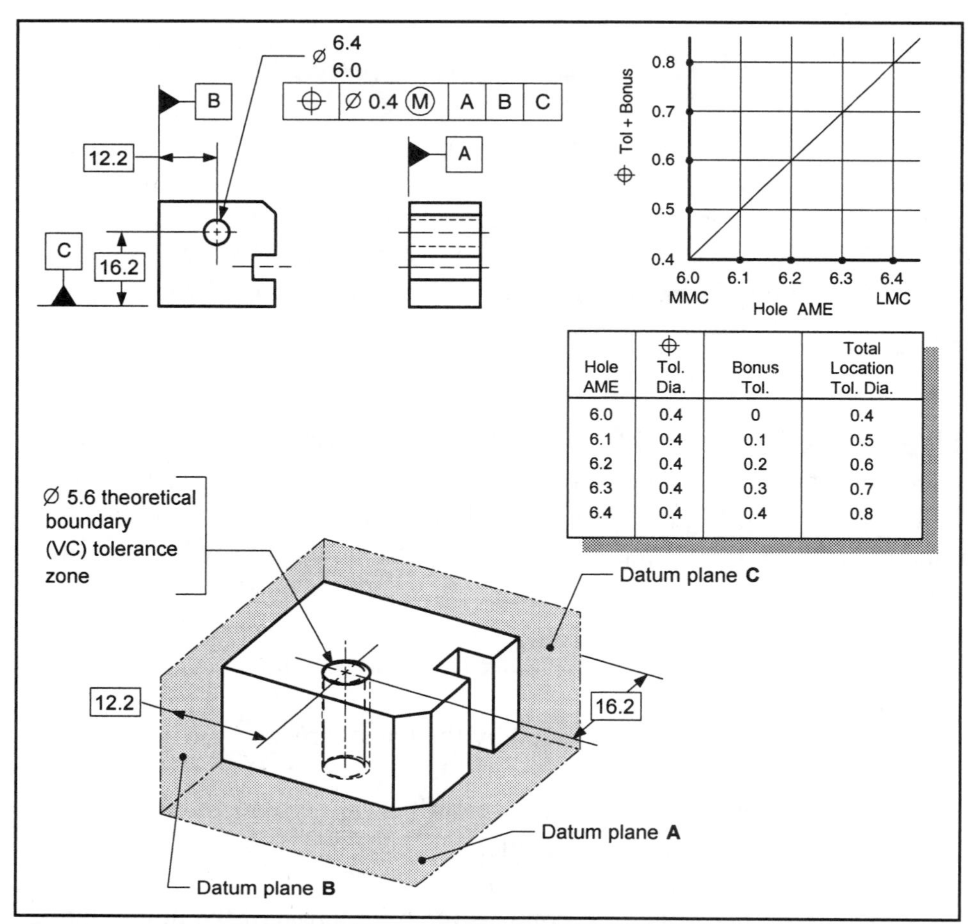

Hole AME	⌖ Tol. Dia.	Bonus Tol.	Total Location Tol. Dia.
6.0	0.4	0	0.4
6.1	0.4	0.1	0.5
6.2	0.4	0.2	0.6
6.3	0.4	0.3	0.7
6.4	0.4	0.4	0.8

FIGURE 8-16 Hole Location Controlled with TOP Using MMC

In Figure 8-16, the maximum allowable bonus is equal to the difference between the MMC and LMC of the AME of the toleranced diameter.

The Location of a Hole Pattern Controlled with Tolerance of Position (MMC)

In certain cases, it may be desired to control the location of a pattern of holes with TOP at MMC. In Figure 8-17, the location of the pattern of holes is being controlled relative to the edges of the part. In this application, the following conditions apply:

- The shapes of the tolerance zones are virtual condition boundaries.
- The tolerance zones are located by the basic dimensions from the datum planes.
- The relationship between the centerlines of the holes and datum plane *A* are implied basic 90°.
- A bonus tolerance is permissible.
- The tolerance zones also control the orientation of the holes relative to the primary datum reference from the TOP callout.
- Rule #1 applies.

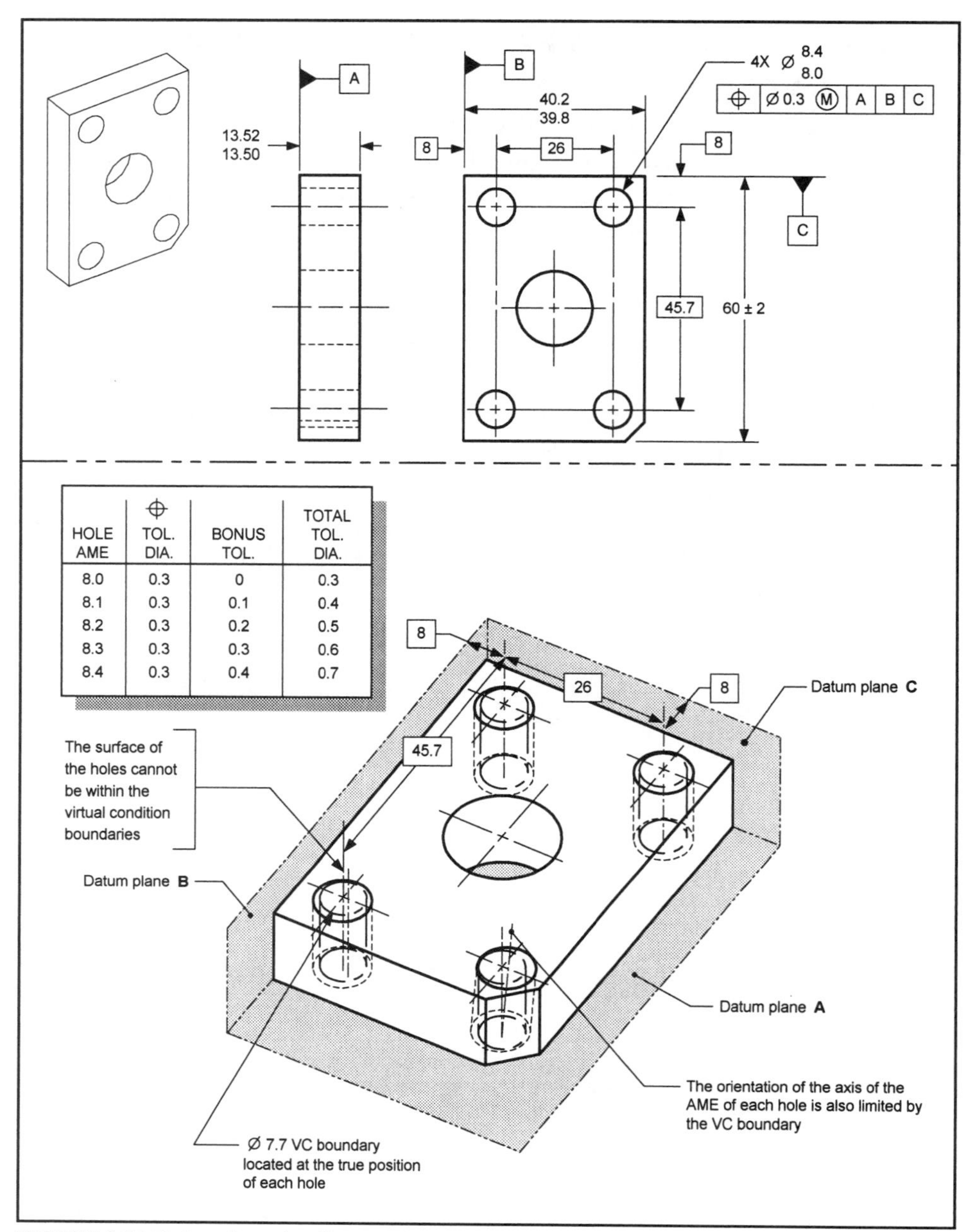

HOLE AME	⌖ TOL. DIA.	BONUS TOL.	TOTAL TOL. DIA.
8.0	0.3	0	0.3
8.1	0.3	0.1	0.4
8.2	0.3	0.2	0.5
8.3	0.3	0.3	0.6
8.4	0.3	0.4	0.7

FIGURE 8-17 Pattern Location Controlled with TOP MMC

In Figure 8-17, the maximum amount of bonus permissible is equal to the difference between the MMC and LMC of the AME of the toleranced FOS.

Coaxial Diameter Applications

Figure 8-18 illustrates the amount of bonus and/or datum shift permissible in coaxial diameter applications. In example one, the toleranced diameter and the datum feature diameter are different diameters. The bonus tolerance permissible comes from the toleranced diameter. The datum shift permissible comes from the datum feature diameter.

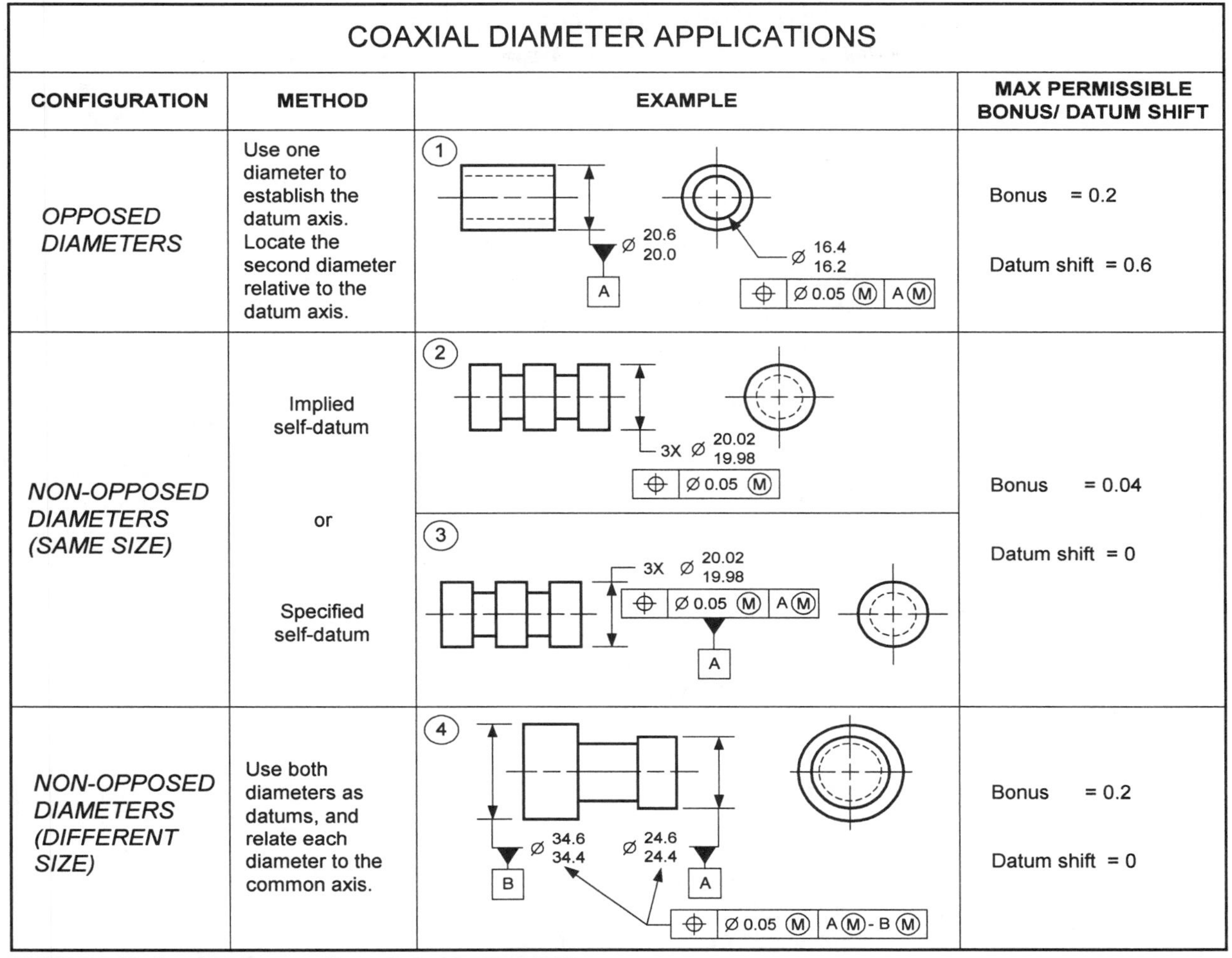

COAXIAL DIAMETER APPLICATIONS

CONFIGURATION	METHOD	EXAMPLE	MAX PERMISSIBLE BONUS/ DATUM SHIFT
OPPOSED DIAMETERS	Use one diameter to establish the datum axis. Locate the second diameter relative to the datum axis.	1: Ø 20.6 / 20.0; A; Ø 16.4 / 16.2; ⌖ Ø 0.05 Ⓜ A Ⓜ	Bonus = 0.2 Datum shift = 0.6
NON-OPPOSED DIAMETERS (SAME SIZE)	Implied self-datum or Specified self-datum	2: 3X Ø 20.02 / 19.98; ⌖ Ø 0.05 Ⓜ 3: 3X Ø 20.02 / 19.98; ⌖ Ø 0.05 Ⓜ A Ⓜ; A	Bonus = 0.04 Datum shift = 0
NON-OPPOSED DIAMETERS (DIFFERENT SIZE)	Use both diameters as datums, and relate each diameter to the common axis.	4: Ø 34.6 / 34.4; B; Ø 24.6 / 24.4; A; ⌖ Ø 0.05 Ⓜ A Ⓜ - B Ⓜ	Bonus = 0.2 Datum shift = 0

FIGURE 8-18 Coaxial Diameters Controlled with TOP MMC

In Figure 8-18, examples two and three show two ways of specifying the relationship of coaxial diameters. They specify the same part requirements with the two dimensioning methods. The gage for the parts in examples two and three would be identical. Example two uses an implied self datum. The toleranced diameters are being located relative to each other. The alignment of the diameters is being controlled. Example three specifies the three diameters as a datum, then relates the diameters back to the datum (or each other). Once again, the alignment of the diameters is being controlled.

Author's Comment

In Figure 8-18, the dimensioning methods shown for non-opposed diameters (same size) are not covered in Y14.5. However, these methods are used on many drawings in industry.

In all four examples, the maximum amount of bonus permissible is equal to the difference between the MMC and LMC of the toleranced FOS. The maximum amount of datum shift available for the part in example one is equal to the difference between the LMC and the virtual condition of the datum FOS. In examples two, three, and four, in the non-opposed diameter applications, there is no datum shift available. The bonus is the datum shift. This is because the toleranced feature is also the datum feature.

Legal Specification Test for TOP

For a TOP control to be a legal specification, it must satisfy the following conditions:

- The TOP must be applied to a FOS.
- Datum references are required. The datum references must ensure repeatable measurements of the toleranced FOS.
- Basic dimensions must be used to establish the true position of the toleranced FOS from the datums referenced (and between features of size in a pattern).

If any of these conditions are not fulfilled, the TOP specification is incorrect or incomplete. Figure 8-19 shows a legal specification flowchart for a TOP specification.

Design Tip
The default condition for a TOP control is RFS, which is very expensive and often not required. For each TOP control, consideration should be given to using the MMC or LMC modifier.

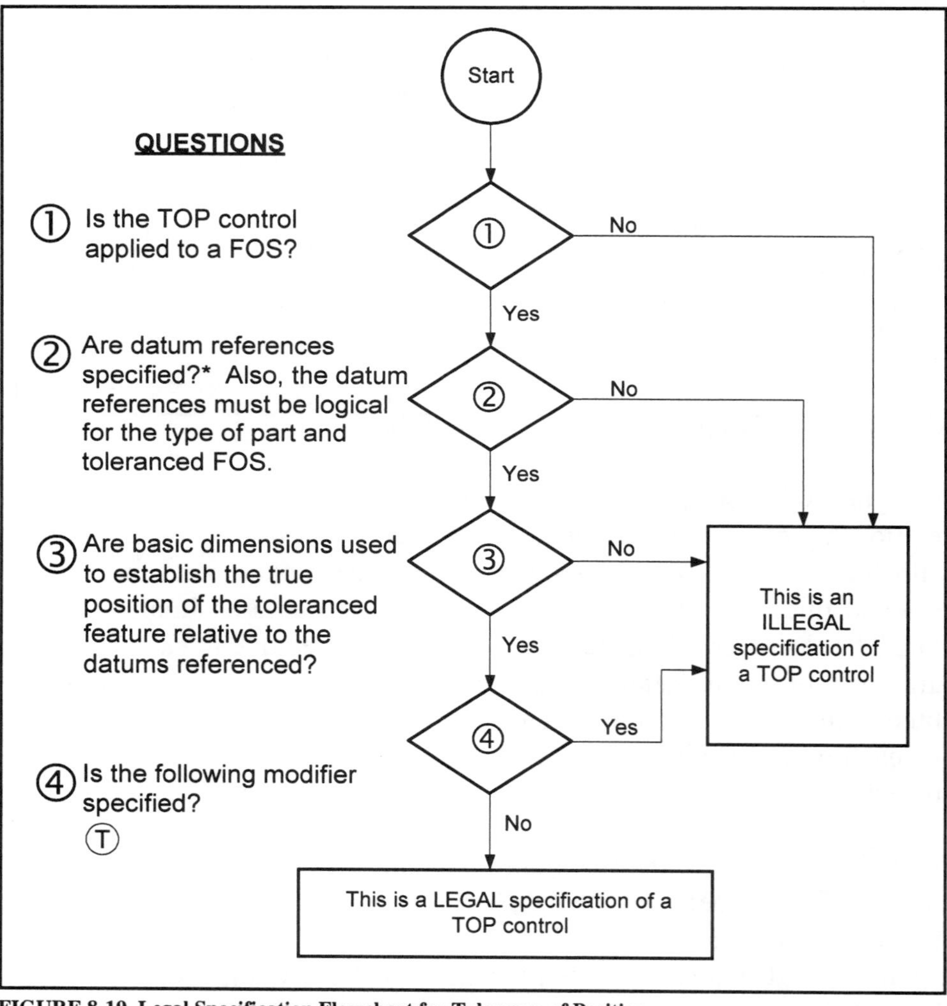

FIGURE 8-19 Legal Specification Flowchart for Tolerance of Position

* Except for coaxial non-opposed diameters, as shown in Figure 8-18.

INSPECTING TOP APPLIED AT MMC

A TOP applied at MMC can be verified in a number of ways. Variable gages, open inspection, CMM, and functional gaging are all common methods of verifying parts dimensioned with TOP. In this text, we will explain the use of functional gaging.

Functional Gage

A ***functional gage*** is a gage that verifies functional requirements of part features as defined by the geometric tolerances. For example, if holes on a part are intended to fit over studs of a mating part, a function of the holes would be to assemble over the studs. To verify the location of the holes, a functional gage that simulates the studs of the mating part could be used. A functional gage does not provide a numerical reading of a part parameter. A functional gage often provides a "pass" or "fail" assessment of a part feature. A functional gage is often referred to as an attribute gage or a fixed gage because it checks attributes of a part FOS (location and orientation).

Author's Comment
Some consideration should be taken to dimension and tolerance part features so that the resulting gage simulates the worst-case boundary of the mating part.

When compared to a variable gage, a functional gage offers several benefits. The list below highlights five benefits of functional gages:

1. The gage represents the worst-case mating part.
2. Parts can be verified quickly.
3. A functional gage is economical to produce.
4. No special skills are required to "read" the gage or interpret the results.
5. In some cases, a functional gage can check several part characteristics simultaneously.

Functional gages are a commonly used method for verifying TOP callout requirements. The functional gage represents the virtual condition of the toleranced FOS. Although functional gages provide many benefits for verifying parts controlled with TOP applied at MMC, their use is not mandatory. A TOP applied at MMC can also be verified with a variable gage or a CMM.

Cartoon Gage

Often, it is desirable to analyze a max. or min. distance on a part in the design stage. Since a functional gage defines the extreme limits of a part FOS, it can be used as a simple method to analyze part distances. Because the functional gage does not exist in the design stage, a cartoon gage is used. A ***cartoon gage*** is a sketch of a functional gage. A cartoon gage defines the same part limits that a functional gage would, but it does not represent the actual gage construction of a functional gage.

The steps for drawing a cartoon gage are shown in Figure 8-20 and are described below:

1. Determine the size of the gage feature. Using the MMC of the toleranced feature, subtract (or add, for an external FOS) the TOP tolerance value to find the virtual condition or gage size of the toleranced FOS.
2. Establish the simulated datums (surfaces or axes) for the datums referenced in the TOP callout.
3. Locate the gage features relative to their respective datums. The basic dimensions from the product drawing are used to locate the gage features relative to the datums.

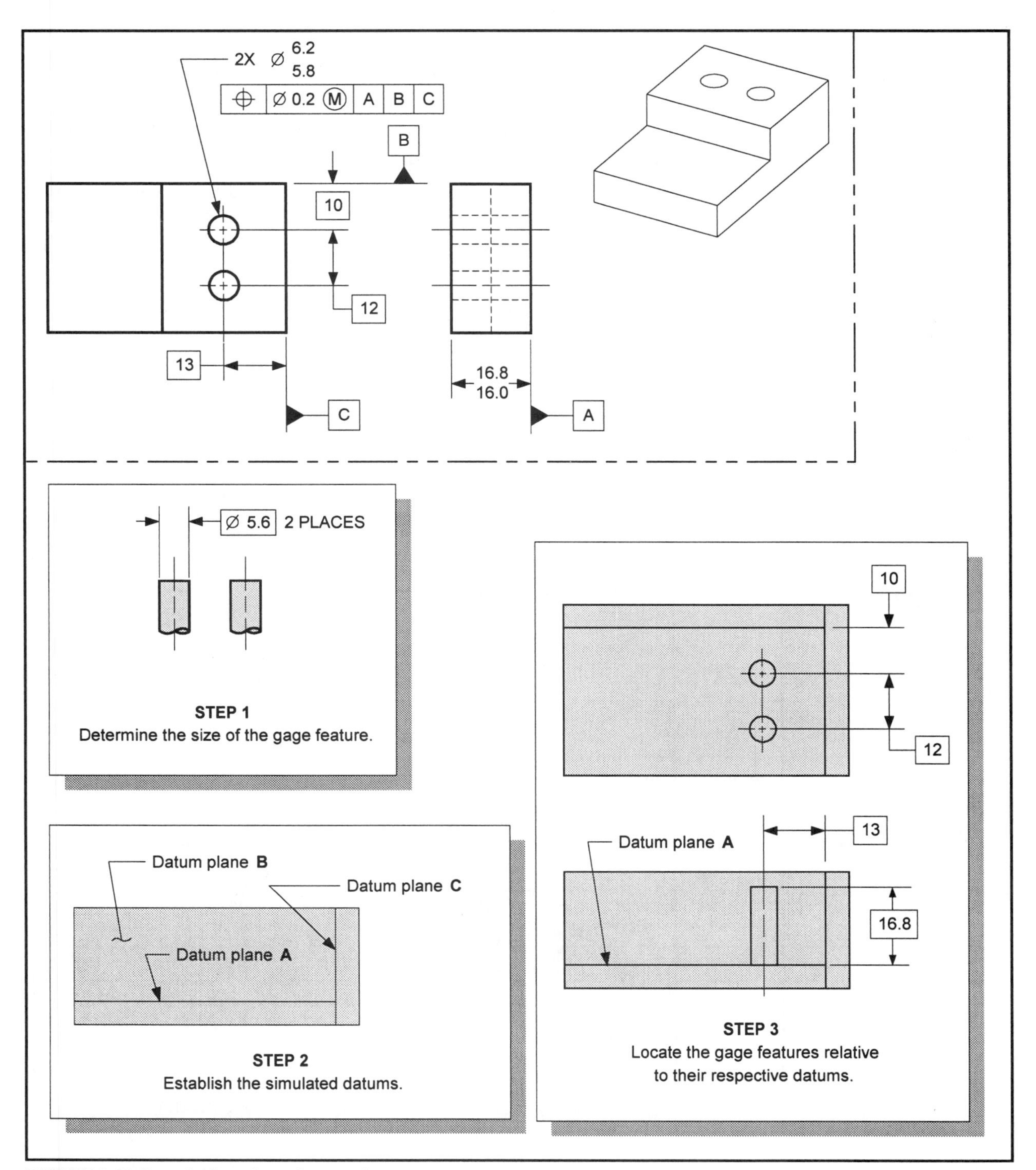

FIGURE 8-20 Steps for Drawing a Cartoon Gage

Figure 8-21 shows a cartoon gage for a coaxial diameter TOP application. When drawing a cartoon gage, the view drawn is based on the designer's judgment. A top, side, isometric, or sectional view may be used, based on the design issue being studied. Note, in this case, external features of size are used. The effects of the TOP callout are added to the MMC of the toleranced diameter to produce the virtual condition (or gage diameter). Also, this gage includes a virtual condition (worst-case boundary) for datum *B*. The gage must be built to the virtual condition of datum feature *B*. This will allow parts to fit on the gage in cases where datum feature *B* is not perfectly square to datum plane *A* (but is still within the perpendicularity tolerance value).

Design Tip

Sketch a cartoon gage for each TOP callout on your drawing. This will help in two ways: it will point out if your TOP control is complete and understandable, and it will help you to visualize the mating part requirement.

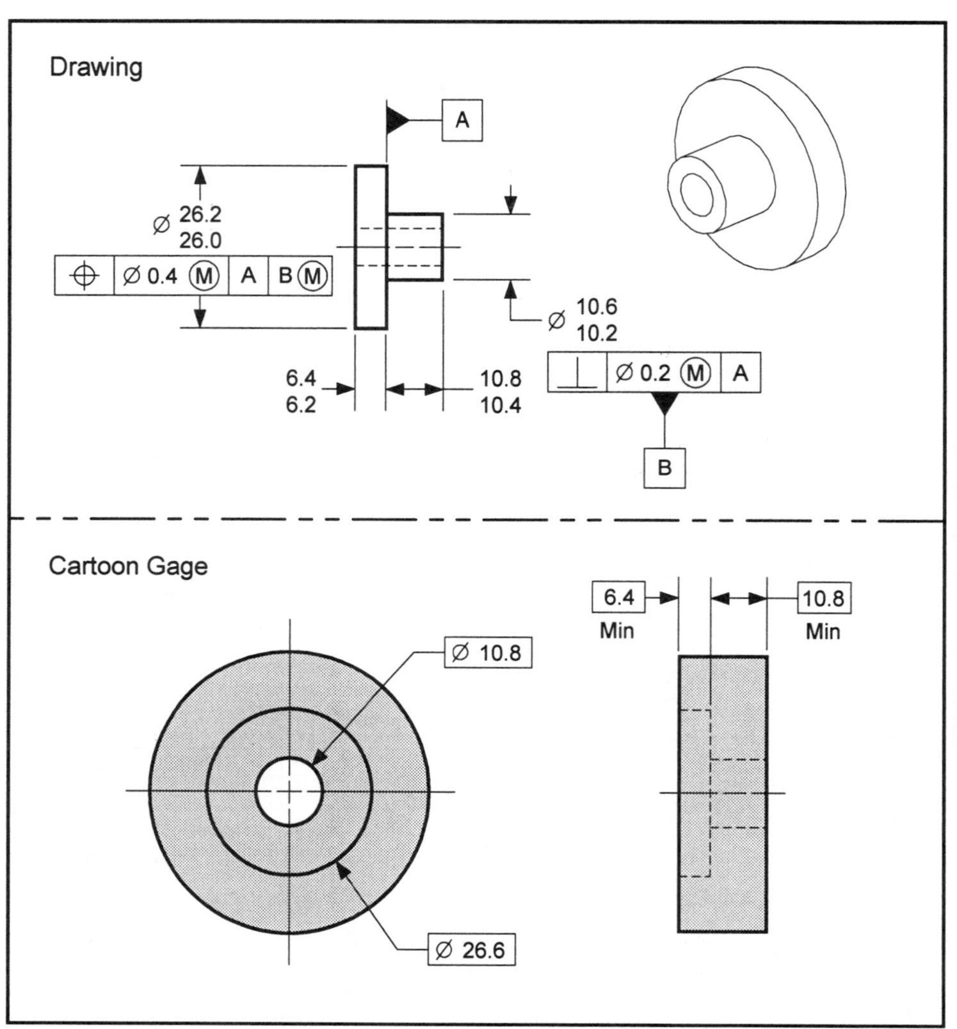

FIGURE 8-21 Cartoon Gage for a Coaxial Diameter TOP Application

This gage only checks the location and orientation of the diameter; the form and the size are additional required checks. In the next chapter, a technique for calculating part distances by using a functional gage is introduced.

Summary

A summarization of TOP information is shown in Figure 8-22.

TOP control	Datum reference required	Can be applied to a		Can affect WCB	Can use Ⓜ or Ⓛ modifier	Can be applied at RFS	Overrides Rule #1
		Surface	FOS				
⌖	Yes*	No	Yes	Yes	Yes	Yes**	No
* Coaxial diameter exception	** Is automatic per Rule #2						

FIGURE 8-22 Summarization of TOP

VOCABULARY LIST

Study Tip
Read each term. If you don't recall its meaning, look it up in the chapter.

New Terms Introduced in this Chapter

Axis theory
Cartoon gage
Functional gage
Implied basic 90° angle
Implied basic zero dimension
Tolerance of position (TOP) control
True position
Variable gage
Virtual condition boundary theory

ADDITIONAL RELATED TOPICS

Author's Comment
These topics, plus advanced coverage of many of the topics introduced in this text, will be covered in my new book on advanced GD&T concepts.

Topic	ASME Y14.5M-1994 Reference
• Simultaneous requirement RFS	Paragraph 5.3.6.1
• Simultaneous requirement MMC	Paragraph 5.3.6.2

QUESTIONS AND PROBLEMS

1. Describe what a tolerance of position control is.

2. Define the term, "true position."

3. List two types of implied basic relationships common with TOP.

4. List six advantages of TOP.

5. List four types of relationships that can be controlled with TOP.

6. Describe the virtual condition boundary theory of TOP.

7. Describe the axis theory of TOP.

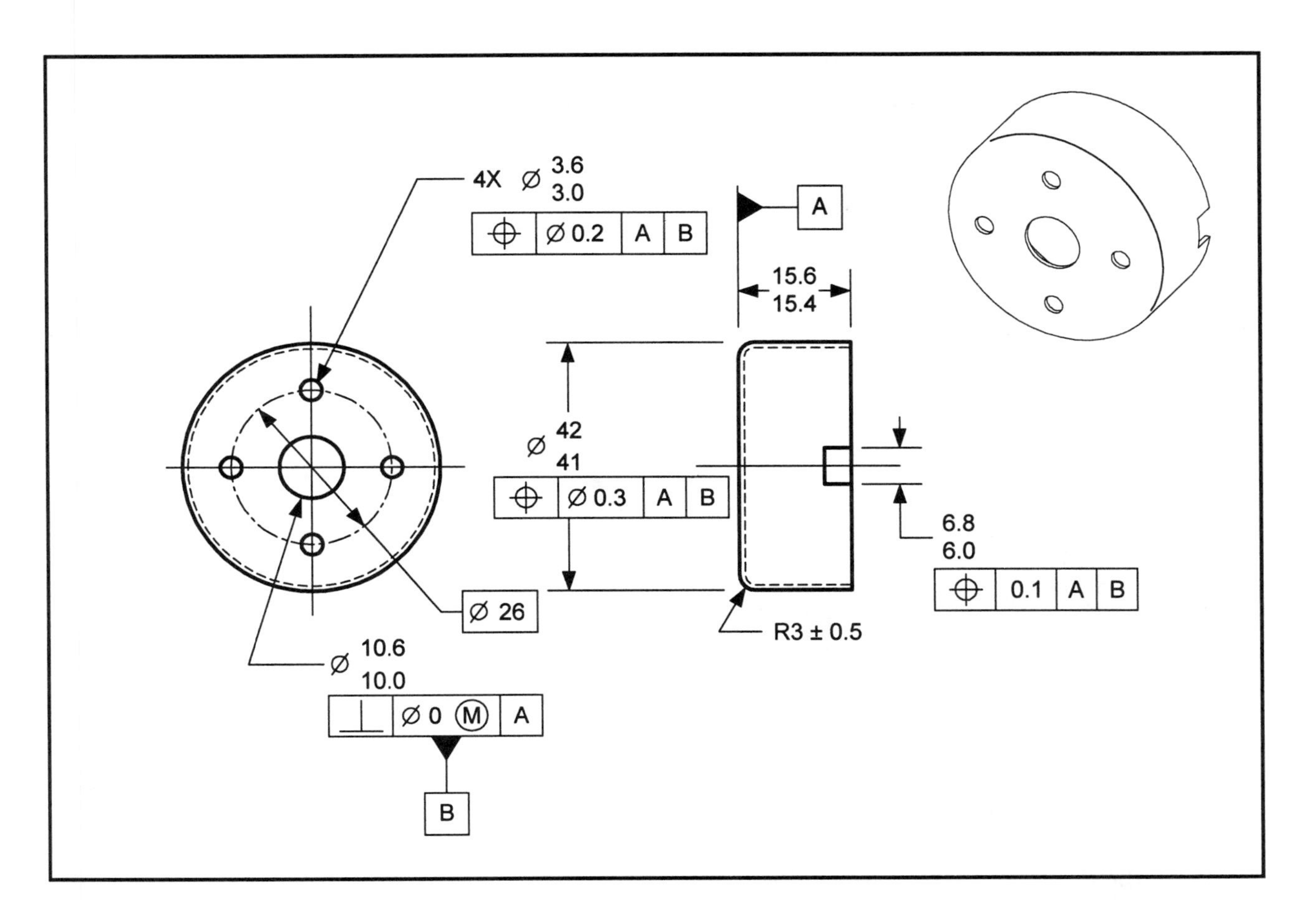

Questions 8-10 refer to the drawing above.

8. Describe the size and shape of the tolerance zones for the location of the 3.0-3.6 dia. holes.

____

9. Describe the size and shape of the tolerance zone for the location of the 6.0-6.8 slot.

____

10. Describe the size and shape of the tolerance zone for the location of the 41-42 diameter.

____

____

11. List three conditions that exist when an MMC modifier is used with a TOP control.

____

____

____

____

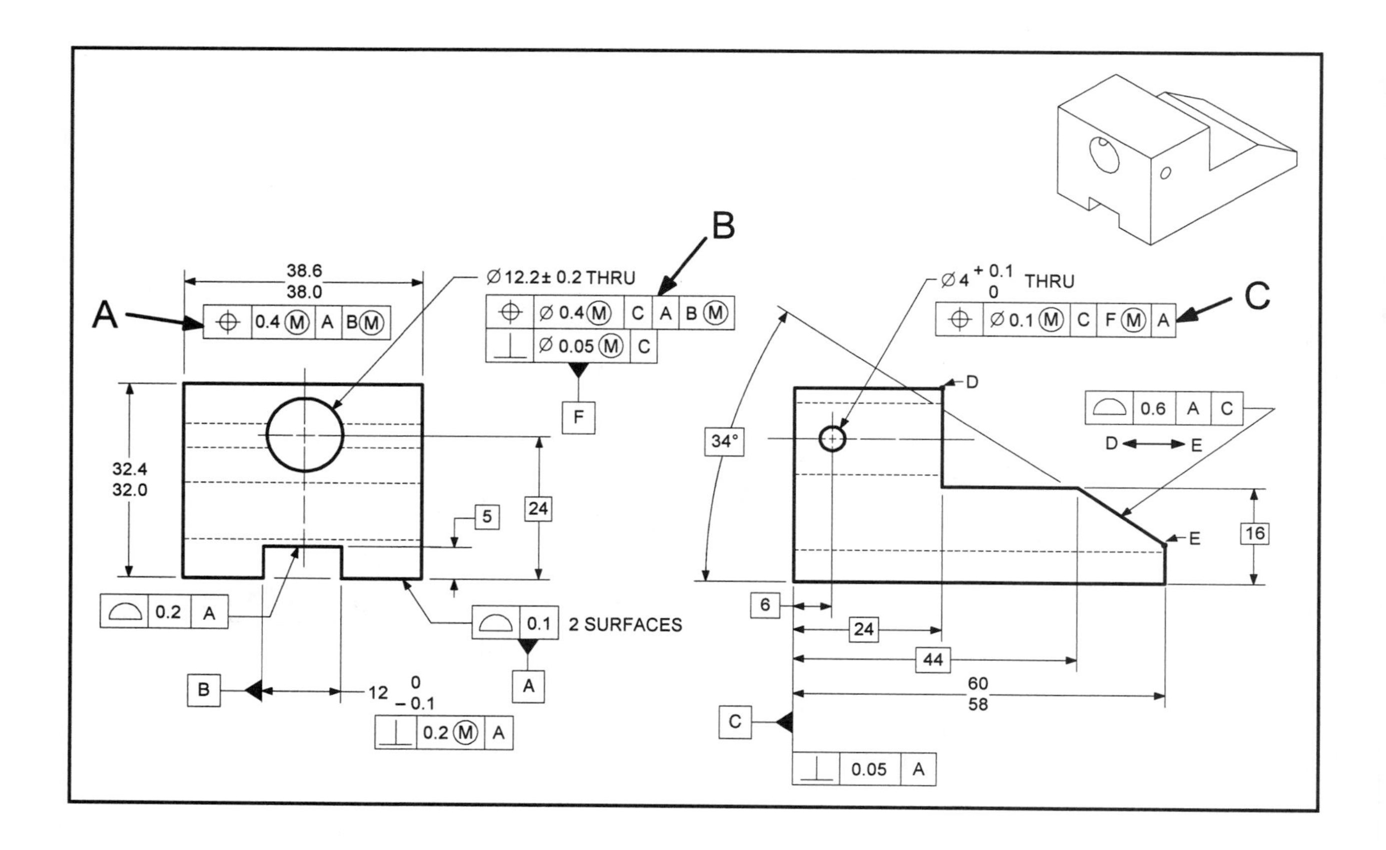

12. Use the drawing above to fill in the chart.

For the TOP callout labeled. . .	The shape of the tolerance zone is. . .	The max permissible bonus is. . .	The max permissible datum shift is. . .
A			
B			
C			

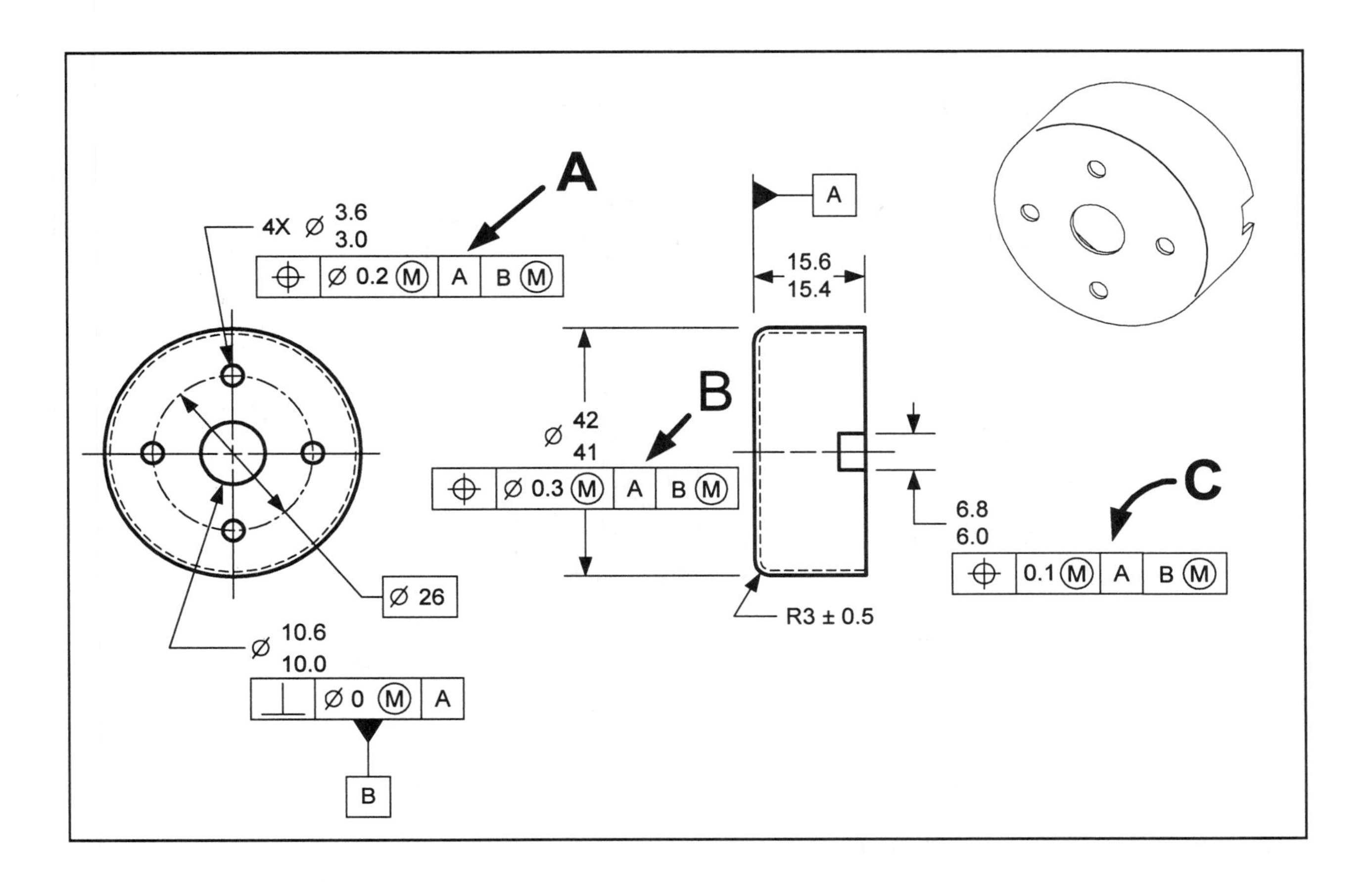

13. Use the drawing above to fill in the chart.

For the TOP callout labeled. . .	The shape of the tolerance zone is. . .	The max permissible bonus is. . .	The max permissible shift is. . .
A			
B			
C			

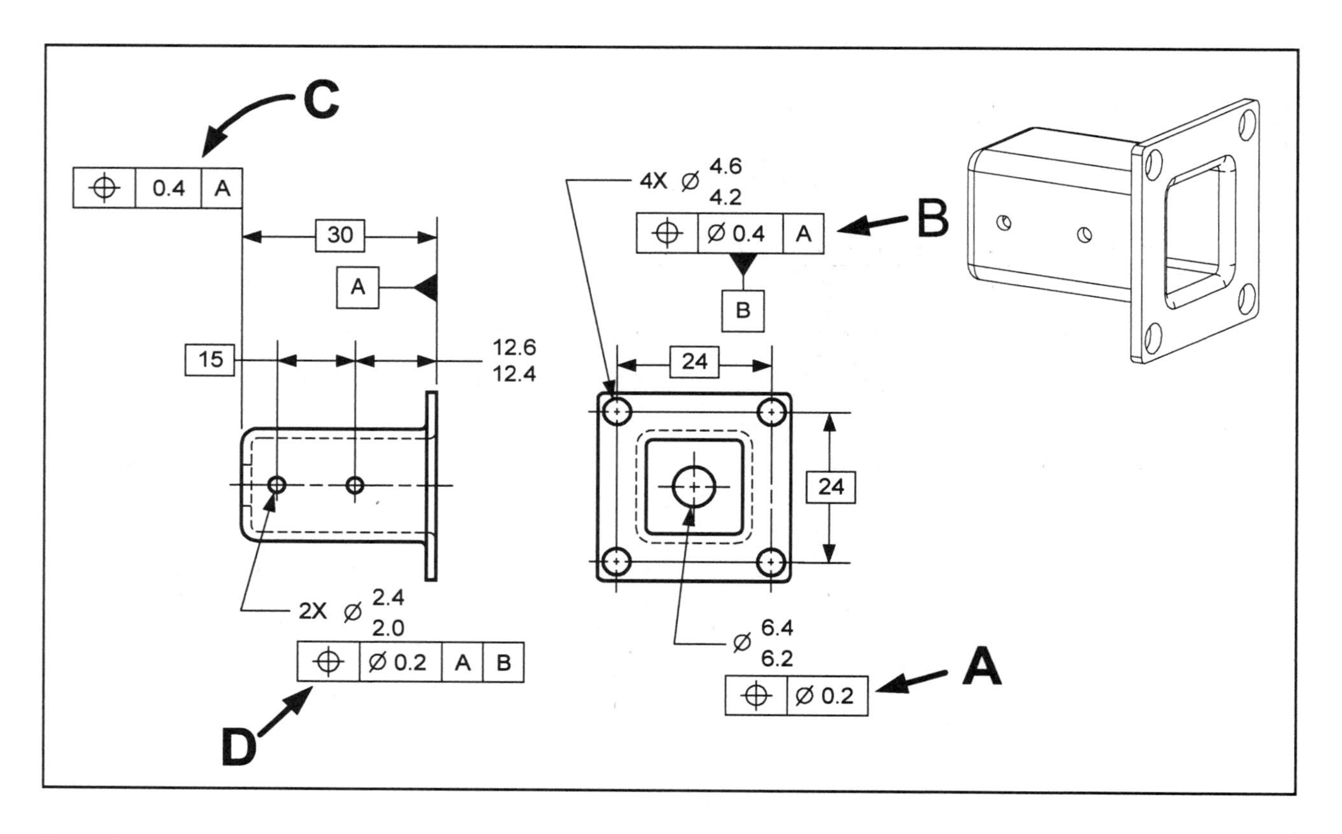

Question 14 refers to the drawing above.

14. For each TOP control listed, indicate if it is a legal specification. If the control is illegal, explain why.
    A ______________________________
    B ______________________________
    C ______________________________
    D ______________________________

15. Define the term, "functional gage."
    ____
    ____

16. List four benefits of a functional gage.

____

____

____

____

17. Define the term, "cartoon gage."

____

____

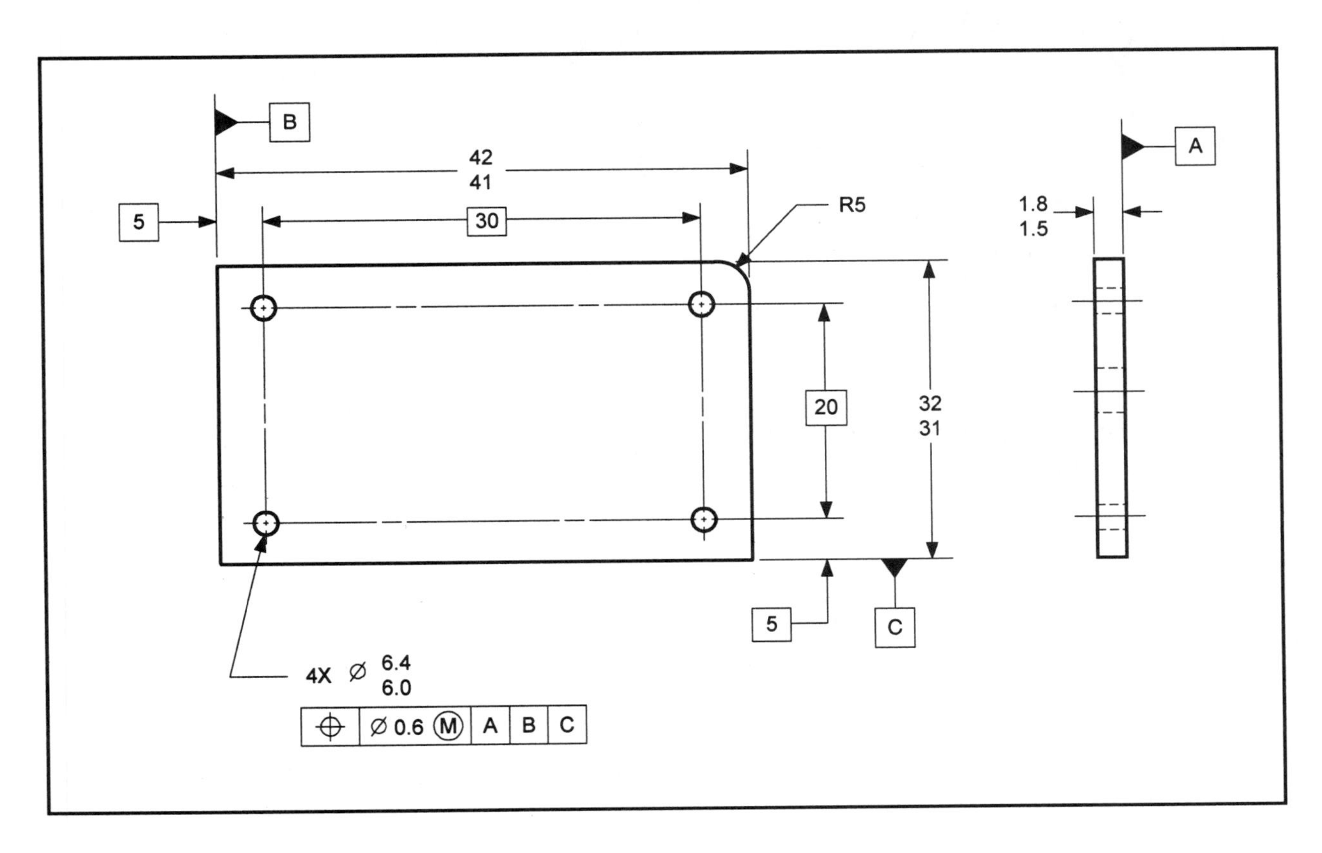

18. Using the drawing above, draw and dimension a cartoon gage for checking the position callout shown.

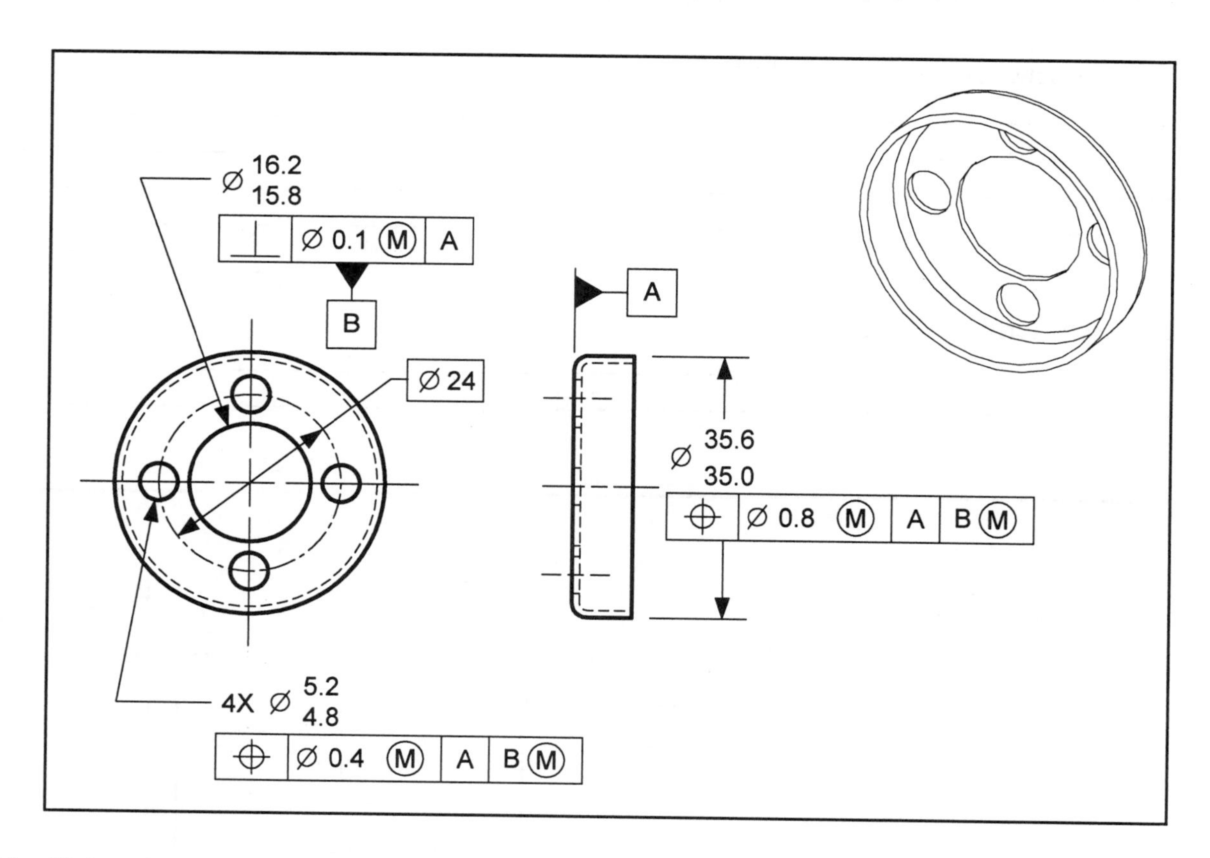

19. Using the drawing above, draw and dimension a cartoon gage for checking each position callout.

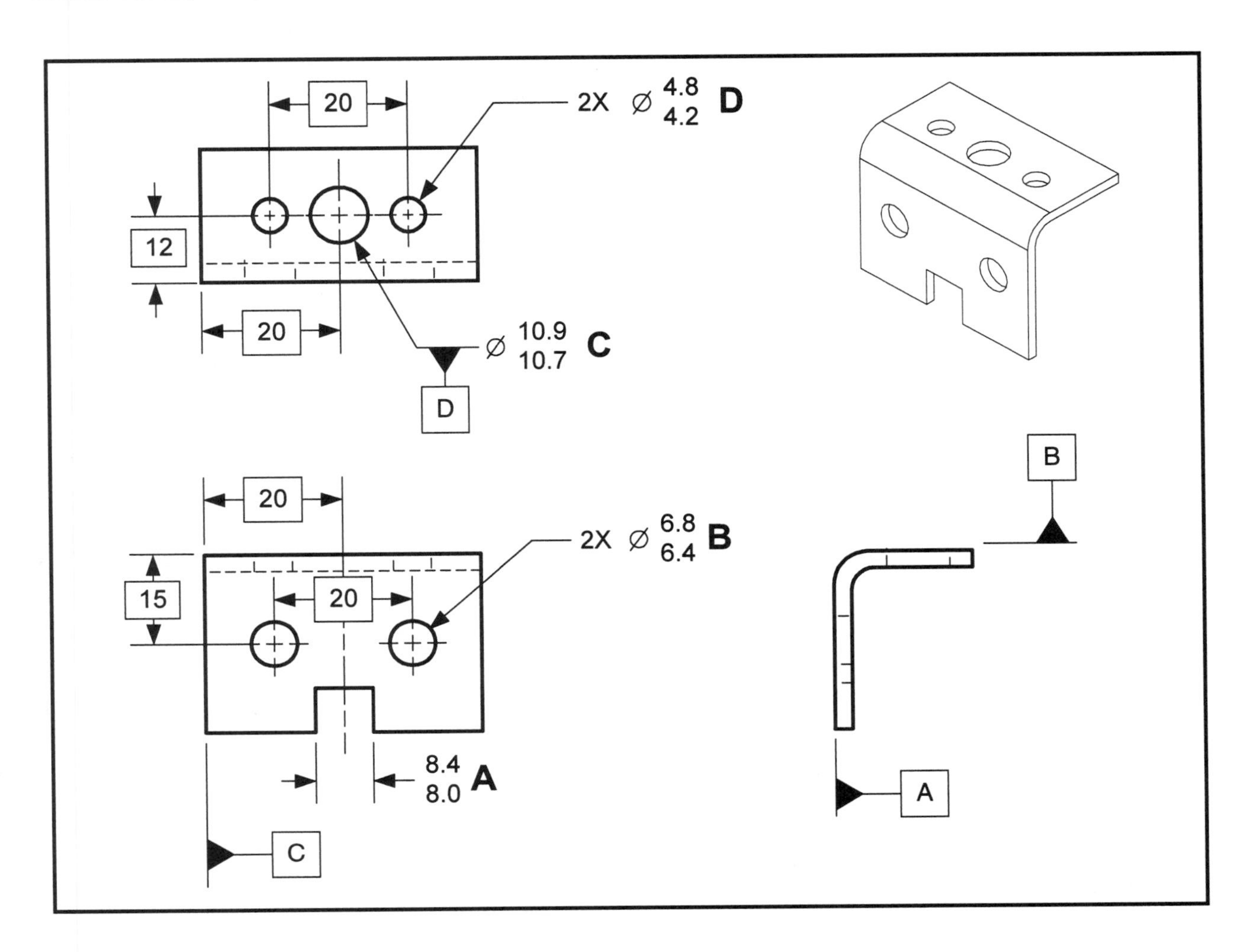

20. Use the instructions to complete the drawing above.

    a. Add a TOP control to the holes labeled *B* that will control their orientation and location relative to datums *A*, *B*, & *C*. The tolerance value should be 0.4 and apply RFS.

    b. Add a TOP control to the slot labeled *A* that will control its orientation and location relative to datums *A*, *B*, & *C*. The tolerance value should be 0.2 and apply RFS.

    c. Add a TOP control to the hole labeled *C* that will control its orientation and location relative to datums *B*, *A*, & *C*. The tolerance value should be 0.1 and apply RFS.

    d. Add a TOP control to the holes labeled *D* that will control their orientation and location relative to datums *B*, *D*, & *A*. The tolerance value should be 0.3 and apply RFS.

Chapter **9**

Tolerance of Position, Part 2

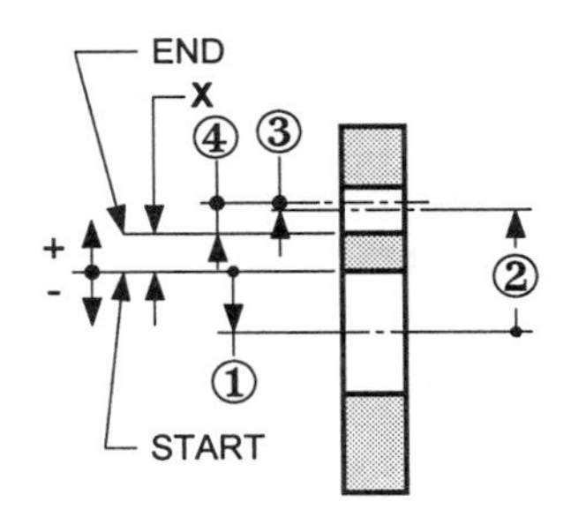

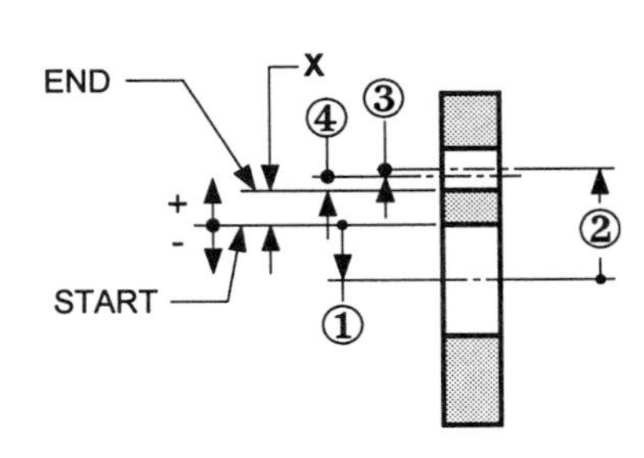

Maximum Distance X

①	- 5.0	MMC hole radius
②	+ 10.0	Basic location
③	+ 0.1	Tol. zone radius
④	- 1.7	MMC hole radius
	+ 3.4	

Minimum Distance X

①	- 5.2	LMC hole radius
②	+ 10.0	Basic location
③	- 0.1	Tol. zone radius
④	- 1.8	LMC hole radius
	+ 2.9	

INTRODUCTION

This chapter covers additional areas on the subject of TOP. First, calculations involving TOP are explained, then the fixed and floating fastener formulas are discussed. Five special applications of TOP are also explained.

CHAPTER GOALS AND OBJECTIVES

There are Three Goals in this Chapter:

9-1. Interpret tolerance of position special applications.
9-2. Calculate distances on a part dimensioned with tolerance of position.
9-3. Calculate tolerance of position tolerance values using the fixed and floating fastener formulas.

Performance Objectives that Demonstrate Mastery of These Goals

Upon completion of this chapter, each student should be able to:

Study Tip
Take a few minutes to fully understand these objectives. When reading this chapter, look for information to help you master these objectives.

Goal 9-1 (pp. 251-261)

- Describe the tolerance zone(s) in a TOP application of holes that are non-parallel and not perpendicular to the datum axis.
- Describe the tolerance zone(s) in a bi-directional TOP application.
- Describe the tolerance zone(s) in a TOP application of an elongated hole.
- Describe when to use a projected tolerance zone modifier.
- Describe the tolerance zone(s) in a TOP application using the projected tolerance zone modifier.
- Describe the tolerance zone(s) in a TOP application used to control a symmetrical relationship.
- Describe when a TOP application should use the LMC modifier.
- Describe how bonus tolerance is calculated in a TOP application that uses the LMC modifier.
- Describe the tolerance zone(s) in a TOP application used to control the spacing and orientation of a hole pattern.
- Describe when a multiple single-segment TOP control should be specified.
- Explain what zero tolerance at MMC dimensioning is.
- Explain three benefits available in a zero tolerance at MMC application.
- Explain the tolerance available in a zero tolerance at MMC application.

Goal 9-2 (pp. 262-265)
- Describe a tolerance stack.
- Calculate distances on a part dimensioned with TOP (RFS).
- Calculate distances on a part dimensioned with TOP (including bonus and shift).

Goal 9-3 (pp. 266-269)
- Describe a fixed fastener assembly.
- Write the formula for fixed fasteners.
- Calculate TOP tolerance values for fixed fastener applications.
- Describe a floating fastener assembly.
- Write the formula for floating fasteners.
- Calculate TOP tolerance values for floating fastener applications.

TOP SPECIAL APPLICATIONS

Tolerance of position is an extremely flexible dimensioning tool. This section explains seven special applications of TOP.

TOP Locating Holes that are Non-Parallel

In certain cases, it may be desired to control the location and orientation of holes that are non-parallel and not perpendicular to the datum axis. This can be accomplished with a TOP control. This type of application is shown in Figure 9-1.

For more info. . .
See Paragraph 5.4.1.6 Y14.5.

In this application, the following conditions apply:
- The shape of the tolerance zones are cylindrical virtual condition boundaries.
- The tolerance zones are located by the basic dimensions relative to the datums referenced.
- The angle of the hole, relative to datum plane *B*, is limited by the TOP.
- Bonus tolerances are permissible.

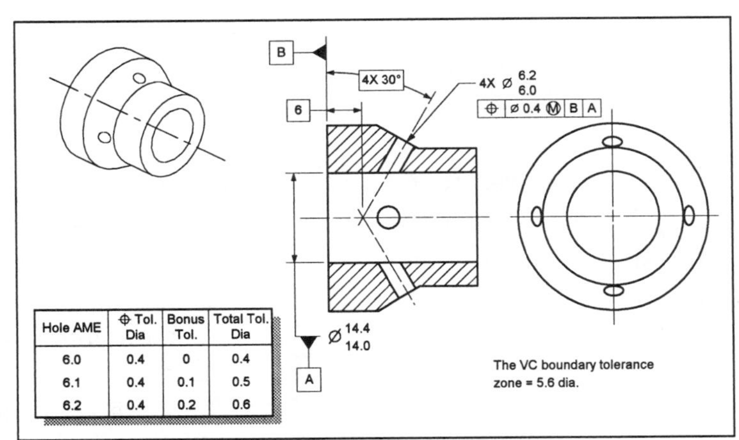

Hole AME	⊕ Tol. Dia	Bonus Tol.	Total Tol. Dia
6.0	0.4	0	0.4
6.1	0.4	0.1	0.5
6.2	0.4	0.2	0.6

FIGURE 9-1 TOP Applied to Holes that are Non-Parallel

Bi-Directional TOP (Locating a Hole in Two Directions)

In certain cases, it may be desired to allow a hole to have more location tolerance in one direction than in another direction. A ***bi-directional control*** is where the location of a hole is controlled to a different tolerance value in two directions. This can be accomplished by using two TOP controls to indicate the direction and magnitude of each positional tolerance relative to the datums specified. The feature control frames are attached to dimension lines. This type of application is shown in Figure 9-2.

For more info. . .
See Paragraph 5.9.1 of Y14.5.

In this application, the following conditions apply:

- The tolerance zones are parallel boundaries in the direction of the TOP control.
- The shape of the tolerance zones is rectangular.
- The tolerance zones are located by the basic dimensions relative to the datums referenced.
- Bonus tolerances are permissible.

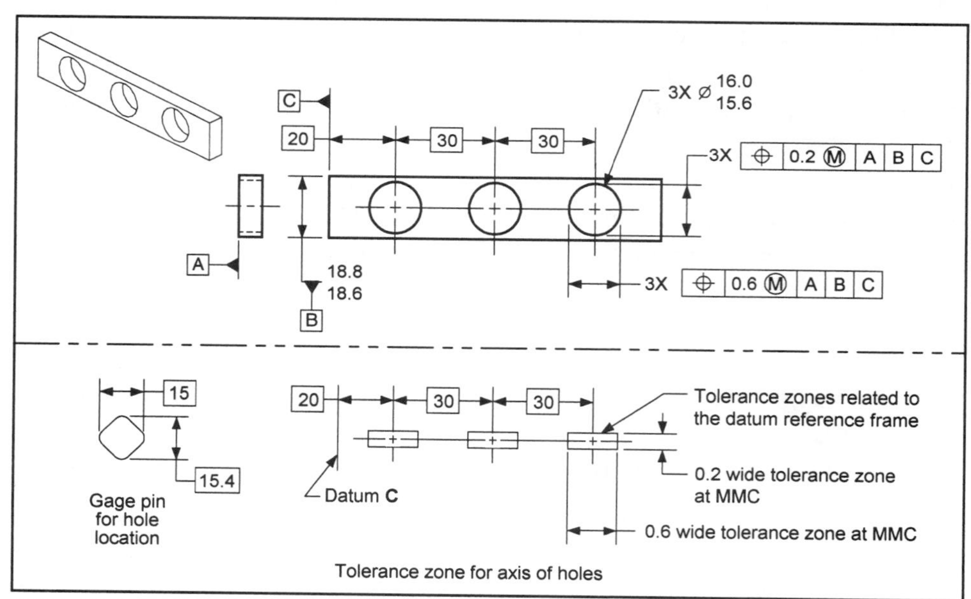

FIGURE 9-2 Bi-Directional TOP Application

Using TOP to Locate an Elongated Hole

In certain cases, it may be desired to allow an elongated hole to have more tolerance in one direction than in another direction. This can be accomplished by using two TOP controls to indicate the direction and magnitude of each positional tolerance relative to the datums specified. The feature control frames are attached to the dimension lines to invoke a bi-directional control. The word "***BOUNDARY***" is placed beneath the feature control frames to invoke a boundary control.

When BOUNDARY is used, the tolerance zone is considered to be a boundary control, and there is no axis interpretation for the application. An example is shown in Figure 9-3.

For more info. . .
See Paragraph 5.10.1 of Y14.5.

TECHNOTE 9-1 Use of the Word "BOUNDARY" Note

When the word "BOUNDARY" is placed beneath a TOP callout, the following conditions apply:

- The tolerance zone shape is a virtual condition boundary.
- There is no axis interpretation for the application.

In this application, the following conditions apply:

- The tolerance zone shape is a boundary of the identical shape as the elongated hole, minus the position tolerance value in each direction.
- There is no axis interpretation.
- The tolerance zones are located by the basic dimensions relative to the datums referenced.
- Bonus tolerances are permissible.
- The elongated hole must also meet its size requirements.

If the same positional tolerance is desired in both directions, a single position tolerance feature control frame may be used. In this instance, the feature control frame is directed to the elongated hole with a leader line.

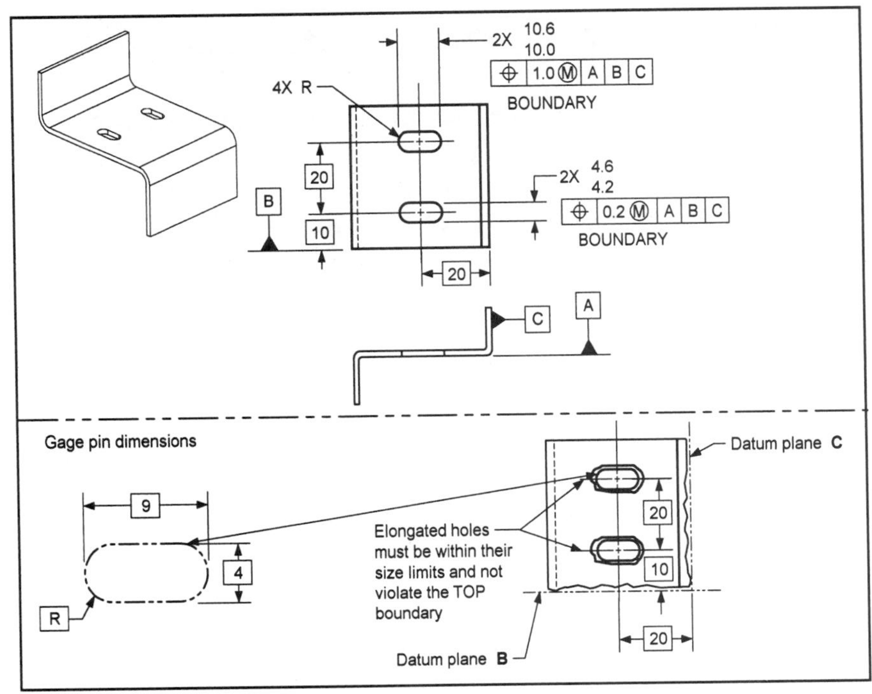

FIGURE 9-3 Elongated Hole Example

Top Using a Projected Tolerance Zone

When dimensioning threaded holes (or press-fit holes), consideration must be given to the variation in perpendicularity of the axis of the hole relative to the mating face of the assembly. The squareness error of the fastener (or press-fit pin) may result in an interference condition with the mating part. Figure 9-4 shows an example. An interference condition can occur where a position tolerance is specified for the hole, and the hole is tipped within the position tolerance zone. When the fastener is placed in the hole, the orientation of the fastener may result in an interference condition near the head of the fastener. This condition is common with fixed fastener applications. Where there is concern that an interference condition may exist, due to the orientation of the fastener, a projected tolerance zone modifier should be used.

Design Tip
A rule of thumb in bolted joint applications: whenever the height of the clearance hole is greater than the depth of the threaded hole, a projected tolerance zone modifier should be specified.

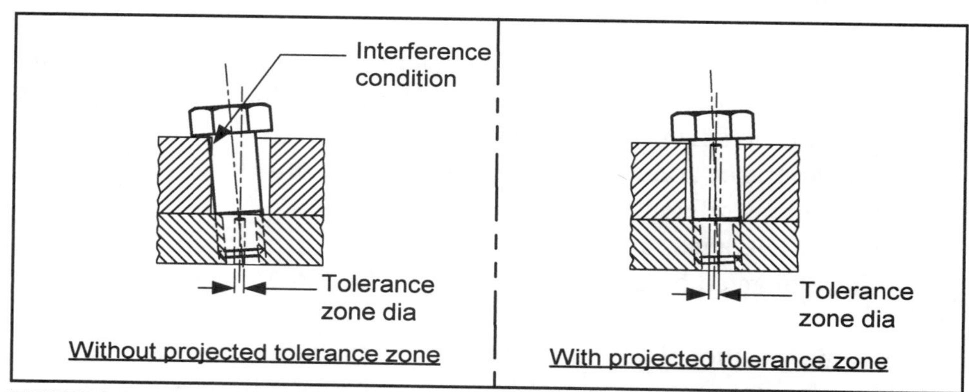

FIGURE 9-4 Fixed Fastener Interference Condition

For more info. . .
See Paragraph 5.5 of ASME Y14.5.

TECHNOTE 9-2 TOP Using A Projected Tolerance Zone

- Where a projected tolerance zone is used, the tolerance zone is projected above the part surface.
- The symbol for the projected tolerance zone modifier is Ⓟ.
- A projected tolerance zone is used to limit the perpendicularity of a hole to ensure assembly with the mating part.

A ***projected tolerance zone*** is a tolerance zone that is projected above the part surface. A projected tolerance zone exists whenever the projected tolerance zone modifier is specified. The projected tolerance zone symbol is a "P" enclosed in a circle. Where a projected tolerance zone is specified, the tolerance zone is projected above the part surface.

Author's Comment
The use of a projected tolerance zone does not permit a bonus tolerance. If bonus is desired, the MMC modifier must also be specified.

Figure 9-5 illustrates the application of a TOP using a projected tolerance zone. The symbol for projected tolerance zone is shown, then the height for the projected tolerance zone is specified. The height for the projected tolerance zone is a minimum and should be equal to the max. thickness of the mating part. The direction and height of the projected tolerance zone are illustrated. By using a projected tolerance zone, the orientation of the fastener is limited, which ensures assembly of the mating part. When using the fixed fastener formula, specifying a projected tolerance zone will ensure that fixed fasteners will not interfere with the clearance holes of mating parts.

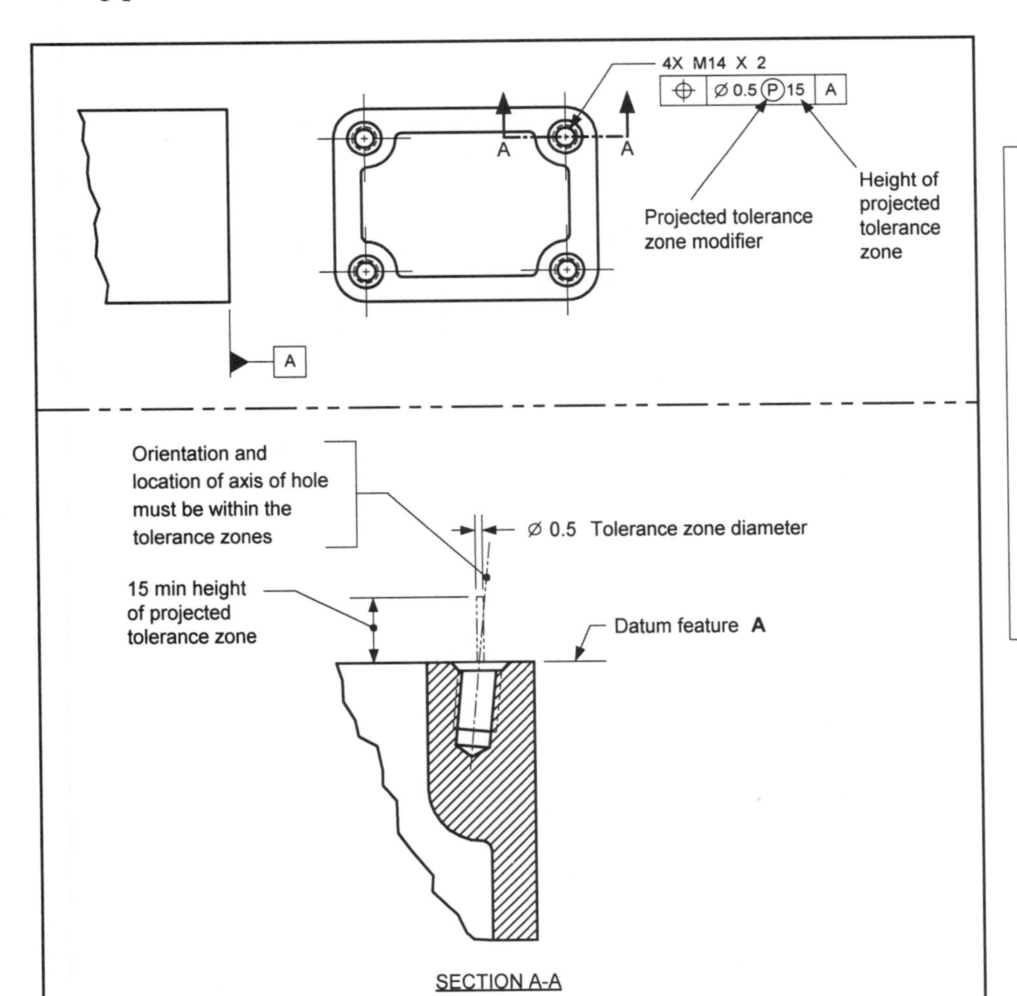

FIGURE 9-5 TOP Using a Projected Tolerance Zone Application

Design Tip
The use of the projected tolerance zone modifier is not necessary when the height of the mating part is thin, as in sheet metal.

Using TOP to Control Symmetrical Relationships

In certain cases, it may be desired to control a symmetrical relationship on a part. When the centerplane of the actual mating envelope of one or more features of size is being controlled with the axis or centerplane of a datum FOS, a TOP may be used. The example shown in Figure 9-6 involves using TOP MMC, which controls a symmetrical relationship to ensure that the part can be assembled.

Design Tip
In most industrial applications, TOP is used to control symmetrical relationships.

In this application, the following conditions apply:

- The tolerance zone shape is two parallel planes.
- The tolerance zone is located by an implied basic zero dimension relative to the datum referenced.
- A bonus tolerance is permissible.
- Datum shift is permissible.

Author's Comment
Controlling symmetrical relationships is not the same as controlling symmetry (as defined by Y14.5).

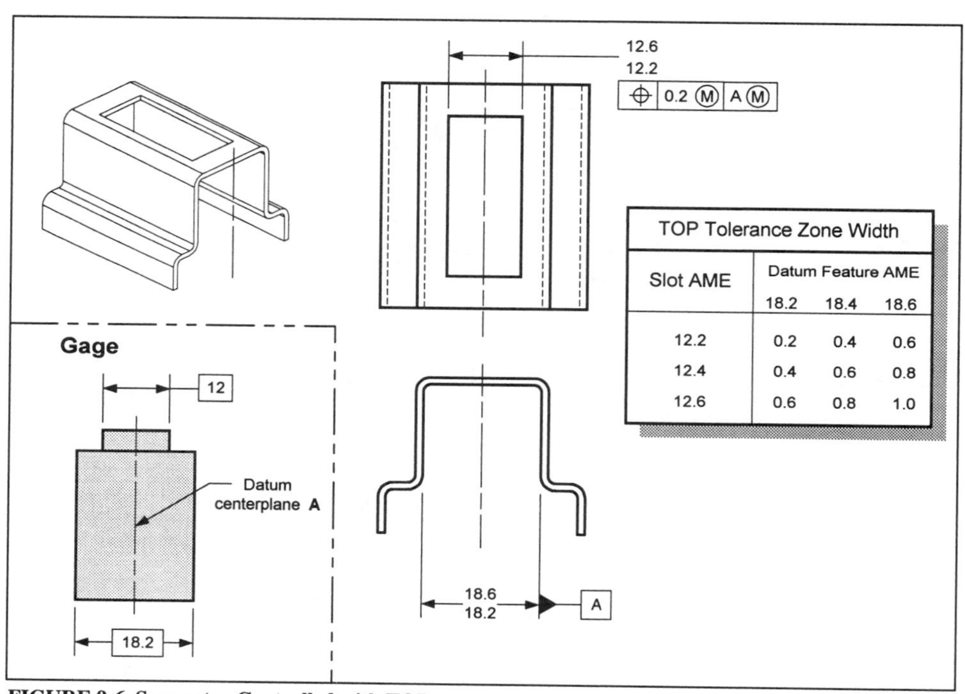

TOP Tolerance Zone Width			
Slot AME	Datum Feature AME		
	18.2	18.4	18.6
12.2	0.2	0.4	0.6
12.4	0.4	0.6	0.8
12.6	0.6	0.8	1.0

FIGURE 9-6 Symmetry Controlled with TOP

The example shown in Figure 9-6 uses a TOP at MMC. The use of TOP to control symmetrical relationships of the actual mating envelope of a FOS can also be applied at RFS or LMC.

TOP with the LMC Modifier

The LMC modifier is used in a TOP control when the functional consideration is to control a minimum distance on a part. The minimum distance can be a minimum wall thickness, a minimum part distance, or minimum machine stock on a casting. When the LMC modifier is used, the TOP applies if the FOS is at LMC. A bonus tolerance is permissible when the toleranced FOS departs from LMC towards MMC. Figure 9-7 illustrates an example of a TOP callout with the LMC modifier to control the minimum wall thickness of a part.

In this application, the following conditions apply:

- The shape of the tolerance zone is a cylindrical boundary.
- The dimension between the centerline of the diameter and the datum axis is an implied basic zero.
- A bonus tolerance is permissible.
- The minimum wall is 1.6 [(24.2 - 20.8 - 0.2) ÷ 2 = 1.6].
- Perfect form at LMC applies (perfect form at MMC is not required).

Author's Comment
Caution: When specifying the LMC modifier, the effects of Rule #1 are reversed. Perfect form at LMC is required. *Study* Paragraph 5.3.5 of Y14.5 before using this concept.

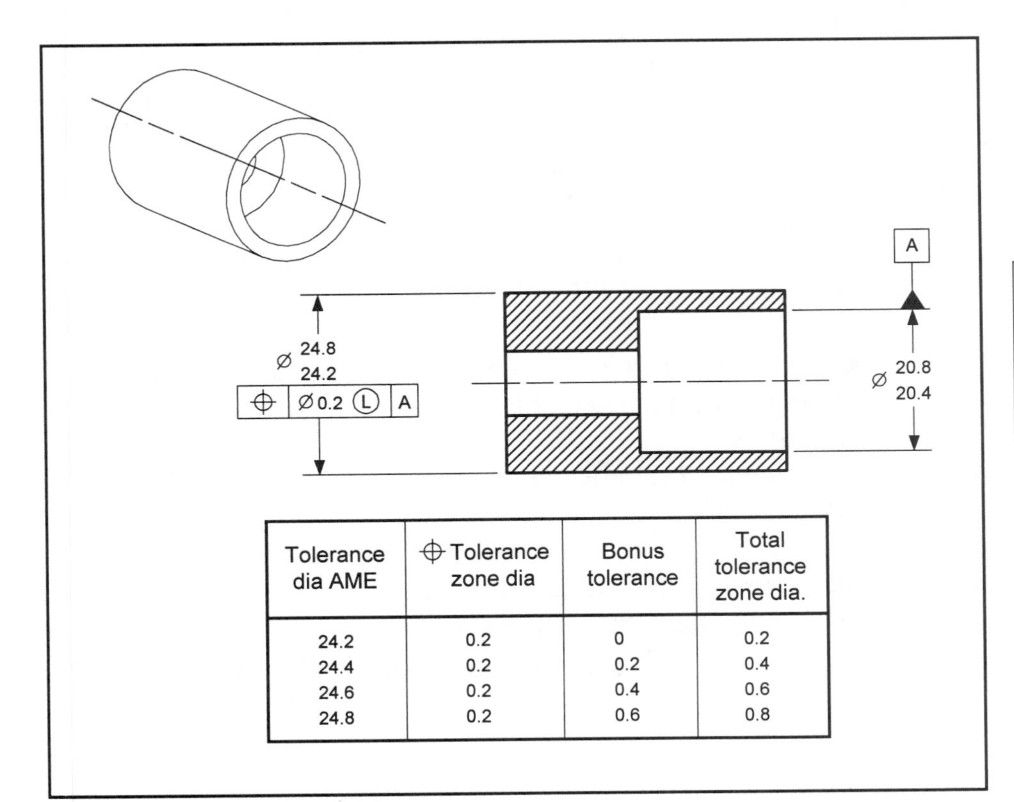

Tolerance dia AME	⌖ Tolerance zone dia	Bonus tolerance	Total tolerance zone dia.
24.2	0.2	0	0.2
24.4	0.2	0.2	0.4
24.6	0.2	0.4	0.6
24.8	0.2	0.6	0.8

FIGURE 9-7 TOP at LMC Application

For more info. . .
See Paragraph 5.3.5 of Y14.5.

Using TOP to Control Spacing and Orientation of a Pattern of Holes

In certain cases, it may be desired to control only the spacing and orientation of holes in a pattern. This can be accomplished by using a TOP control with a single datum reference. Figure 9-8 shows an example. In this figure, the TOP control limits the spacing between the holes and the squareness of the holes relative to datum plane *A*, but the TOP control does not control the location of the hole pattern. Notice the gage has only one datum simulator for datum *A*. The four gage pins limit the perpendicularity and spacing of the holes.

Author's Comment
Figure 9-8 is a partial drawing. A relationship between the outsides edges of the part and the four holes needs to be established.

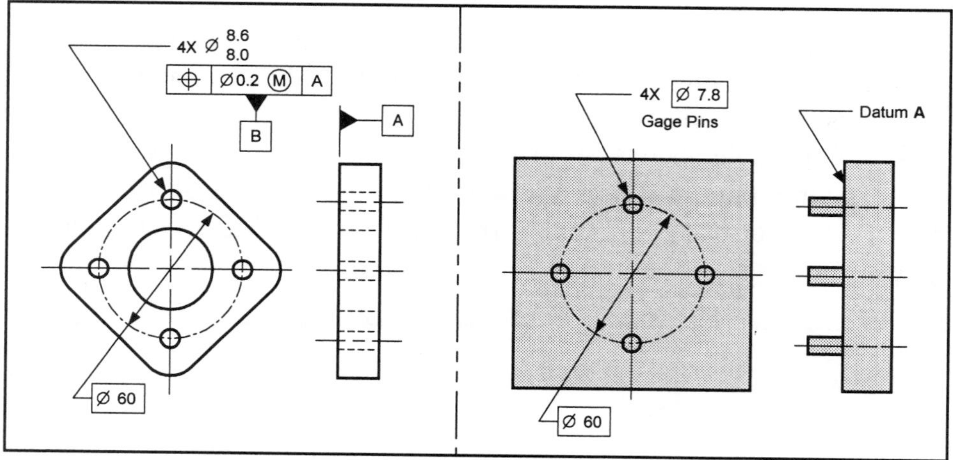

FIGURE 9-8 TOP with a Single Datum Reference Applied to a Hole Pattern

The designer typically uses a TOP control on a hole pattern with a single datum reference in two cases:

1. When the hole pattern is used as a datum feature, and only the spacing and perpendicularity need to be defined, as shown in Figure 9-8
2. When a hole pattern is toleranced with a multiple single-segment TOP control

When a hole pattern is used as a datum feature, it does not have to be located from the outside edges of the part. The outside edges of the part can be defined from the hole pattern and toleranced with a profile control.

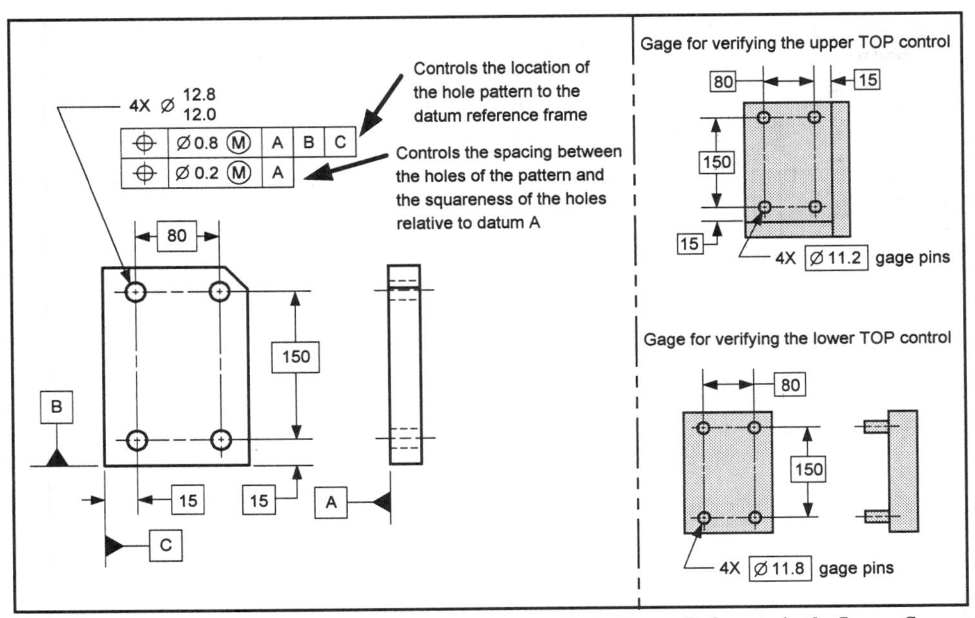

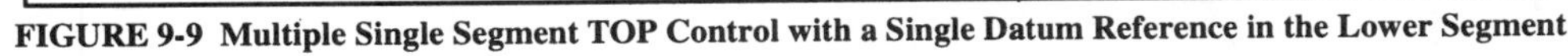

FIGURE 9-9 Multiple Single Segment TOP Control with a Single Datum Reference in the Lower Segment

Design Tip
The use of TOP to control a hole pattern is only introduced in this text. Y14.5 contains a method called "composite tolerancing" for dimensioning hole patterns with complex functional relationships.

Multiple Single-Segment TOP Controls

A ***multiple single-segment TOP control*** is when two (or more) single-segment TOP callouts are used to define the location, spacing, and orientation of a pattern of features of size. In Figure 9-9, the upper segment of the TOP control provides tolerance for the location of the pattern of holes relative to the outside of the part. The lower segment of the TOP control provides tolerance for the spacing between the holes and the orientation of the holes relative to datum *A*. Each segment of the TOP control is an independent requirement.

A designer uses this type of control when a hole pattern can have a large tolerance with respect to the outside edges of the part, but requires a tighter tolerance for squareness and/or spacing within the hole pattern.

Author's Comment
Multiple single-segment TOP controls may be specified with two or three datum references in the lower segment. The lower segment may reference different datums than the upper segment.

TOP with Zero Tolerance at MMC

A significant function of the design of a part is how it is dimensioned. Tight tolerances do not guarantee a quality part, only an expensive one. A dimensioning method that can help to reduce cost is called "zero tolerance (ZT) at MMC." ZT at MMC appears restrictive to anyone not familiar with the benefits and capabilities of GD&T. However, once the ZT method is understood, it becomes apparent that ZT at MMC protects the functional requirements of a design, while offering maximum flexibility to manufacturing.

ZT at MMC is a method of tolerancing part features that includes the geometric tolerance value with the FOS tolerance and states a zero tolerance at MMC in the feature control frame.

For more info. . .
See Paragraph 5.3.3 of Y14.5.

Design Tip
ZT at MMC should be considered whenever the function of a FOS is assembly.

Figure 9-10*A* shows a part with a conventional TOP of 0.3 at MMC. Figure 9-10*B* shows the same part dimensioned with the ZT at MMC method. The 0.3 tolerance has been removed from the TOP callout and included in the size tolerance. Note that the functional parameter—the virtual condition of the hole—is the same for both parts. With ZT at MMC, all of the hole location tolerance is derived from the bonus tolerance, so manufacturing can divide the available tolerance between size and location to best suit the process for the part.

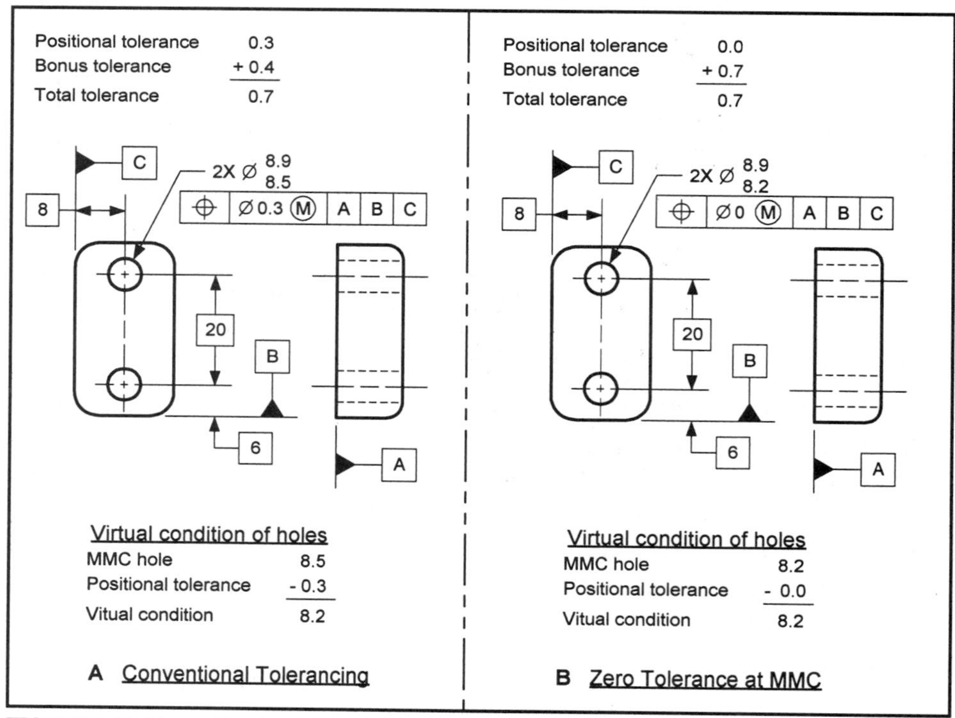

FIGURE 9-10 Conventional and Zero Tolerance at MMC Comparison

Author's Comment
All the benefits of the ZT at MMC method of tolerancing are also available with straightness, parallelism, perpendicularity, and angularity.

There are three primary benefits to ZT at MMC:

1. It provides flexibility for manufacturing.
2. It prevents the rejection of usable parts.
3. It reduces manufacturing costs.

The effects of ZT at MMC can be demonstrated through the use of a tolerance analysis chart. A ***tolerance analysis chart*** is a means of graphically displaying the limits of a part as defined by the print specifications. The tolerance analysis chart in Figure 9-11*A* describes the parameters for parts from Figure 9-10*A*. On the vertical scale, the allowable positional tolerance values are listed. The horizontal scale shows the virtual condition and hole sizes for the part.

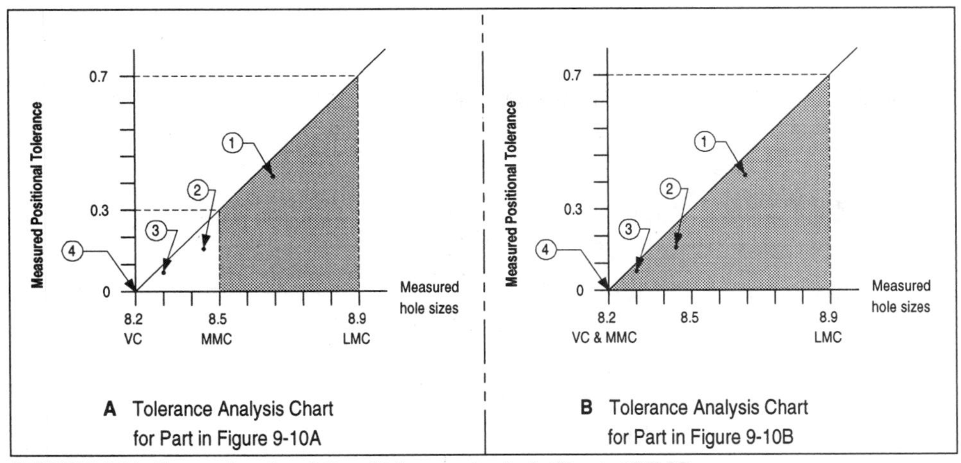

FIGURE 9-11 Conventional and Zero Tolerance Analysis Chart at MMC

The shaded area of the chart represents the acceptable parts according to the print specifications. The dots labeled "1, 2, 3, and 4" represent the actual hole size and location for four parts. Three of the four parts would be rejected on the basis of the hole size not being to print specifications. It is important to note that the virtual condition of the hole size between the MMC and the virtual condition may be a functional part if it is located properly. Conventional TOP dimensioning can result in functional parts being rejected.

Figure 9-11*B* shows a tolerance analysis chart for the parts from Figure 9-10*B*. The shaded area of the chart represents the acceptable parts according to the print specifications from Figure 9-10*B*. The dots labeled, "1, 2, 3, and 4" represent the actual hole size and location of the same four parts from Figure 9-11*A*. Note that all four parts are now considered acceptable. ZT at MMC increases the zone of acceptable parts by making the MMC and virtual condition of the hole equal. As a bonus, any hole that meets its virtual condition requirement (the functional requirement), will also meet its size requirement.

When dimensioning holes for the function of assembly, the designer should consider ZT at MMC. With this method of dimensioning, functional parts are not rejected, and more flexible manufacturing results in additional tolerance to produce parts.

TECHNOTE 9-3 Zero Tolerance at MMC

Zero tolerance at MMC is a method of tolerancing that restates a geometric tolerance so it is included with the FOS tolerance. Three benefits of zero tolerance at MMC are:

1. It provides flexibility for manufacturing.
2. It prevents the rejection of usable parts.
3. It reduces manufacturing costs.

TOP CALCULATIONS

In industry, it is very common to make a calculation to find a max. or min. distance on a part. These calculations are referred to as tolerance stacks. A ***tolerance stack*** is a calculation used to find an extreme max. or min. distance on a part. When calculating the extreme max. or min., all of the part tolerances must be included. This section explains how to use TOP in tolerance stacks.

Design Tip
Use tolerance stacks to predict how parts will assemble or function before hardware is made.

Tolerance Stacks Using TOP at RFS

Tolerance stacks on a part that involves TOP applied RFS are straightforward. In this section, a simple method used to calculate the maximum and minimum distance between the edges of two holes is explained. The process involves using the basic dimension between the holes, the TOP tolerance value, and the MMC or LMC hole size.

Figure 9-12 shows an example of a part calculation to find the max. and min. distance between two holes on a part. When reading the steps below, refer to Figure 9-12.

The steps for making a tolerance stack to calculate a max. and min. part distance are as follows:

1. Label the start and end points of the distance to be calculated. On the start point of the calculation, draw a double-ended arrow. Label the arrow that points towards the end point of the calculation as positive (+). Label the other arrow as negative (-). Each time a distance that is in the direction of the positive arrow is used in the calculation, the distance will be a positive value. When a distance is used in the negative direction, it will be a negative value.

2. Establish a loop of part dimensions or gage distances (as in Figure 9-12) from the start point to the end point of the calculation.

3. Calculate the answer.

When solving for a min. distance, half the TOP tolerance value is subtracted from the calculation. When solving for a max. distance, half the TOP tolerance value is added in the calculation.

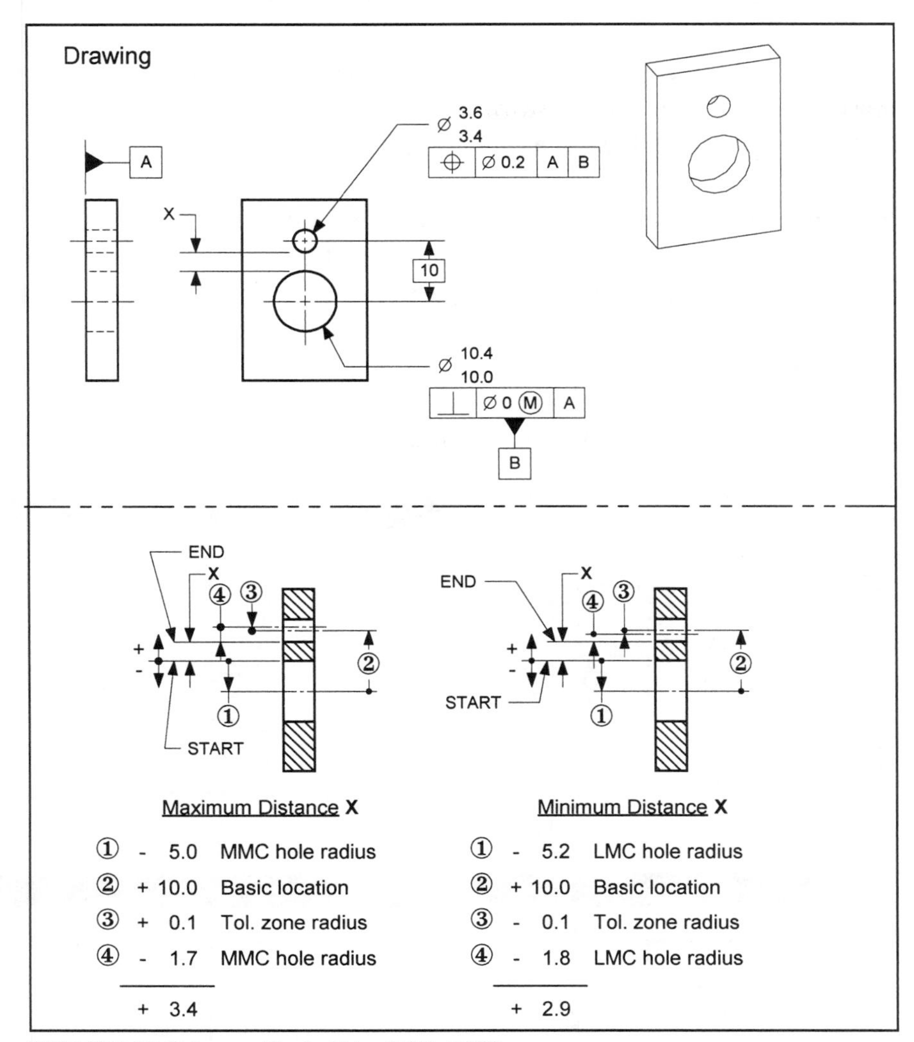

FIGURE 9-12 Tolerance Stacks Using TOP at RFS

Tolerance Stacks Using TOP at MMC

Tolerance stacks on a part that involves TOP applied at MMC can be best learned by using the gage method. The gage method involves using the cartoon gage to calculate part distance. A major advantage of the gage method is that the effects of bonus tolerances or datum shift are automatically included in the analysis. The gage method involves five steps:

1. Draw the cartoon gage.
2. Draw the part on the gage in the position that gives the extreme condition being calculated.
3. Label the start and end points of the distance being calculated. (The start and end points are always on the part.)
4. Establish a path of continuous known distances (either the part or the gage can be used) from the start point to the end point of the calculation.
5. Calculate the answer.

The five steps for the gage method are demonstrated in Figure 9-13. On this part, the max. and min. distance "X" is to be calculated. In step one, the cartoon gage shows the gage pins for both of the toleranced holes.

In steps two, three, and four, the part is drawn on the cartoon gage. The part is shown in the position that creates the max. or min. condition. An "X" is drawn on each plate where a part surface touches a gage surface. The stack indicator is shown on the start point of the stack. The stack path is labeled.

Step five shows the calculation for the max. or min. condition.

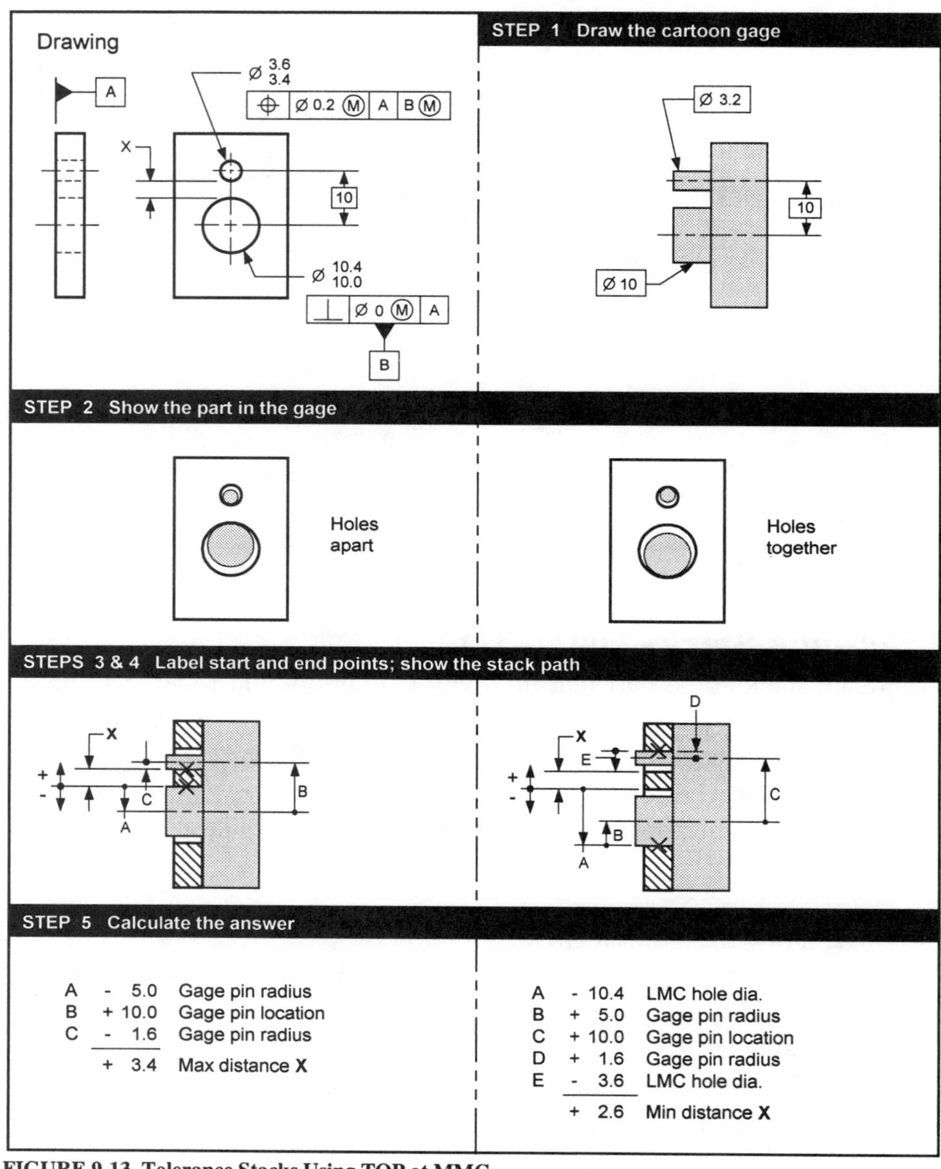

FIGURE 9-13 Tolerance Stacks Using TOP at MMC

Figure 9-14 shows another example of a part distance calculation on a part that involves TOP. On this part, the max. and min. "X" is to be calculated. The gage method was used to find the distances. Note that in the max. X calculation, the upper limit of the width dimension (70.4) was used in the calculation because it results in distance X becoming greater. In the min. X calculation, the min. of the width dimension (69.6) and the LMC hole size were used because they result in distance X becoming smaller.

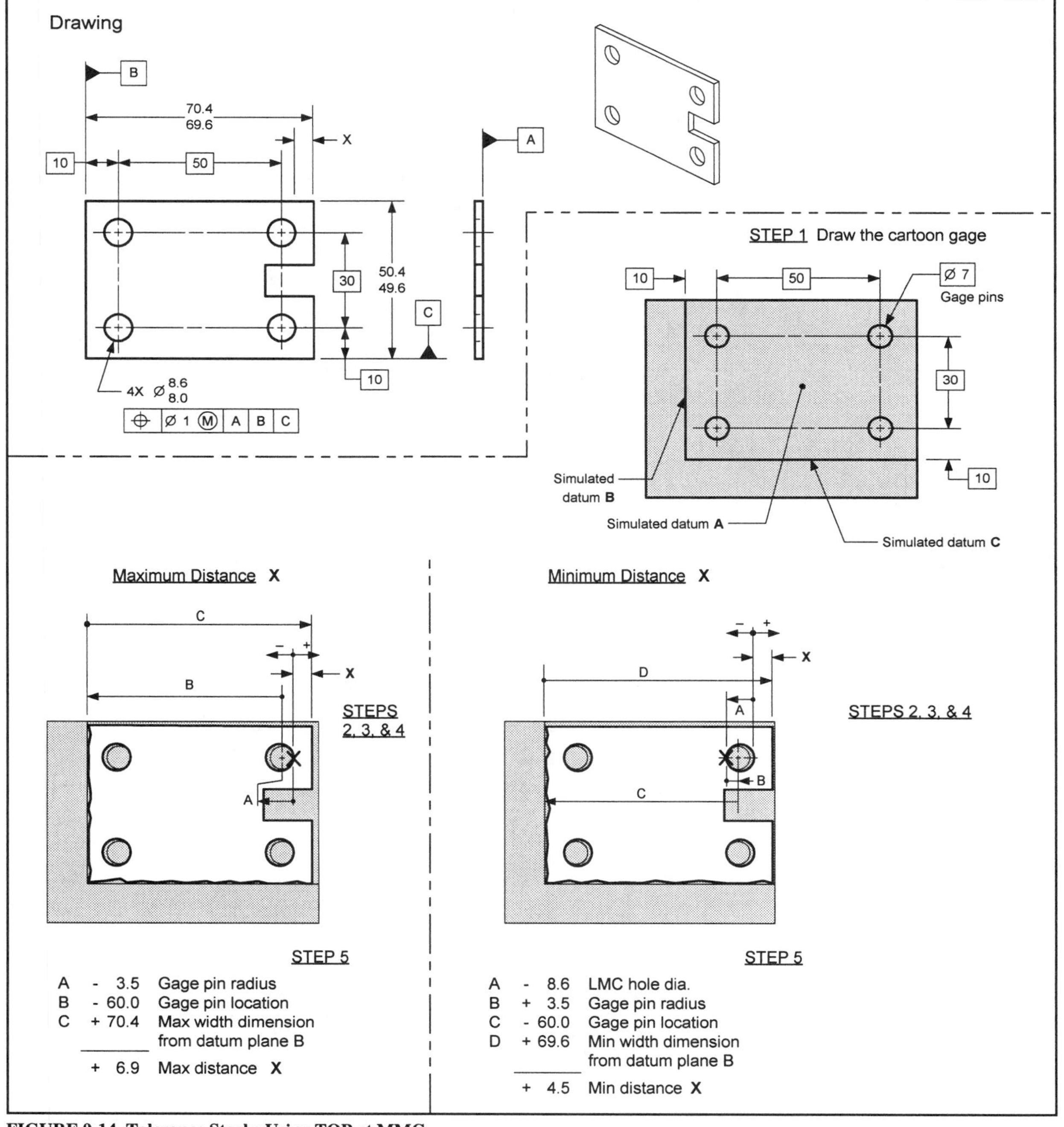

FIGURE 9-14 Tolerance Stacks Using TOP at MMC

Author's Comment
The most economical method for dimensioning holes in a bolted joint application is ZT at MMC, as described on pages 259-261. The fixed and floating fastener formulas are convenient, but are more expensive requirements for manufacturing.

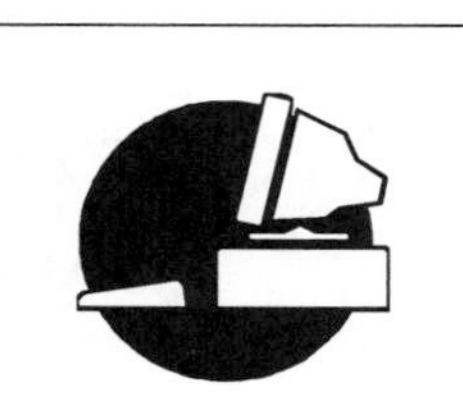

For more info. . .
See Appendix *B* of Y14.5.

FIXED AND FLOATING FASTENER CALCULATIONS

When designing products with fasteners, the fixed and floating fastener formulas are convenient design tools. They allow a designer to quickly determine the TOP tolerance values for the mating parts involved. The fixed and floating fastener formulas can be used on all types of hole patterns. This section provides an introduction on how to use these formulas.

Fixed Fastener Assemblies

A ***fixed fastener assembly*** is where the fastener is held in place (restrained) into one of the components of the assembly. Often, the holes in one component of the assembly are clearance holes, and the holes in other component are threaded holes (or a press fit, like a dowel pin). This type of assembly is called a fixed fastener assembly because the fastener is "fixed" in the assembly. An example of a fixed fastener assembly is shown in Figure 9-15. The components are assembled with four M14 screws. The cover has four clearance holes. The housing has four M14 X2 threaded holes. Both hole patterns are dimensioned with TOP.

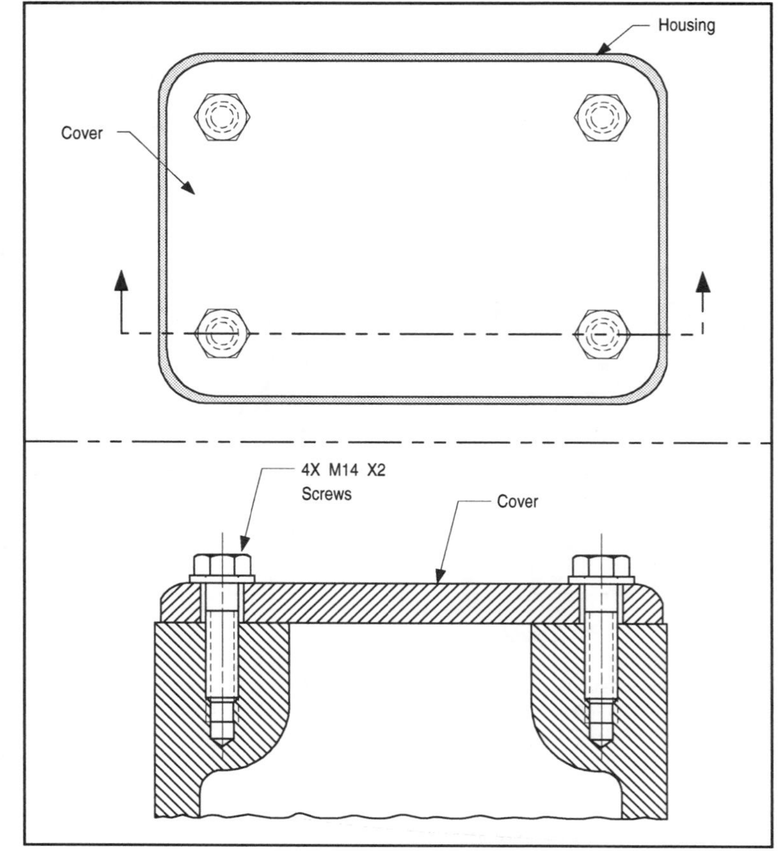

FIGURE 9-15 Example of a Fixed Fastener Assembly

The procedure for determining the amount of tolerance for fixed fastener applications is a simple process. (The formula in this text applies when the projected tolerance zone modifier is used on the threaded hole.)

The ***fixed fastener formula*** is:

$$H = F + 2T \quad \text{or} \quad T = \frac{H - F}{2}$$

Where:

T = position tolerance diameter
H = MMC of the clearance hole
F = MMC of the fastener

Design Tip
The projected tolerance zone modifier is often used on the threaded hole when the fixed fastener formula is used. See page 255.

Since the function of the holes is assembly, the MMC modifier is used in the tolerance portion of the TOP callout. This allows additional position tolerance as the holes depart from MMC.

Figure 9-16 shows an example of using the fixed fastener formula for determining the TOP tolerance values. The cover and housing are from the assembly shown in Figure 9-15. Using the fixed fastener formula, the total tolerance for both parts is 0.4. The tolerance for each part is 0.2.

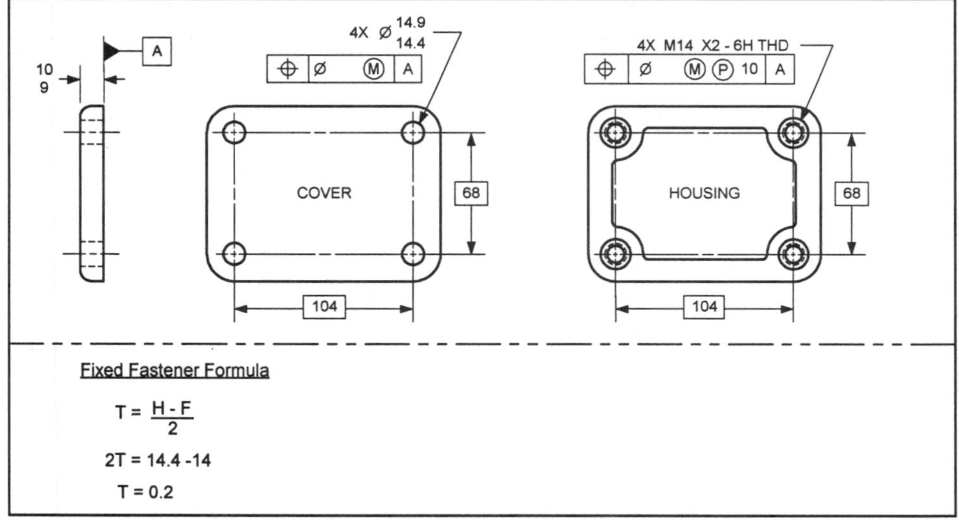

FIGURE 9-16 Using the Fixed Fastener Formula to Determine TOP Tolerance Values

The fixed fastener formula is used to determine the position tolerance values, which ensures that the parts will assemble. This results in a "no interference, no clearance" condition when the components are at MMC and located at their extreme position.

For more info. . .
The formulas that apply when the projected tolerance zone modifier is not specified are shown in Y14.5, Appendix B.

The examples shown in this text specify the projected tolerance zone modifier. If the projected tolerance zone is not specified, different formulas apply.

TECHNOTE #9-4 Fixed Fastener Assemblies

- A fixed fastener assembly is held in place (fixed) into one of the components of the assembly.

- The fixed fastener formula is $T = \frac{H - F}{2}$ or $H = F + 2T$

Floating Fastener Assemblies

Floating fastener assembly is where two (or more) components are held together with fasteners (such as bolts and nuts), and both components have clearance holes for the fasteners. This type of assembly is called a floating fastener assembly because the fasteners can "float" (move) in the holes of each part. An example of a floating fastener assembly is shown in Figure 9-17.

Design Tip
When using the fixed and floating fastener formulas, you should also consider shank to thread runout and the straightness of the fastener.

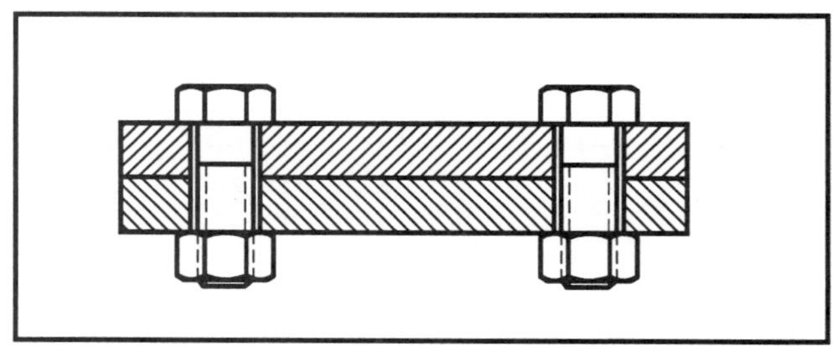

FIGURE 9-17 Examples of a Floating Fastener Assembly

The plates are assembled with four M14 bolts and nuts. Both plates have the same diameter bolt clearance holes and use TOP to dimension the hole locations.

When determining the amount of tolerance for the position callouts, the floating fastener formula may be used. The ***floating fastener formula*** is:

$$T = H - F$$

Where:

T = position tolerance diameter (for each part)
H = MMC of the clearance hole
F = MMC of the fastener

Once the position tolerance is determined, it applies to each part in the assembly (that has the hole size used in the formula). Since the function is assembly, the MMC modifier is used in the tolerance portion of the TOP callout. This allows additional position tolerance as the holes depart from MMC. Figure 9-18 shows an example of using the floating fastener formula for determining the TOP tolerance values.

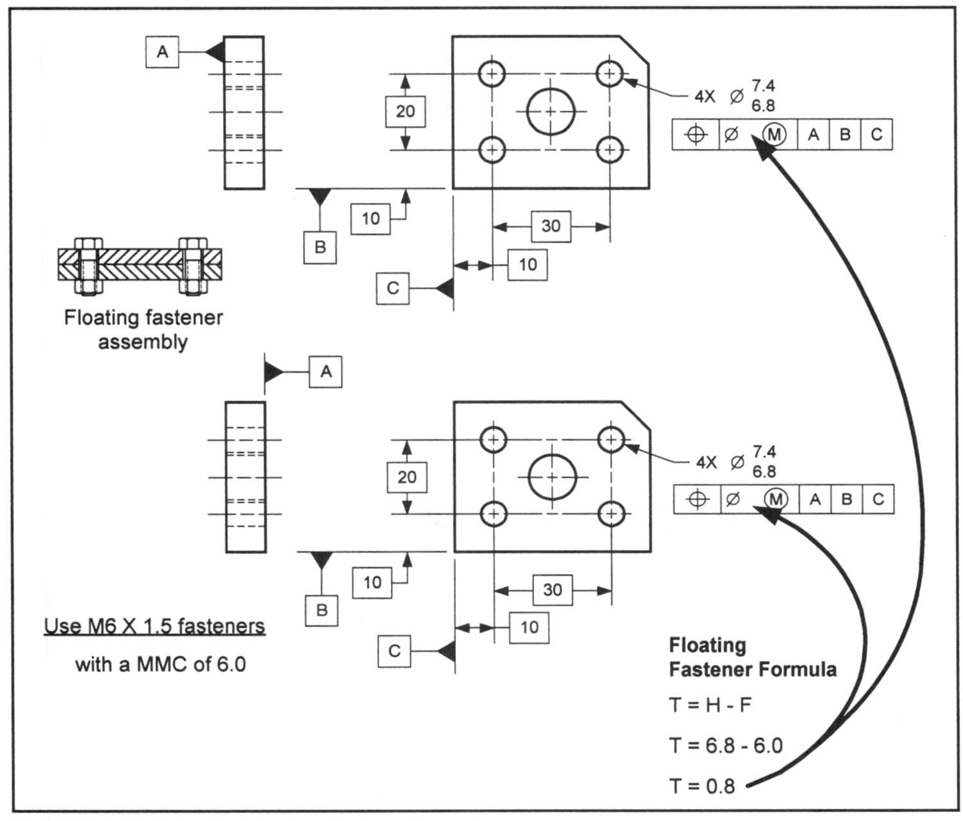

FIGURE 9-18 Using the Floating Fastener Formula

The floating fastener formula is used to determine the position tolerance values, which ensures that the parts will assemble. It results in a "no interference, no clearance" condition when the components are at MMC and are located at their extreme position. Consideration must be given to additional geometric conditions that are not accounted for in the floating fastener formula.

TECHNOTE 9-5 Floating Fastener Assemblies

- A floating fastener assembly is where two or more components are held together with fasteners (such as bolts and nuts) and all the components have clearance holes for the fasteners.

- The floating fastener formula is: T = H - F

VOCABULARY LIST

Study Tip
Read each term. If you don't recall the meaning of a term, look it up in the chapter.

New Terms Introduced in this Chapter

Bi-directional control
BOUNDARY
Fixed fastener assembly
Fixed fastener formula
Floating fastener assembly
Floating fastener formula
Multiple single-segment TOP control
Projected tolerance zone
Tolerance analysis chart
Tolerance stack
Zero tolerance (ZT) at MMC

Author's Comment
These topics, plus advanced coverage of many of the topics introduced in this text, will be covered in my new book on advanced GD&T concepts.

ADDITIONAL RELATED TOPICS

Topic	ASME Y14.5M-1994 Reference
• Simultaneous TOP requirements	Paragraph 5.3.6.2
• Composite positional tolerancing	Paragraph 5.4.1

QUESTIONS AND PROBLEMS

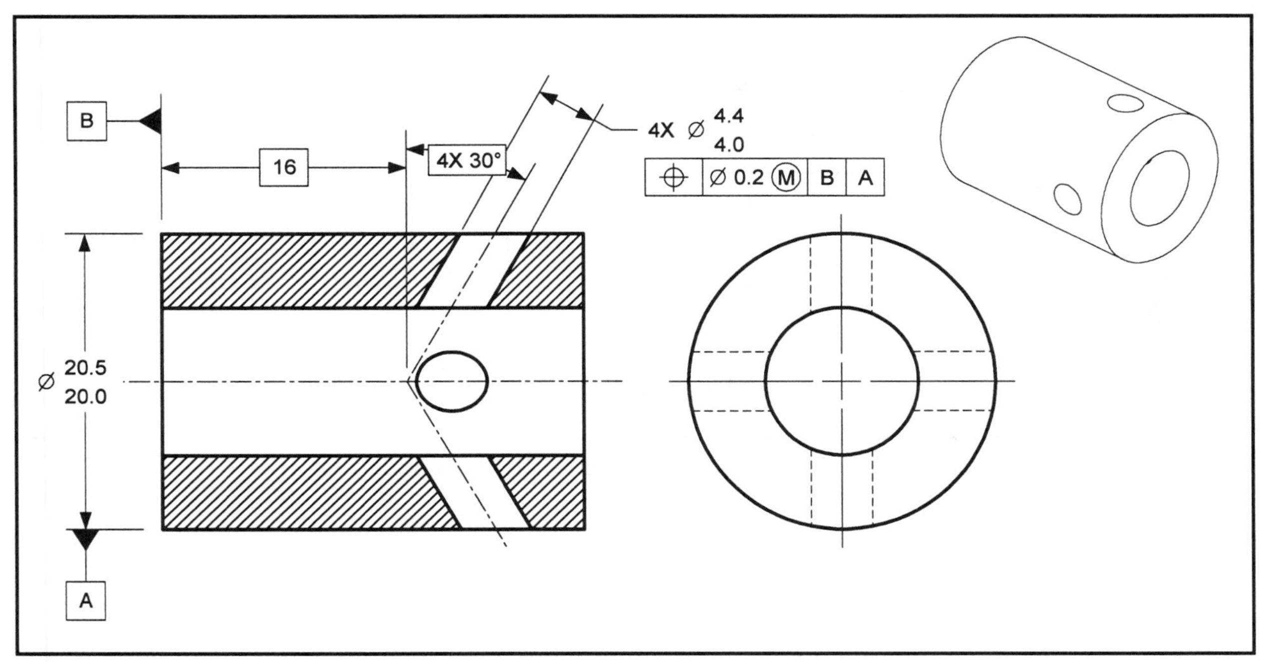

Questions 1-3 refer to the figure above.

1. What is the shape of the tolerance zone for the holes?
____

2. What limits the angular relationship of the holes to datum plane *B*?____________________
____

3. Fill in the chart below.

Toleranced Hole AME	⌖ Tolerance Zone Diameter
4.0	
4.1	
4.2	
4.3	
4.4	

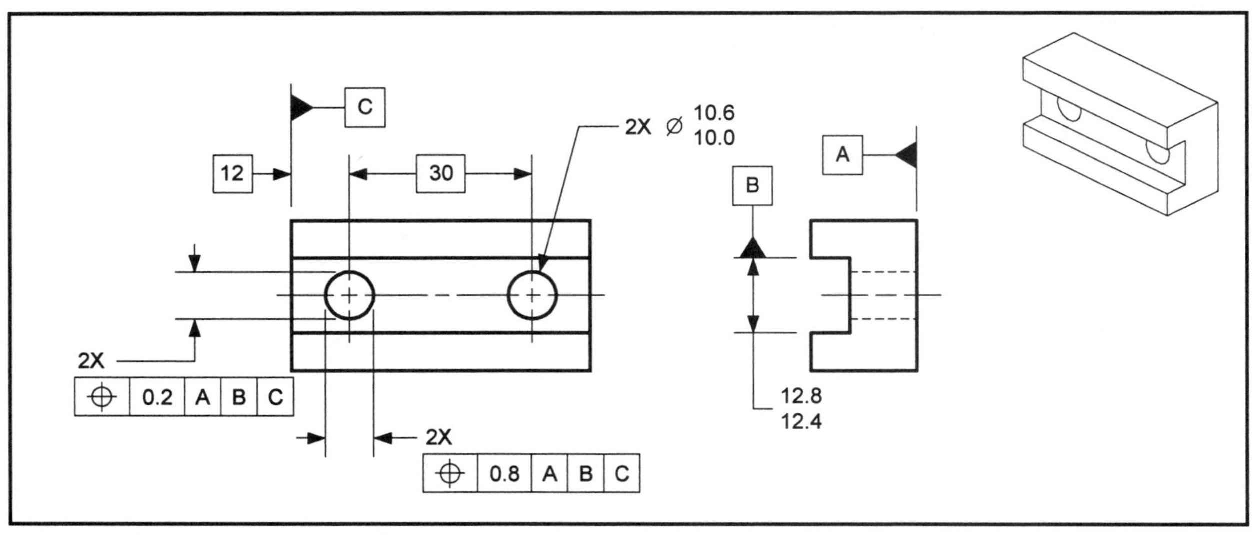

4. Using the figure above, draw the tolerance zones for the hole locations.

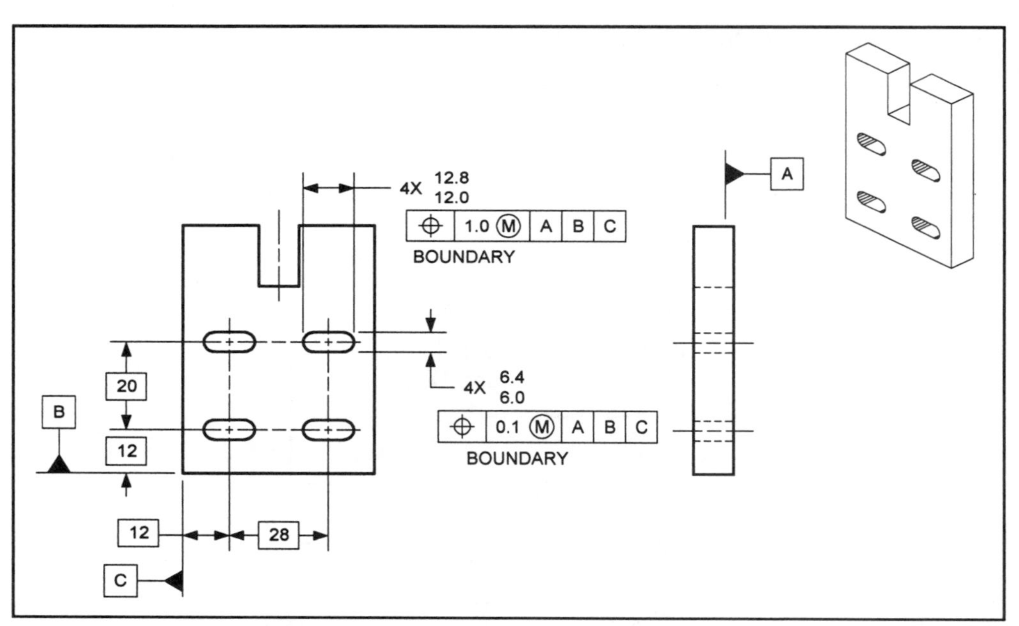

Questions 5 and 6 refer to the figure above.

5. In the space below, draw a gage pin for verifying the locations of the elongated holes.

6. In the space below, draw the tolerance zone for the location of the axis of the elongated holes.

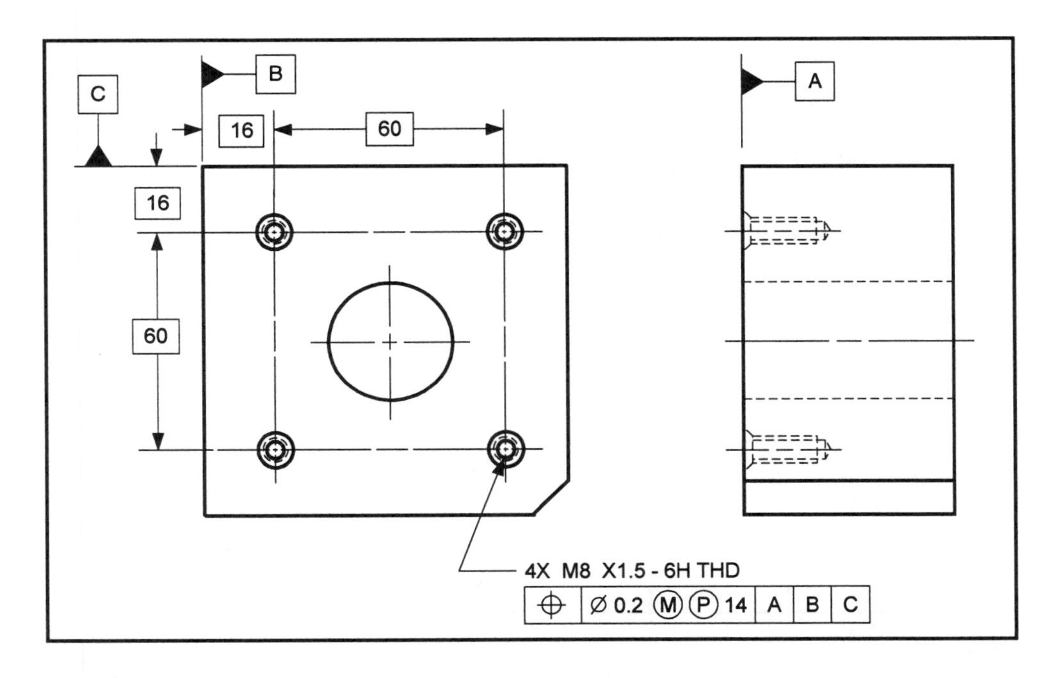

Questions 7 - 9 refer to the figure above.

7. The Ⓟ stands for ______________________________

8. The 14 stands for______________________________

9. Draw the TOP tolerance zone on the figure.

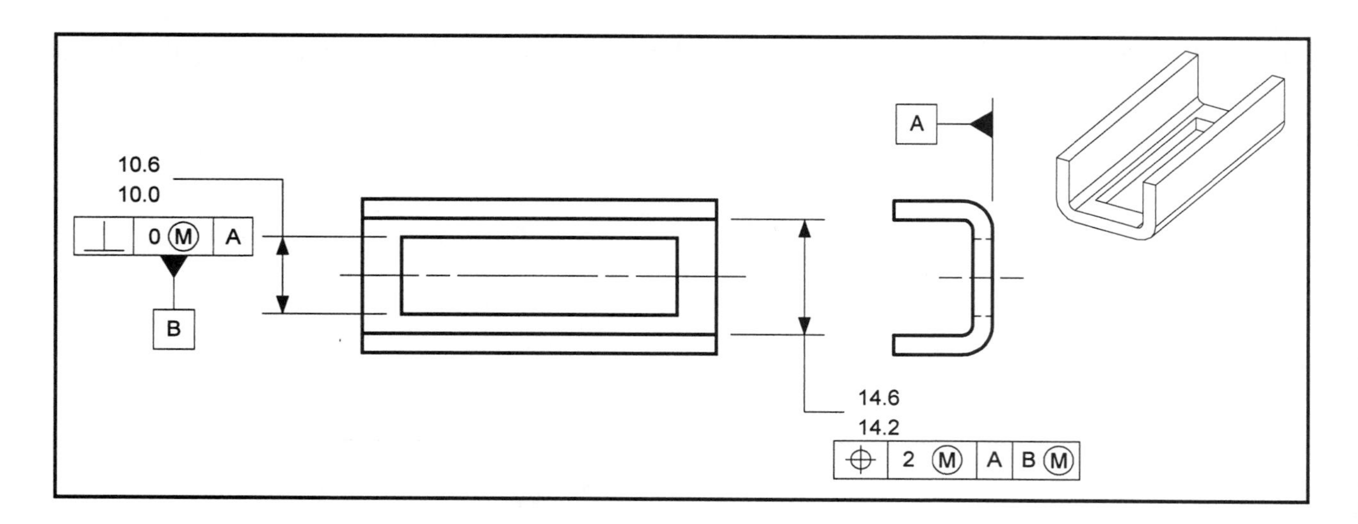

Questions 10 and 11 refer to the drawing above.

10. What is the shape of the tolerance zone of the TOP control?

____

11. Fill in the chart below.

Position Tolerance Zone Width at Centerplane				
Toleranced Feature AME	Datum Feature AME			
	10.0	10.2	10.4	10.6
14.2				
14.4				
14.6				

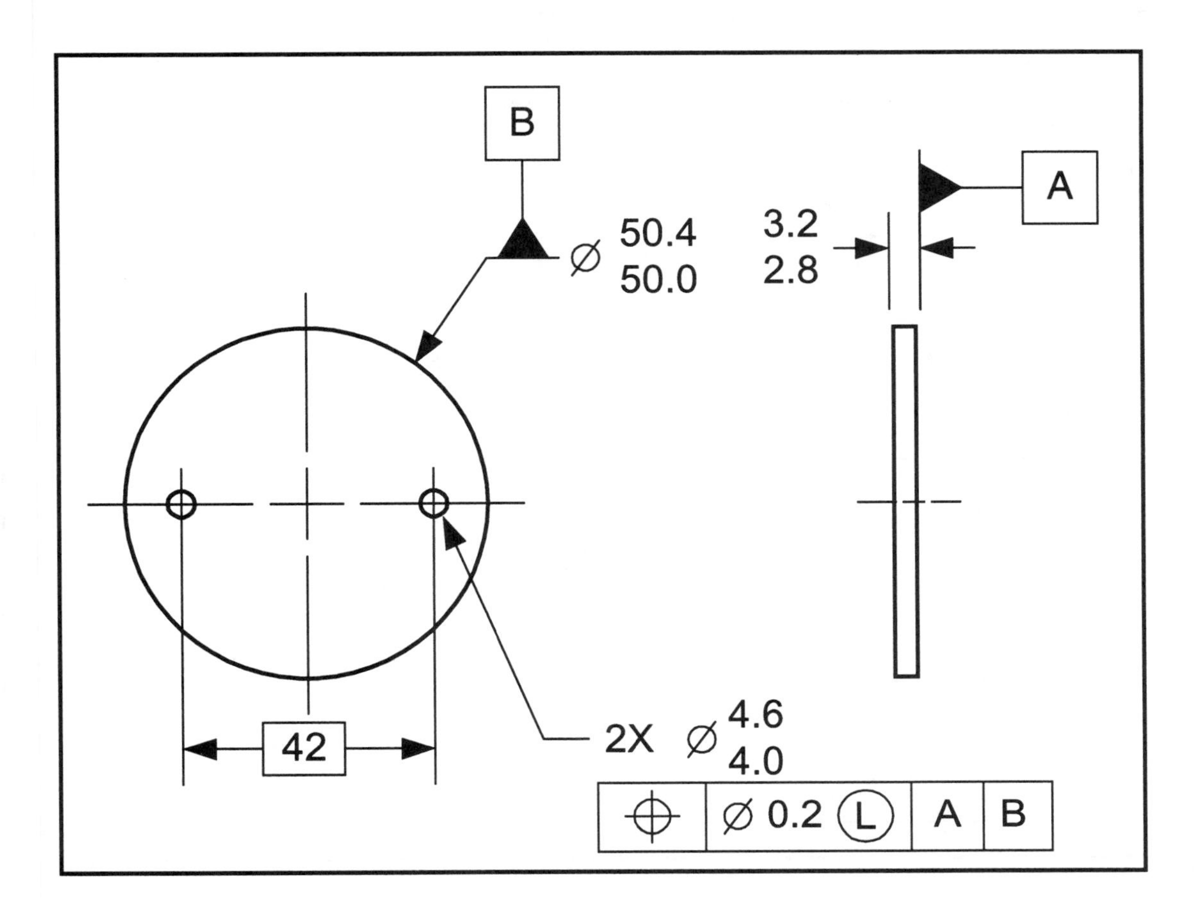

Questions 12 and 13 refer to the figure above.

12. How much bonus tolerance is permissible for the TOP callout?________________________

13. Fill in the chart below.

Toleranced Hole AME	⌖ Tolerance Zone Diameter
4.6	
4.4	
4.2	
4.0	

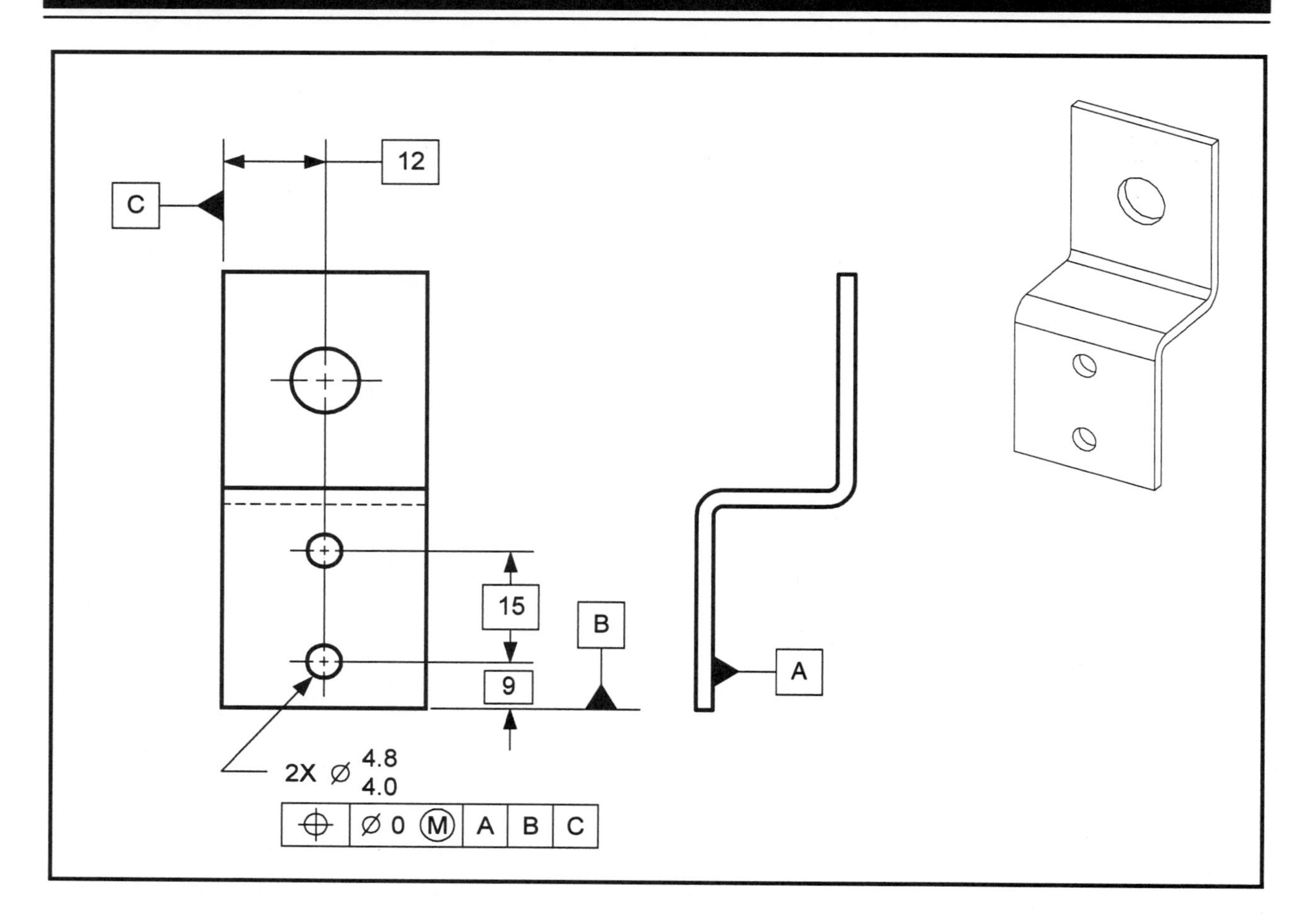

Questions 14 - 16 refer to the figure above.

14. What is the virtual condition of the holes?_______________

15. What is the gage pin dia. for verifying the hole locations?____________________

16. Fill in the chart below.

Hole AME	⌖ Tol. Dia.	Bonus Tol.	Total Location Tol. Dia.
4.0			
4.2			
4.4			
4.6			
4.8			

17. Write the definition of a tolerance stack.

____

____

18. Using the figure below, calculate the max. and min. distance *X*.

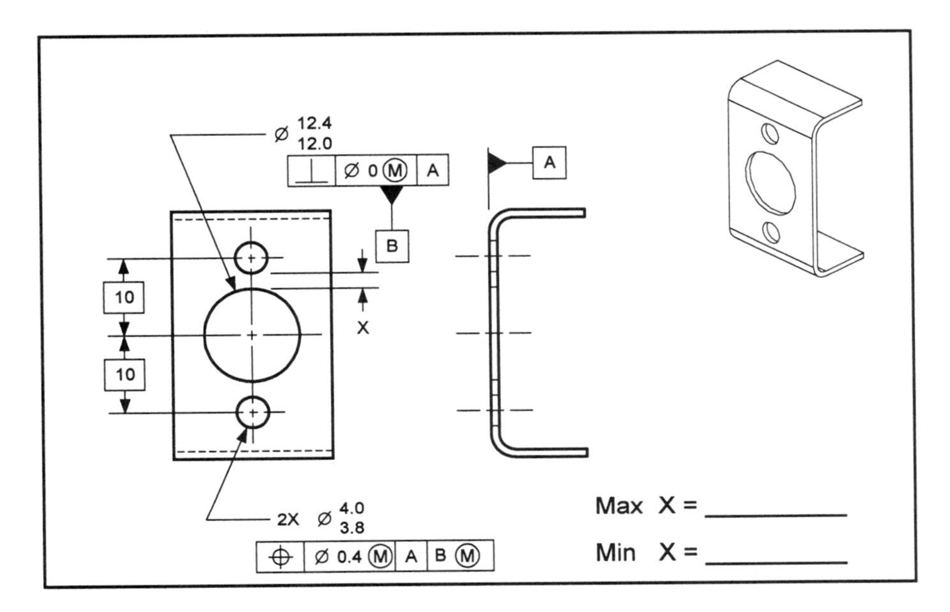

19. Using the figure below, calculate the max. and min. distance *X*.

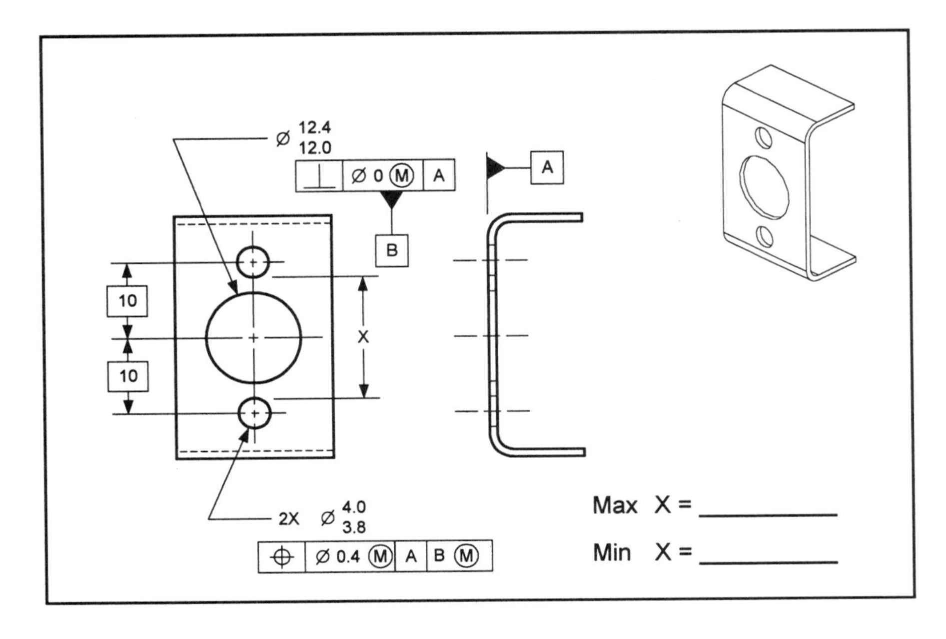

20. In the space below, write the fixed and floating fastener formulas.

    Fixed Fastener Formula:

    Floating Fastener Formula:

21. Using the fixed fastener formula, calculate the position tolerance values for the clearance holes and the tapped holes.

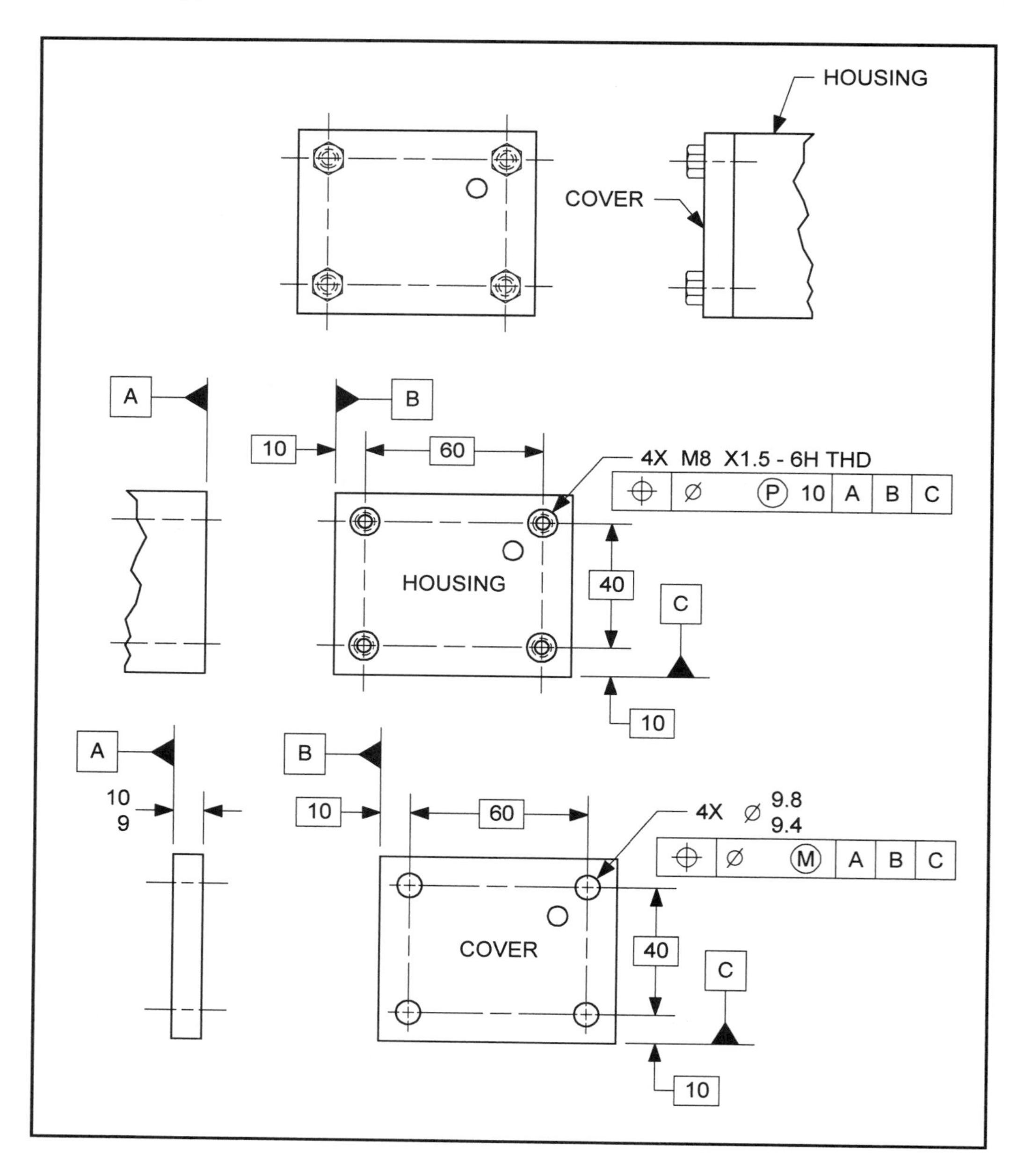

22. Using the floating fastener formula, calculate the position tolerance values for the clearance holes in both parts. (Assume the M8 fasteners to be perfect.)

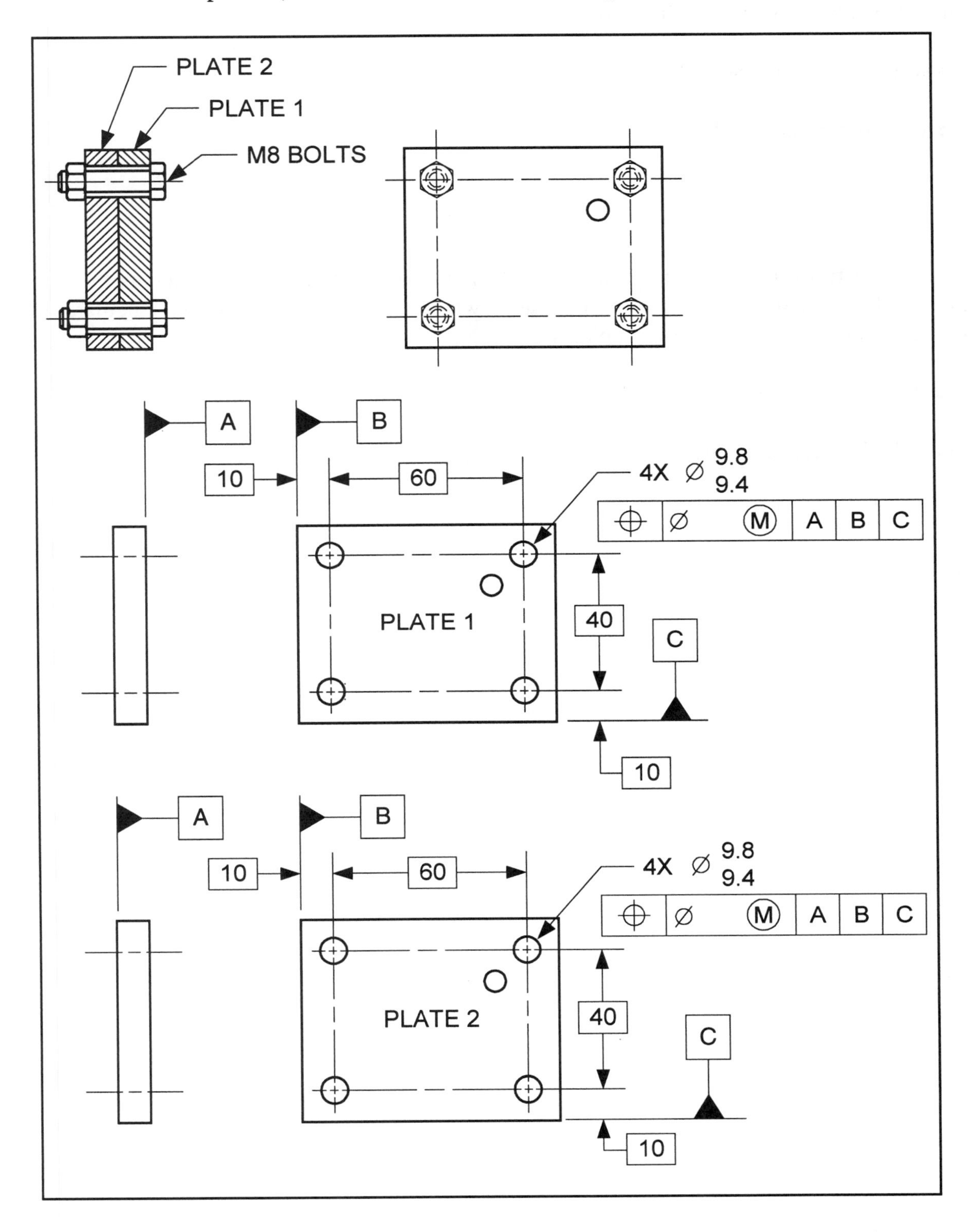

23. Use the part from the figure in question 21. With all dimensions and requirements the same (except the clearance holes on the cover, which are 8.6-9.0 diameter) calculate the positional tolerance values for the housing and the cover.

    Housing TOP tolerance value: _______________

    Cover TOP tolerance value: _______________

24. Use the part from the figure in question 22. With all dimensions and requirements the same (except the clearance holes on both plates, which are 8.3-8.6 diameter) calculate the positional tolerance values for both plates.

    Plate 1 TOP tolerance value: ____________________

    Plate 2 TOP tolerance value: ____________________

Chapter 10

Concentricity and Symmetry Controls

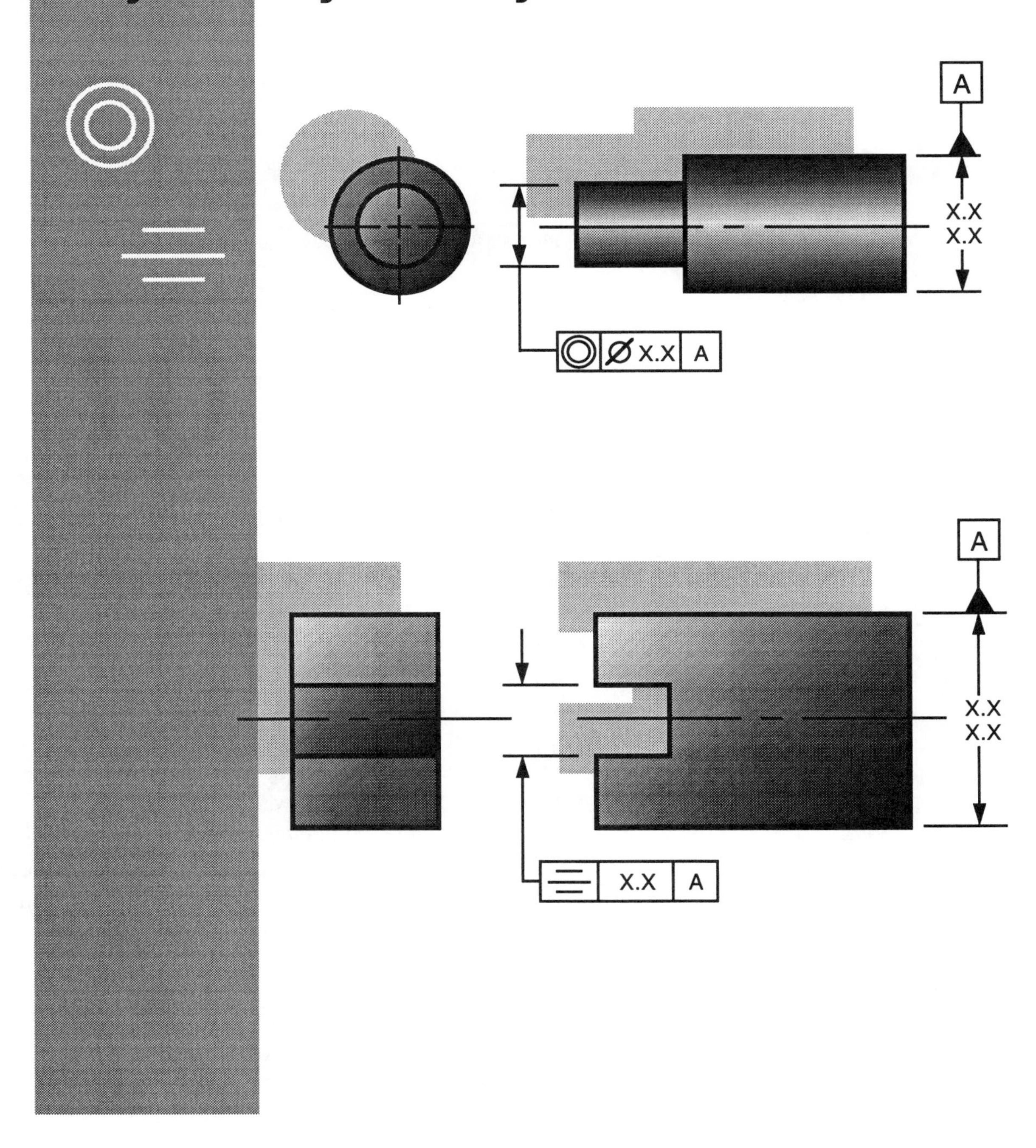

INTRODUCTION

This chapter will help you to read and understand the concentricity and symmetry controls. These controls are a type of location control. A symmetry control limits the location of median points of a symmetrical planar FOS relative to a datum centerplane. The symbols for these controls are shown in Figure 10-1.

FIGURE 10-1 Concentricity and Symmetry Controls

CHAPTER GOALS AND OBJECTIVES

There are Two Goals in this Chapter:

10-1. Interpret the concentricity control.
10-2. Interpret the symmetry control.

Performance Objectives that Demonstrate Mastery of These Goals

Upon completion of this chapter, each student should be able to:

Study Tip
Take a few minutes to fully understand these objectives. When reading this chapter, look for information to help you master these objectives.

Goal 10-1 (pp. 283-287)

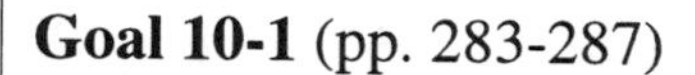

- Describe what concentricity is.
- Describe a median point.
- Describe what a concentricity control tolerance zone applies to.
- Describe the tolerance zone for a concentricity control.
- Describe two differences between concentricity and total runout.
- Describe one difference between concentricity and TOP (RFS).
- Determine if a concentricity control specification is legal.
- Describe how a concentricity control can be inspected.

Goal 10-2 (pp. 288-293)

- Describe what symmetry is.
- Describe the tolerance zone for a symmetry control.
- Describe what a symmetry control tolerance zone applies to.
- Describe the differences between symmetry and TOP.
- Determine if a symmetry control specification is legal.
- Describe how a symmetry control can be inspected.

CONCENTRICITY CONTROL

Definition

Concentricity is the condition where the median points of all diametrically opposed elements of a cylinder (or a surface of revolution) are congruent with the axis of a datum feature. A ***median point*** is the midpoint of a two-point measurement.

For more info. . .
See Paragraph 5.12 of Y14.5.

A ***concentricity control*** is a geometric tolerance that limits the concentricity error of a part feature. The tolerance zone for a concentricity control is three-dimensional; it is a cylinder that is coaxial with the datum axis. The diameter of the cylinder is equal to the concentricity control tolerance value. The median points of correspondingly located elements of the feature being controlled, regardless of feature size, must lie within the cylindrical tolerance zone. When using a concentricity control, the specified tolerance and the datum references always apply on an RFS basis.

Author's Comment
Concentricity may also use a spherical tolerance zone. This condition is rare and beyond the scope of this text.

An example of a concentricity tolerance zone is shown in Figure 10-2.

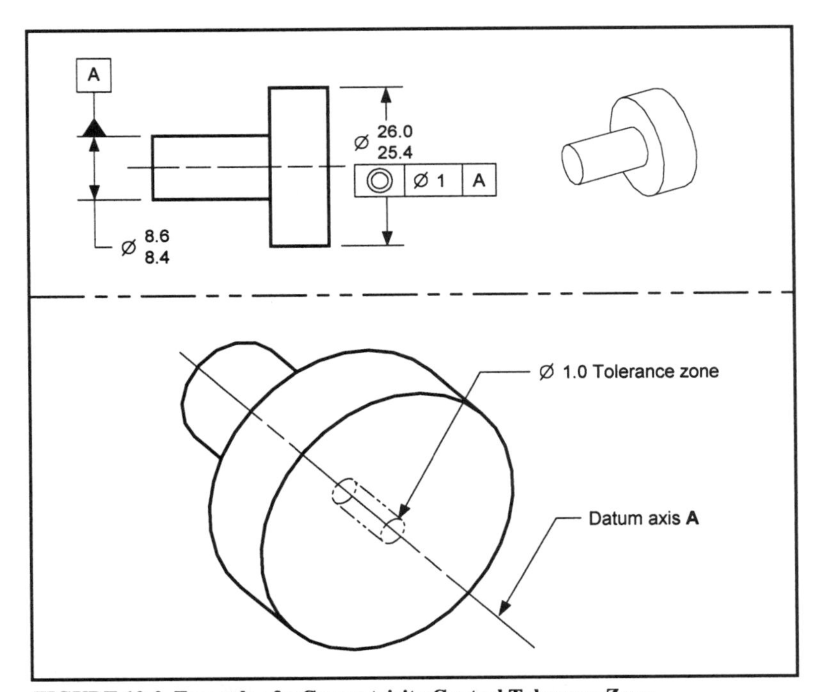

FIGURE 10-2 Example of a Concentricity Control Tolerance Zone

TECHNOTE 10-1 Concentricity

- The tolerance zone is a cylinder centered about the datum axis.
- The median points of the toleranced feature must be within the tolerance zone.

Concentricity Application

In industry concentricity controls are only used in a few unique applications. Concentricity is used when a primary consideration is precise balance of the part, equal wall thickness, or another functional requirement that calls for equal distribution of mass. The toleranced FOS may contain flats or be lobed and still be perfectly concentric. Before using a concentricity control, the use of tolerance of position or runout should be considered. When specifying concentricity, the form of the toleranced diameter is allowed to vary to a greater extent than if a runout control was used. In Figure 10-3, a concentricity control is applied to a diameter.

Design Tip
First consider using TOP at MMC to define a coaxial relationship on a part. TOP is less expensive to produce and to inspect.

When concentricity is applied to a diameter, the following conditions apply:

- The diameter must meet its size and Rule #1 requirements.
- The concentricity control tolerance zone is a cylinder that is coaxial with a datum axis.
- The tolerance value defines the diameter of the tolerance zone.
- All median points of the toleranced diameter must be within the tolerance zone.

The maximum possible distance between the median points of the toleranced diameter and the datum axis is half the concentricity tolerance value.

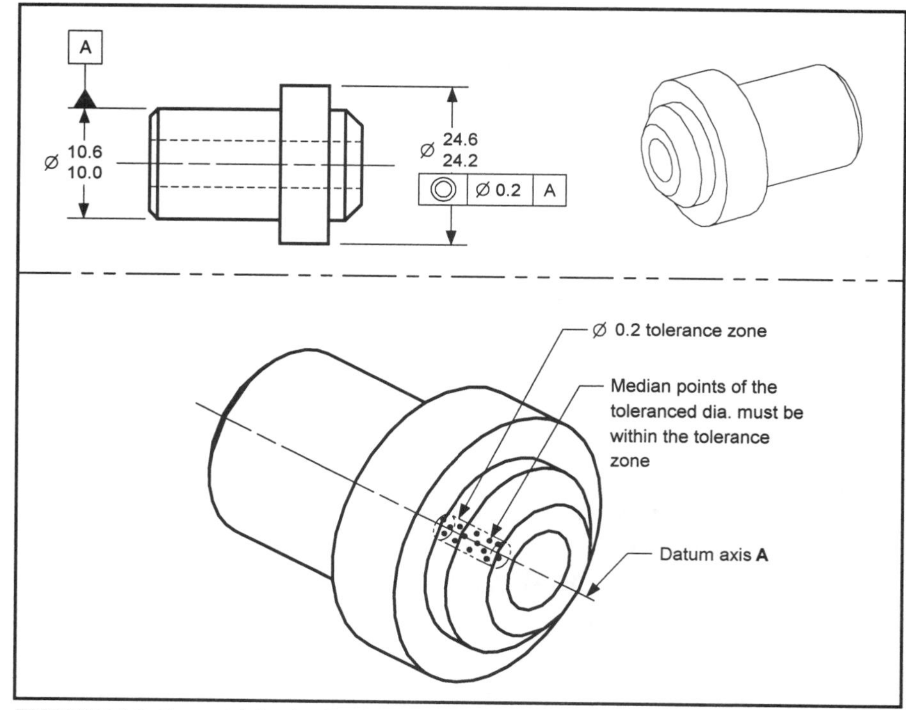

FIGURE 10-3 Concentricity Application

Differences Between Concentricity, Runout, and TOP (RFS)

When dimensioning coaxial diameters, several geometric controls can be used. On a part that rotates around an axis, three geometric controls are common. The designer can choose between concentricity, total runout and TOP (RFS). The chart in Figure 10-4 shows a comparison between these controls.

CONCEPT	GEOMETRIC CONTROL		
	CONCENTRICITY	TOTAL RUNOUT	TOP (RFS)
Tolerance zone	Cylinder	Two coaxial cylinders	Cylinder
Tolerance zone applies to . . .	Median points of toleranced diameter	Surface elements of a toleranced diameter	Axis of the AME of the toleranced diameter
Relative cost to produce	$$	$$$	$
Relative cost to inspect	$$$	$$	$
Part characteristics being controlled	Location and orientation	Location, orientation, and form	Location and orientation

FIGURE 10-4 Differences Between Concentricity, Runout, and TOP

Design Tip

As a rule of thumb, runout and concentricity should only be considered on parts that rotate.

TECHNOTE 10-2 Differences Between Concentricity and Runout

Two differences between runout and concentricity are:

1. The shape of the tolerance zone
2. Runout affects form

One difference between TOP (RFS) and concentricity is:

- With TOP the axis of the AME must be within the tolerance zone. With concentricity, the median points of the toleranced diameter must be within the tolerance zone.

Legal Specification Test for a Concentricity Control

For a concentricity control to be a legal specification, it must satisfy the following conditions:

- The feature control frame must be applied to a surface of revolution that is coaxial to the datum axis.
- Datum references are required. The datum references must ensure that a legal datum axis is established.
- The ∅ symbol must be shown in the tolerance portion of the feature control frame.
- The Ⓜ Ⓛ Ⓣ Ⓟ modifiers may not be used in the feature control frame.

If any of these conditions are not fulfilled, the concentricity specification is incorrect or incomplete. Figure 10-5 shows a legal specification flowchart for a concentricity control.

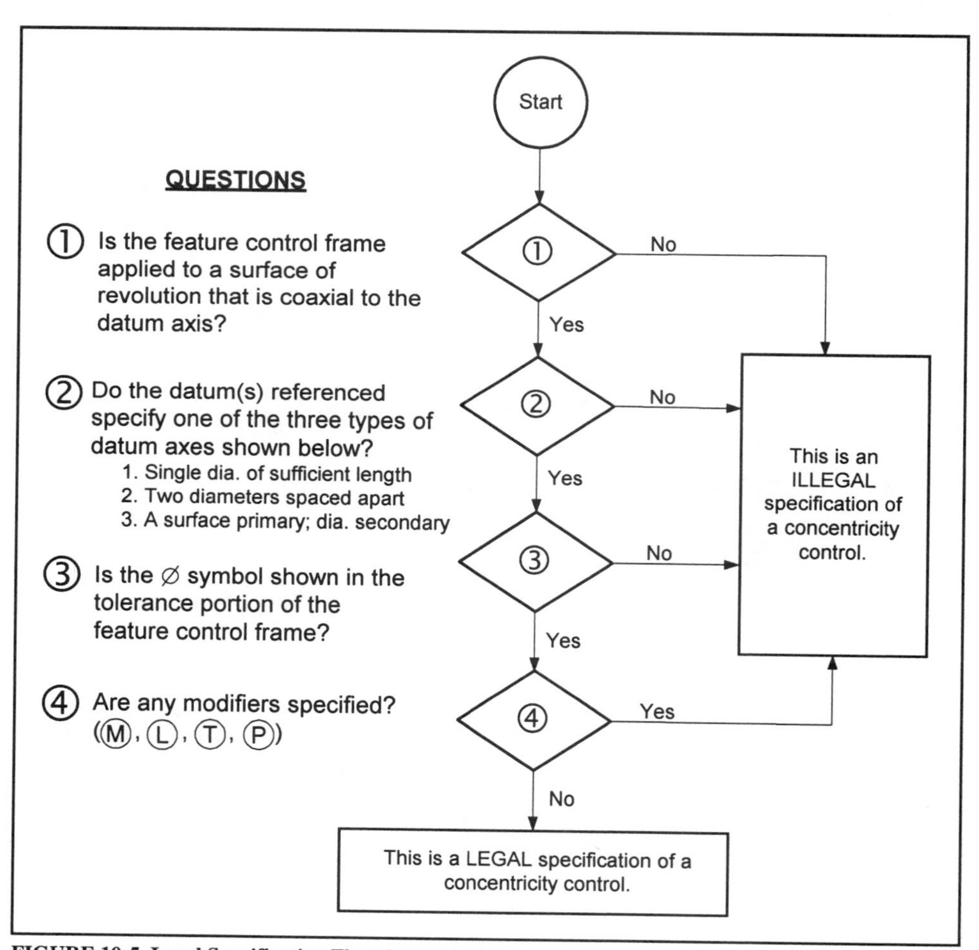

FIGURE 10-5 Legal Specification Flowchart for a Concentricity Control

Inspecting Concentricity

Figure 10-6 shows a part with a concentricity control. When inspecting the large diameter of this part, three separate checks are required: the size of the diameter, its Rule #1 boundary, and the concentricity requirement. Chapter 2 explains how to check the size and Rule #1 boundary; now we will look at how to inspect the concentricity requirement.

Inspecting concentricity is different from inspecting runout or position of coaxial diameters. Inspecting concentricity requires the establishment and verification of the location of a feature's median points.

One way to inspect a concentricity control is shown in Figure 10-6.

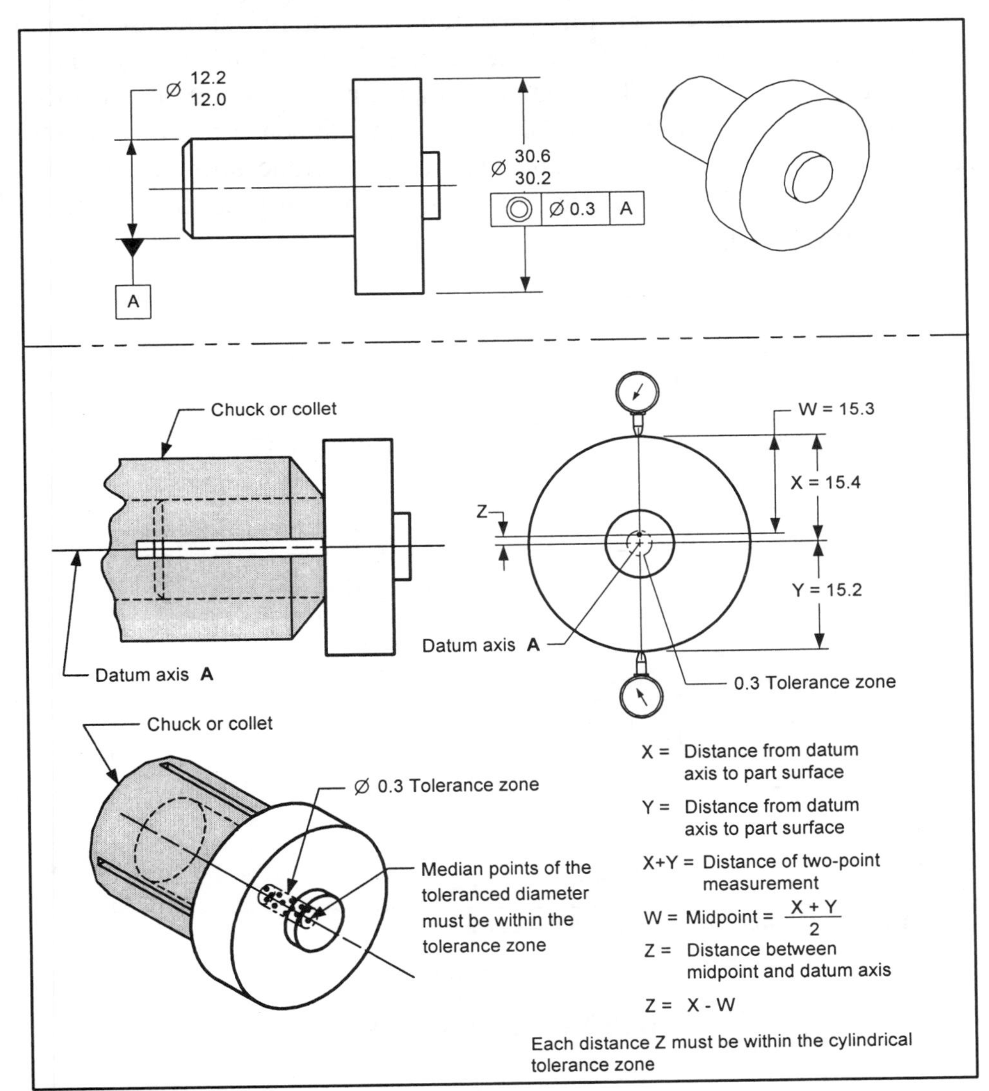

FIGURE 10-6 Inspecting Concentricity

SYMMETRY CONTROL

Symmetry is similar to concentricity. The difference is that, while concentricity is used on surface of revolution, symmetry is used on planar features of size.

For more info. . .
See Paragraph 5.14 of Y14.5.

Definition

Symmetry is the condition where the median points of all opposed elements of two or more feature surfaces are congruent with the axis or centerplane of a datum feature. A ***symmetry control*** is a geometric tolerance that limits the symmetry error of a part feature. A symmetry control may only be applied to part features that are shown symmetrical to the datum centerplane. The tolerance zone is centered about the datum centerplane. The width between the planes is equal to the symmetry control tolerance value. The median points must lie within the parallel plane tolerance zone, regardless of feature size. When using a symmetry control, the specified tolerance and the datum references must always be applied on an RFS basis. An example of a symmetry control tolerance zone is shown in Figure 10-7.

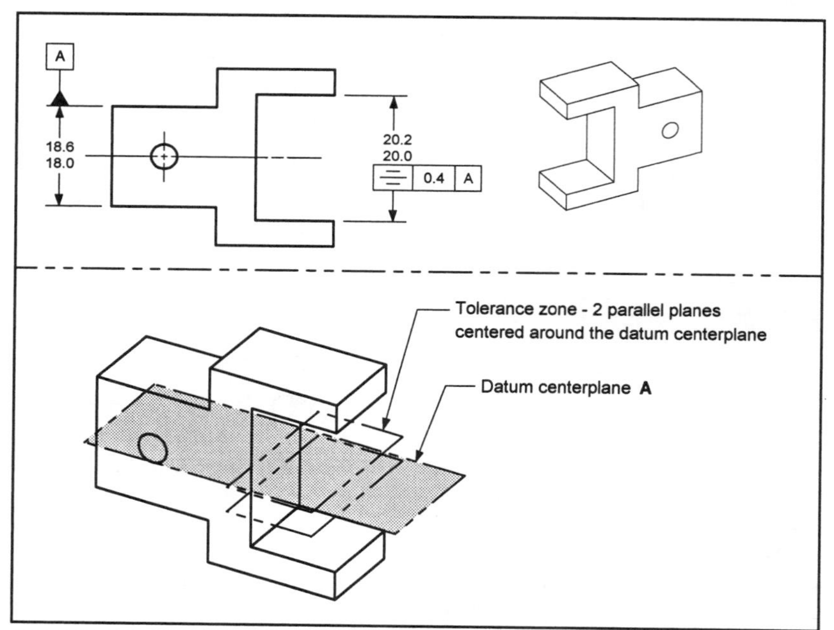

FIGURE 10-7 Symmetry Control Tolerance Zone

TECHNOTE 10-3 Symmetry Control

- The tolerance zone is two parallel planes centered about a datum axis or centerplane.
- The median points of the toleranced feature must be within the tolerance zone.

Symmetry Application

Symmetry controls are only used in a few unique applications in industry. Symmetry is used when a primary consideration of symmetrical features is precise balance of the part, equal wall thickness, or another functional requirement that calls for equal distribution of part mass. Otherwise, TOP is recommended to control symmetrical relationships. In Figure 10-8, a symmetry control is applied to a slot. When symmetry is applied to a slot, the following conditions apply:

- The slot must meet its size and Rule #1 requirements.
- The symmetry control tolerance zone is two parallel planes that are centered about the datum centerplane.
- The tolerance value of the symmetry control defines the distance between the parallel planes.
- All the median points of the toleranced slot must be within the tolerance zone.

Design Tip
First consider using TOP at MMC to define a symmetrical relationship on a part. TOP is less expensive to produce and to inspect.

The maximum possible distance between the median points of the toleranced feature and the datum centerplane is half the symmetry tolerance value.

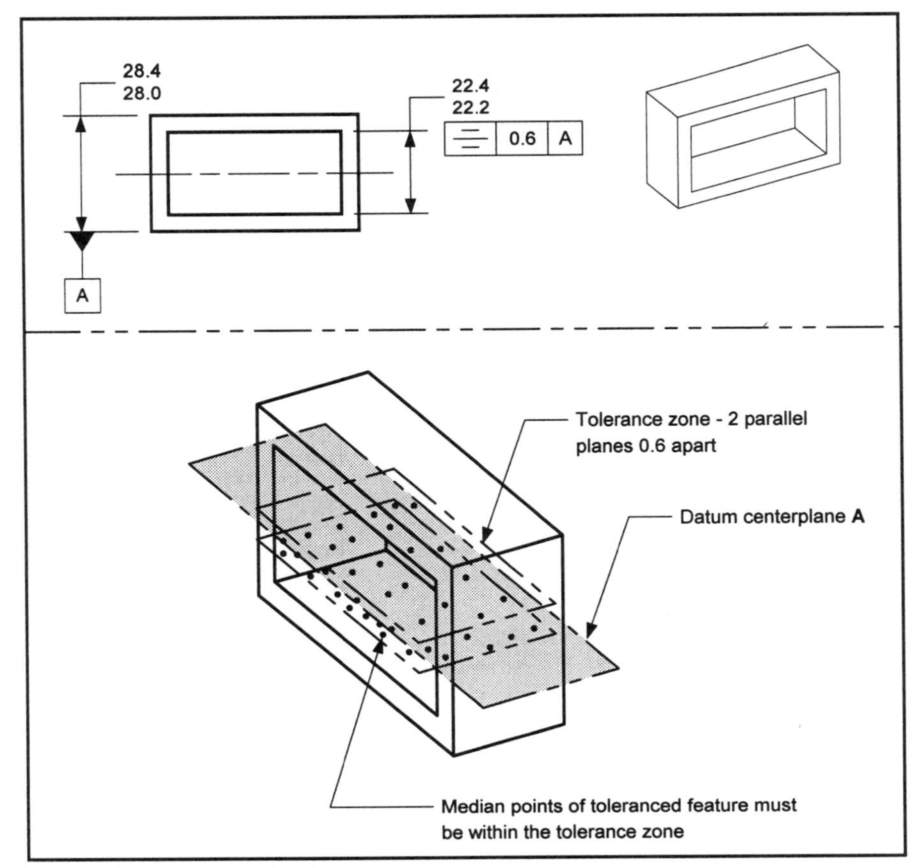

FIGURE 10-8 Symmetry Applications

Differences Between Symmetry and TOP (RFS)

Symmetry and tolerance of position (RFS) are two geometric controls that can be used to tolerance symmetrical part features. Often confusion exists over which control is best to use in a given situation. Understanding differences between these controls will help to eliminate the confusion over choosing a symbol in a part application.

Symmetry controls the location of the median points of a part feature. TOP controls the location of the centerplane of the actual mating envelope of a part feature. In general, TOP is considered a more economical tolerance to produce and to verify. Figure 10-9 shows a comparison between symmetry and TOP (RFS).

CONCEPT	GEOMETRIC CONTROL	
	SYMMETRY	TOP (RFS)
Tolerance zone	Two parallel planes	Two parallel planes
Tolerance zone applies to . . .	Median points of toleranced FOS	The centerplane of the AME
Types of part characteristics being controlled	Orientation and location	Orientation and location
Relative cost to produce	$$$	$$
Relative cost to inspect	$$$	$$

FIGURE 10-9 Differences Between Symmetry and TOP

Legal Specification Test for a Symmetry Control

For a symmetry control to be a legal specification, it must satisfy the following conditions:

- The feature control frame must be applied to a planar FOS that is symmetrical about the datum centerplane.
- Datum references are required. The datum references must ensure that a legal datum centerplane is established.
- No modifiers may be used in the feature control frame.

If any of these conditions are not fulfilled, the symmetry specification is incorrect or incomplete. Figure 10-10 shows a legal specification flowchart for a symmetry control.

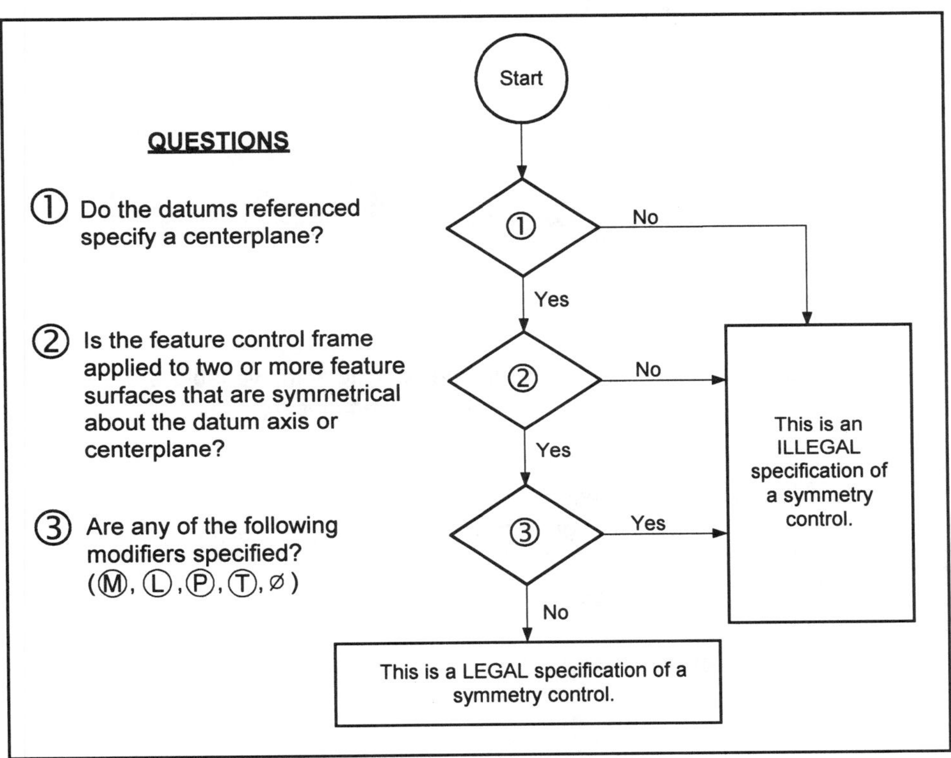

FIGURE 10-10 Legal Specification Flowchart for a Symmetry Control

Inspecting Symmetry

Figure 10-7 shows a part with a symmetry control. When inspecting this part, three separate checks are required: the size of the slot, its Rule #1 boundary, and its symmetry. Symmetry requires the establishment and verification of the location of a feature's median points.

One way to inspect the symmetry control for the part in Figure 10-7 is shown in Figure 10-11.

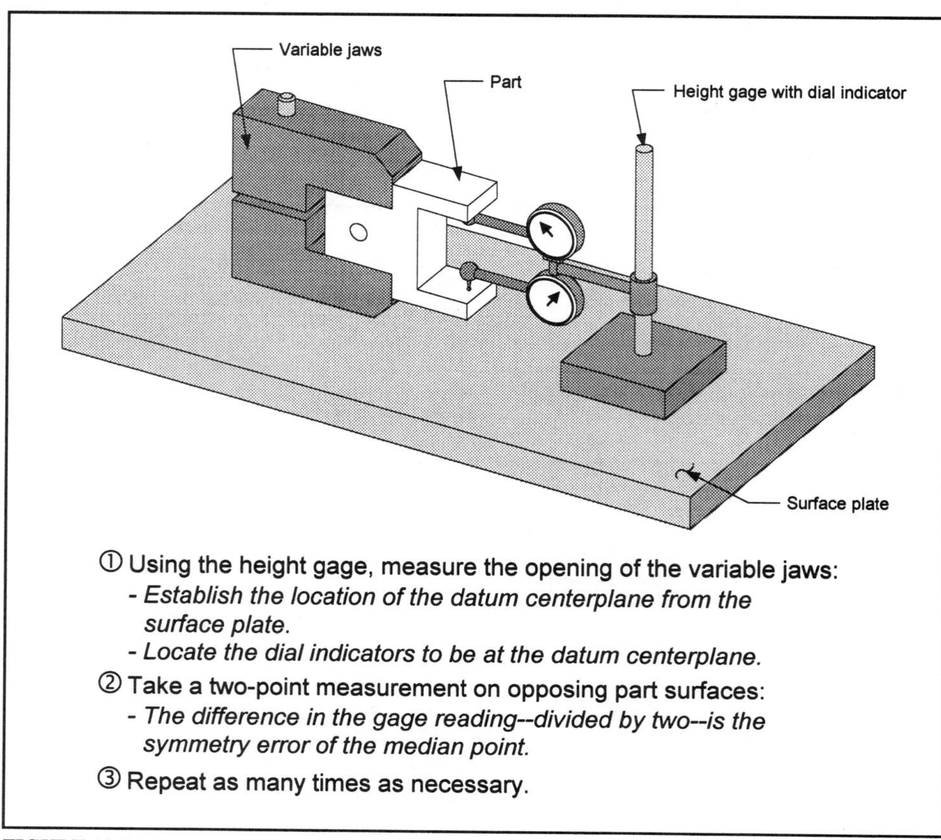

FIGURE 10-11 Inspecting Symmetry

Summary

A summarization of concentricity and symmetry control information is shown in Figure 10-12.

Symbol	Datum reference required	Can be applied to a		Can affect WCB	Can use Ⓜ Ⓛ Ⓣ Ⓟ modifier	Can override Rule #1	Tolerance zone shape
		Two or more planar feature surfaces	Cylindrical or surface of revolution				
◎	Yes	No	Yes	Yes	No	No	Cylindrical
⌯	Yes	Yes	No	Yes	No	No	Parallel planes

FIGURE 10-12 Summarization of Concentricity and Symmetry Controls

VOCABULARY LIST

New Terms Introduced in this Chapter

Concentricity
Concentricity control
Median point
Symmetry
Symmetry control

Study Tip
Read each term. If you don't recall its meaning, look it up in the chapter.

ADDITIONAL RELATED TOPICS

Topic	ASME Y14.5M-1994 Reference
• Spherical tolerance zone	Paragraph 5.12.1

Author's Comment
These topics, plus advanced coverage of many of the topics introduced in this text, will be covered in my new book on advanced GD&T concepts.

QUESTIONS AND PROBLEMS

1. Describe the tolerance zone for concentricity.

____

____

2. Describe the tolerance zone for symmetry.

____

____

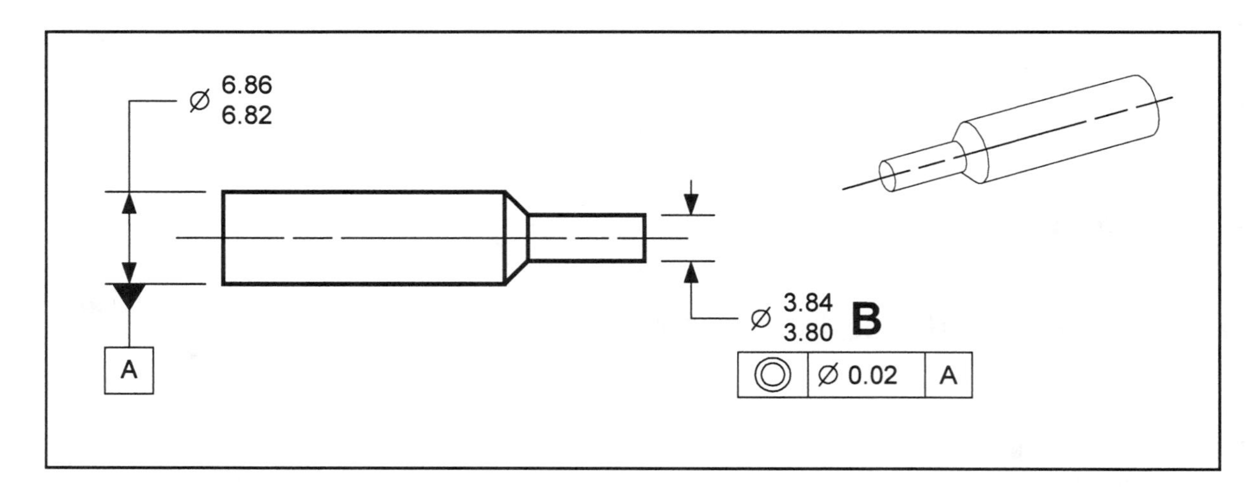

Questions 3-7 refer to the drawing above.

3. What controls the location of diameter *B*?

____

4. What are the shape and size of the tolerance zone for the concentricity callout?

____

5. Does Rule #1 apply to diameter *B*? ______________________________

6. The __________ of diameter *B* must be within the concentricity tolerance zone.
   A. median points
   B. axis of the AME
   C. TIR
   D. runout

7. The maximum possible offset between a median point of diameter *B* and the datum axis is ______________.

8. Fill in the chart below.

CONCEPT	CONCENTRICITY	TOTAL RUNOUT	TOP (RFS) (applied to a dia)
Describe the shape of the tolerance zone.			
What characteristic of the toleranced feature must be within the tolerance zone?			
Does Rule #1 still apply to the toleranced feature?			
What type of characteristics of the toleranced feature are being controlled?			

9. For each symbol shown below, indicate if it is a legal specification. If a control is illegal, explain why.

A. | ◎ | 0.5 | A | ______________________________

B. | ◎ | ⌀0.5 Ⓟ 10 | A | ______________________________

C. | ◎ | ⌀ 0.5 Ⓢ | A Ⓢ | ______________________________

D. | ◎ | ⌀ 0.5 | A | ______________________________

E. | ◎ | ⌀ 0.5 Ⓜ | A | ______________________________

10. Describe a median point.

____

____

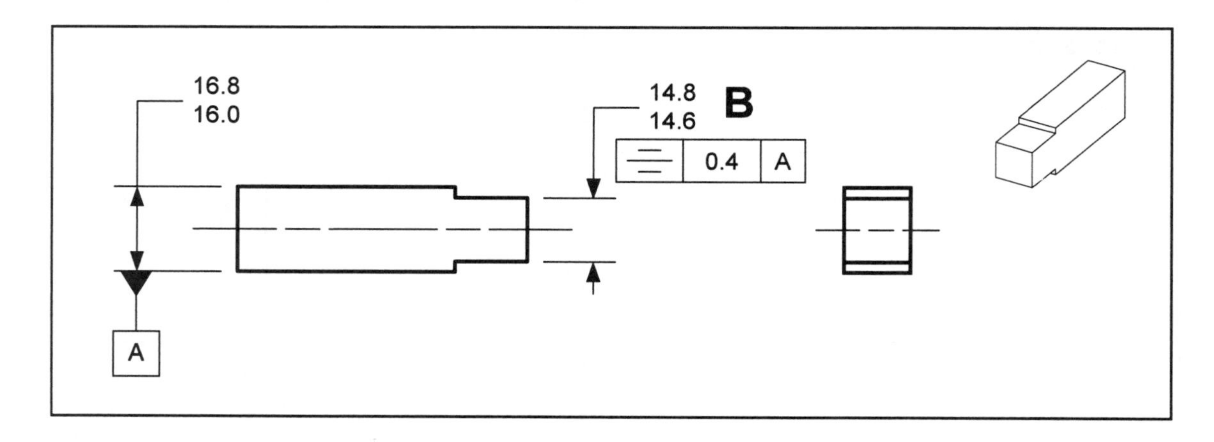

Questions 11-14 refer to the drawing above.

11. What is the shape and size of the tolerance zone for the symmetry callout?

____

12. Does Rule #1 apply to tab *B*? ______________________________

13. The symmetry callout controls the ________________ of tab *B*?
    A. centerplane of the AME
    B. median points
    C. straightness
    D. runout

14. The maximum offset of the centerplane of tab *B* relative to datum centerplane *A* is ____.

15. Fill in the chart below.

CONCEPT	SYMMETRY	POSITION (RFS) (applied to a planar FOS)
Tolerance zone shape		
What characteristic of the toleranced feature must be within the tolerance zone?		
Does Rule #1 apply to the toleranced feature?		
What type of characteristics of the toleranced feature are being controlled?		

16. For each symbol shown below, indicate if it is a legal specification. If a control is illegal, explain why.

A. | ⌯ | 0.5 | A | ______________________________

B. | ⌯ | 0.5 Ⓜ | A | ______________________________

C. | ⌯ | 0.5 Ⓢ | A Ⓢ | ______________________________

D. | ⌯ | 0.5 Ⓛ | A Ⓛ | ______________________________

E. | ⌯ | 0.5 Ⓟ 10 | A | ______________________________

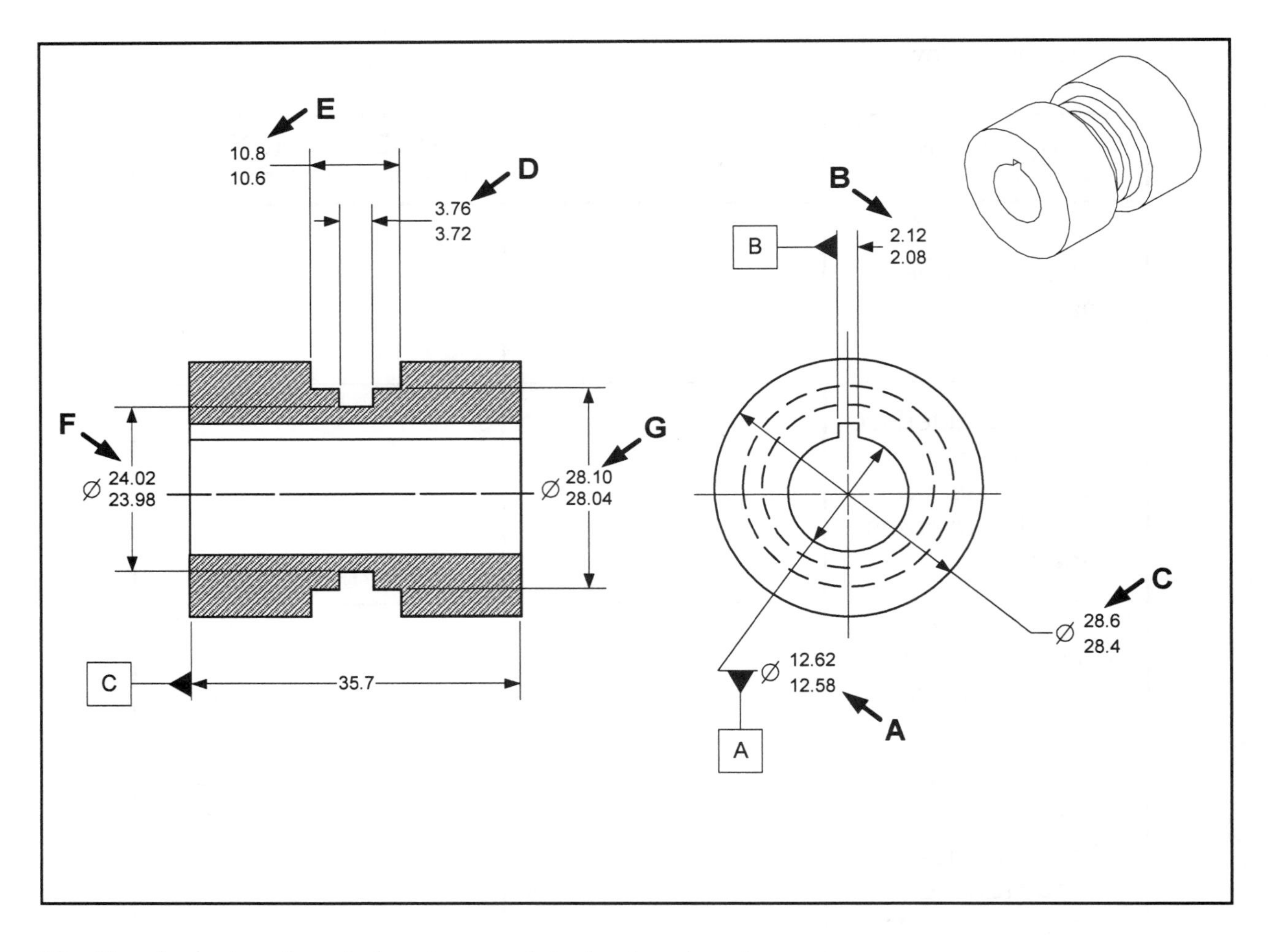

17. Use the instructions below to complete the drawing above.

   a. Add a symmetry control to the keyway labeled *B* that will control its symmetry relative to datum axis *A*. The tolerance value should be 0.06.

   b. Add a concentricity control to the diameter labeled *C* that will control its concentricity relative to datum axis *A*. The tolerance value should be 0.1.

   c. Add a symmetry control to the groove labeled *D* that will control its symmetry relative to datum centerplane *C*. The tolerance value should be 0.12.

   d. Add a symmetry control to the groove labeled *E* that will control its symmetry relative to datum centerplane *C*. The tolerance value should be 0.12.

   e. Add a concentricity control to the diameter labeled *F* that will control its concentricity relative to datum axis A. The tolerance value should be 0.06.

   f. Add a concentricity control to the diameter labeled *G* that will control its concentricity relative to datum axis A. The tolerance value should be 0.06.

Chapter **11**

Runout Controls

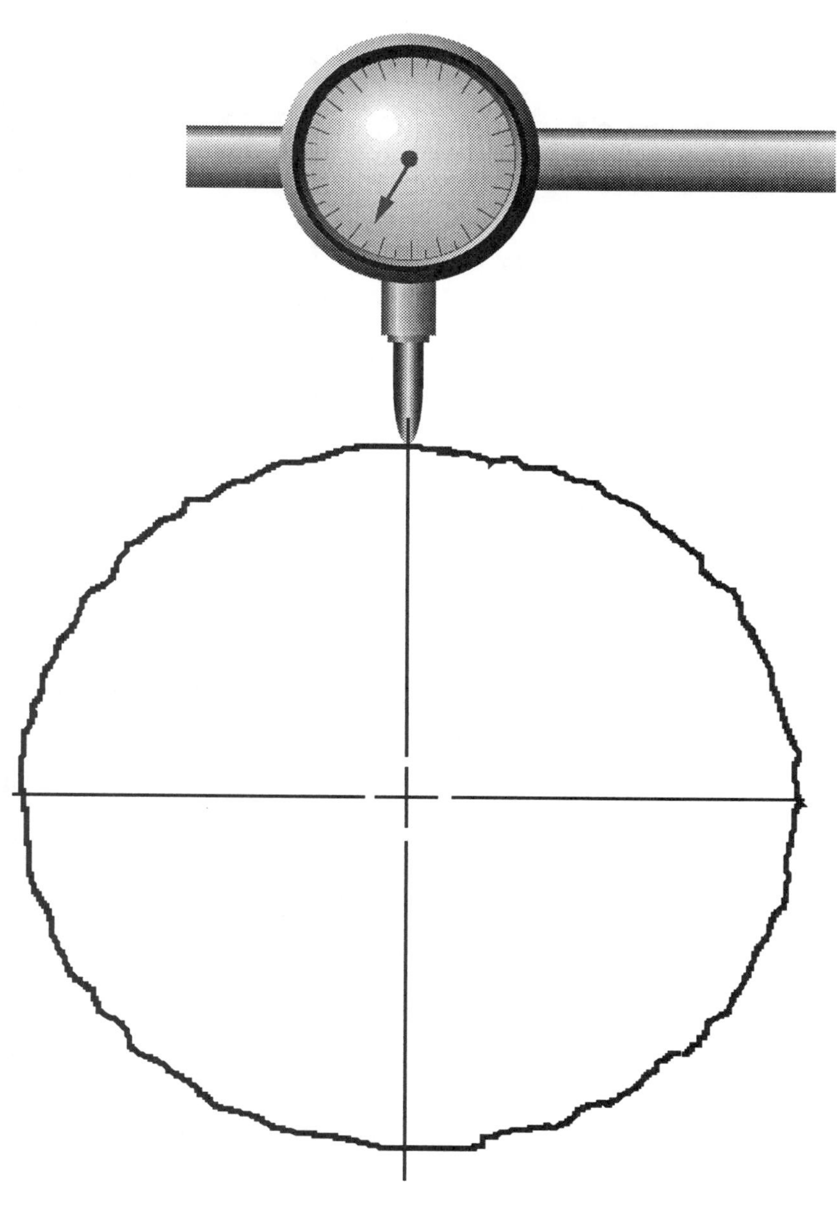

INTRODUCTION

This chapter explains the concepts involved with the runout controls. Runout is a composite control. A ***composite control*** controls the form, location, and orientation of a part feature simultaneously (in a single gage reading). Runout controls are often used to control the coaxiality of diameters. A runout control always requires a datum axis. There are two types of runout controls: circular runout and total runout. The symbols for the runout controls are shown in Figure 11-1.

Author's Comment
Older drawings used terms like FIM (Full Indicator Movement), TIR (Total Indicator Reading), or TIM (Total Indicator Movement) to indicate runout.

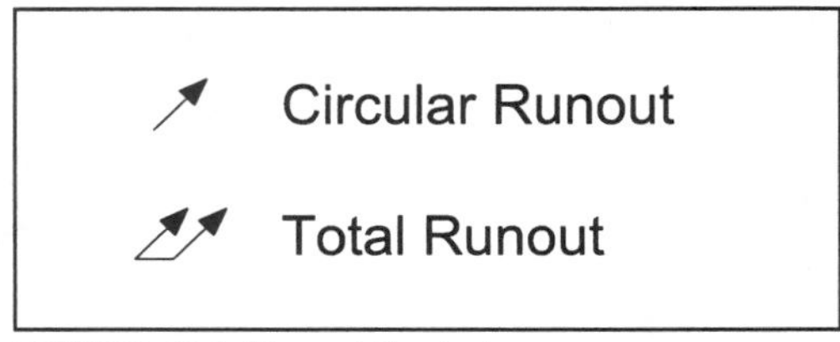

FIGURE 11-1 Runout Controls

A runout control can be applied to any part feature that surrounds, or is intersected by, the datum axis. A runout tolerance value specified in a feature control frame indicates the maximum permissible indicator reading (gage travel) of the considered feature, when the part is rotated 360° about its datum axis.

CHAPTER GOALS AND OBJECTIVES

There are Two Goals in this Chapter:

11-1. Interpret the circular runout control.
11-2. Interpret the total runout control.

Performance Objectives that Demonstrate Mastery of These Goals

Upon completion of this chapter, each student should be able to:

Goal 11-1 (pp. 301-309)

- List three ways a datum axis can be specified for a runout application.
- Describe what circular runout is.
- Describe the tolerance zone for a circular runout control (applied to a diameter).
- Describe how circular runout is a composite control.
- Determine the maximum amount of axis offset from a circular runout callout.
- Calculate the WCB in a circular runout application.
- Determine if a circular runout control specification is legal.
- Describe a method of inspection for circular runout.

Goal 11-2 (pp. 310-319)
- Describe what total runout is.
- Describe the tolerance zone for a total runout control (applied to a diameter).
- Describe how total runout is a composite control.
- Determine the max. amount of axis offset from a total runout callout.
- Calculate the WCB in a total runout application.
- Determine if a total runout control specification is legal.
- Describe a method of inspection for total runout.
- Describe two differences between circular and total runout.
- Calculate distances on a part that uses runout.

Study Tip
Take a few minutes to fully understand these objectives. When reading this chapter, look for information to help you master these objectives.

CIRCULAR RUNOUT

Establishing a Datum Axis for Runout

There are three ways to establish a datum axis for a runout specification. They are:

1. A single diameter of sufficient length
2. Two coaxial diameters a sufficient distance apart to create a single datum axis
3. A surface and a diameter at right angles

The three ways to establish a datum axis are illustrated in Figure 11-2. Functional design requirements and part shape are considerations for selecting one of these methods to establish a datum axis. Usually, the features used for the datum axis are the same features that locate the part in the assembly. A single diameter is used when the diameter is long enough to orient the part. Two coaxial diameters are used when they equally establish the orientation of the part. A surface primary, diameter secondary is used when the surface orients the part. When a surface is primary, the diameter should be very short.

TECHNOTE 11-1 Establishing A Datum Axis For Runout

There are three ways to establish a datum axis for a runout control. They are:

1. A single diameter of sufficient length
2. Two coaxial diameters a sufficient distance apart to create a single datum axis
3. A surface and a diameter at right angles

Design Tip
If a diameter is too short to establish an axis for inspection, the diameter will not serve well as a primary datum feature in the function of the part.

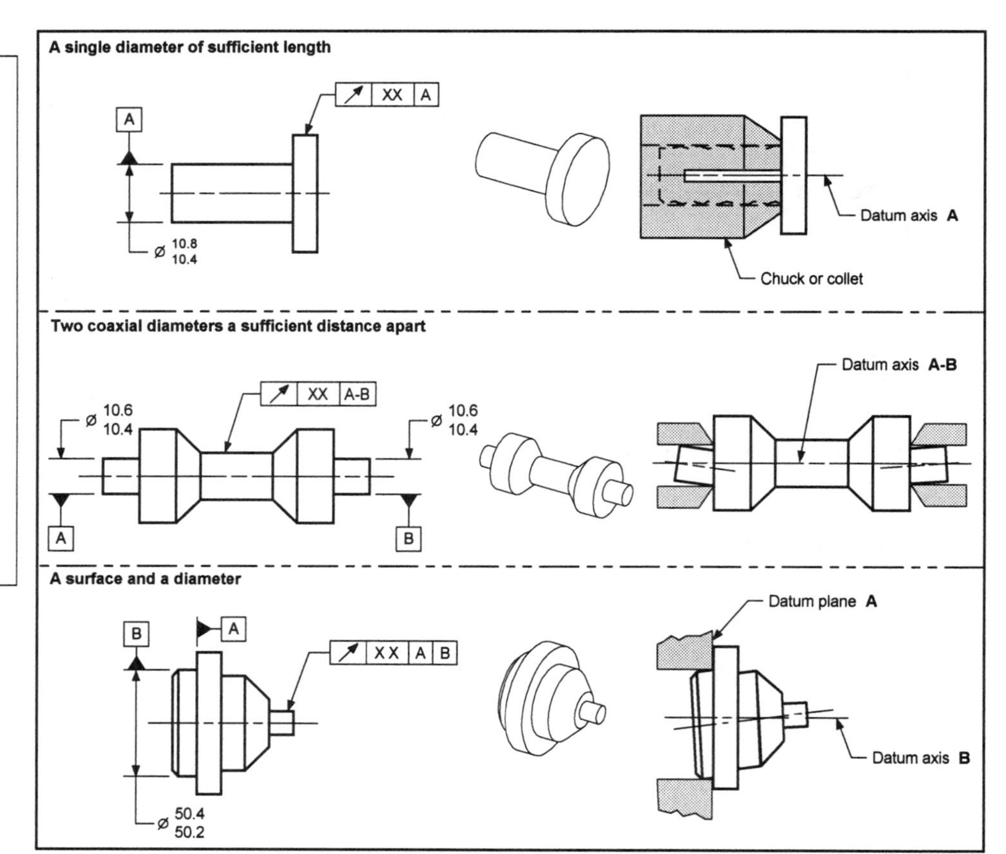

FIGURE 11-2 Establishing a Datum Axis for Runout

For more info. . .
See Paragraph 6.7.1.2.1 of Y14.5.

Definition of Circular Runout

Circular runout is a composite control that affects the form, orientation, and location of circular elements (individually) of a part feature relative to a datum axis. A ***circular runout control*** is a geometric tolerance that limits the amount of circular runout of a part surface. Circular runout applies independently to each circular element of a diameter. It is referred to as a ***composite control*** because it controls the form, location, and orientation of a part feature simultaneously (in a single gage reading). Circular runout is frequently used to control the location of circular elements of a diameter. When applied to a diameter, it controls the form (circularity) and location of the diameter to a datum axis.

When the tolerance zone shape for a circular runout control is applied to a diameter, it is easily visualized; it is two coaxial circles whose centers are located on the datum axis. The radial distance between the circles is equal to the runout tolerance value. Figure 11-3 illustrates the tolerance zone for circular runout.

The size of the larger circle of the tolerance zone is established by the radius to the surface element that is farthest from the datum axis. The inner circle of the tolerance zone is offset from the larger circle by the runout tolerance value. Figure 11-3 illustrates the size of a circular runout tolerance zone. When verifying a diameter controlled with circular runout, a dial indicator is placed perpendicular to the surface being verified. The part is rotated 360° and the indicator measures the distance between the circles.

Author's Comment
If the runout tolerance value is smaller than the size tolerance of the diameter, the roundness of the diameter will be limited by the runout control.

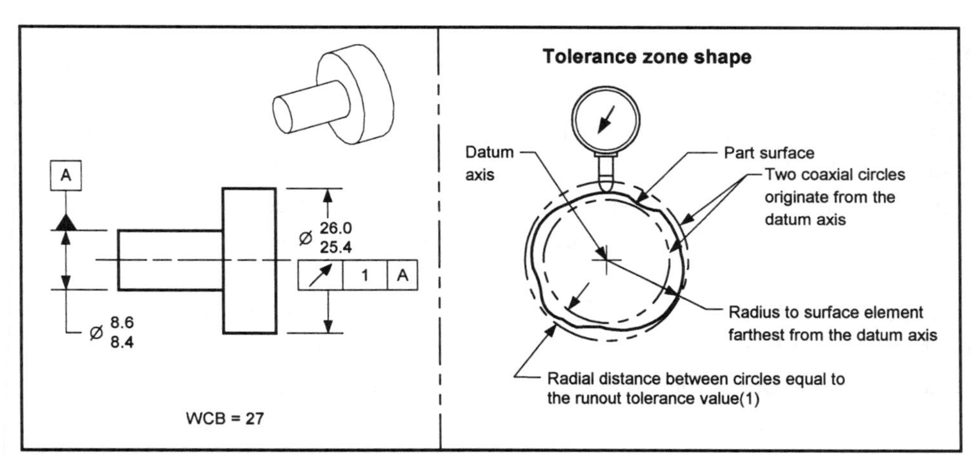

FIGURE 11-3 Circular Runout Tolerance Zone

TECHNOTE 11-2 Circular Runout Tolerance Zone

The tolerance zone for circular runout is two coaxial circles that are located on the datum axis.

Circular Runout as a Composite Control

Circular runout is a composite control. It limits the circularity, orientation, and axis offset of a diameter. When runout is applied to a diameter, it affects the WCB of the diameter. When verifying runout, the dial indicator reading includes several types of part errors, including form, orientation, and location errors. Figure 11-4 illustrates these errors.

In Figure 11-4*A*, a part is dimensioned with a circular runout control. The figure demonstrates how various indicator readings could occur.

In Figure 11-4*B*, a part is chosen, and the circular element being verified is perfectly round and perfectly coaxial with the datum axis. As the part is rotated 360° about the datum axis, the gage reading (runout value) will be zero.

Author's Comment
This condition is the worst-case condition. It is used to calculate tolerance stacks.

In Figure 11-4*C*, a part is chosen and the circular element being verified is out-of-round (lobed). The out-of-round element is still within the size limits of the diameter. The axis of the circular element is still perfectly coaxial with the datum axis. As this part is rotated 360° about the datum axis, an indicator reading—or runout error—is obtained. The indicator will read the out-of-round of the circular element as a runout error.

In Figure 11-4*D*, a part is chosen, and the circular element being verified is perfectly round, but its axis is offset from the datum axis. As this part is rotated 360° around the datum axis, an indicator reading (or runout error) is obtained. This time, the entire runout error is produced from the axis offset. Note that an axis offset of 0.15 produces a runout indicator reading of 0.3. The maximum possible axis offset is equal to one-half the runout tolerance value. Whenever any roundness error exists in the toleranced feature, the allowable axis offset will be reduced by the amount of the roundness error. However, the indicator reading does not separate the roundness error from the axis offset.

Circular runout is also an indirect orientation control. In Figure 11-4*D*, each end of the axis of the toleranced diameter could be mislocated in the opposite directions by one-half the runout tolerance value. This would result in a parallelism control (or in some cases, a perpendicularity control) control equal to the runout tolerance value.

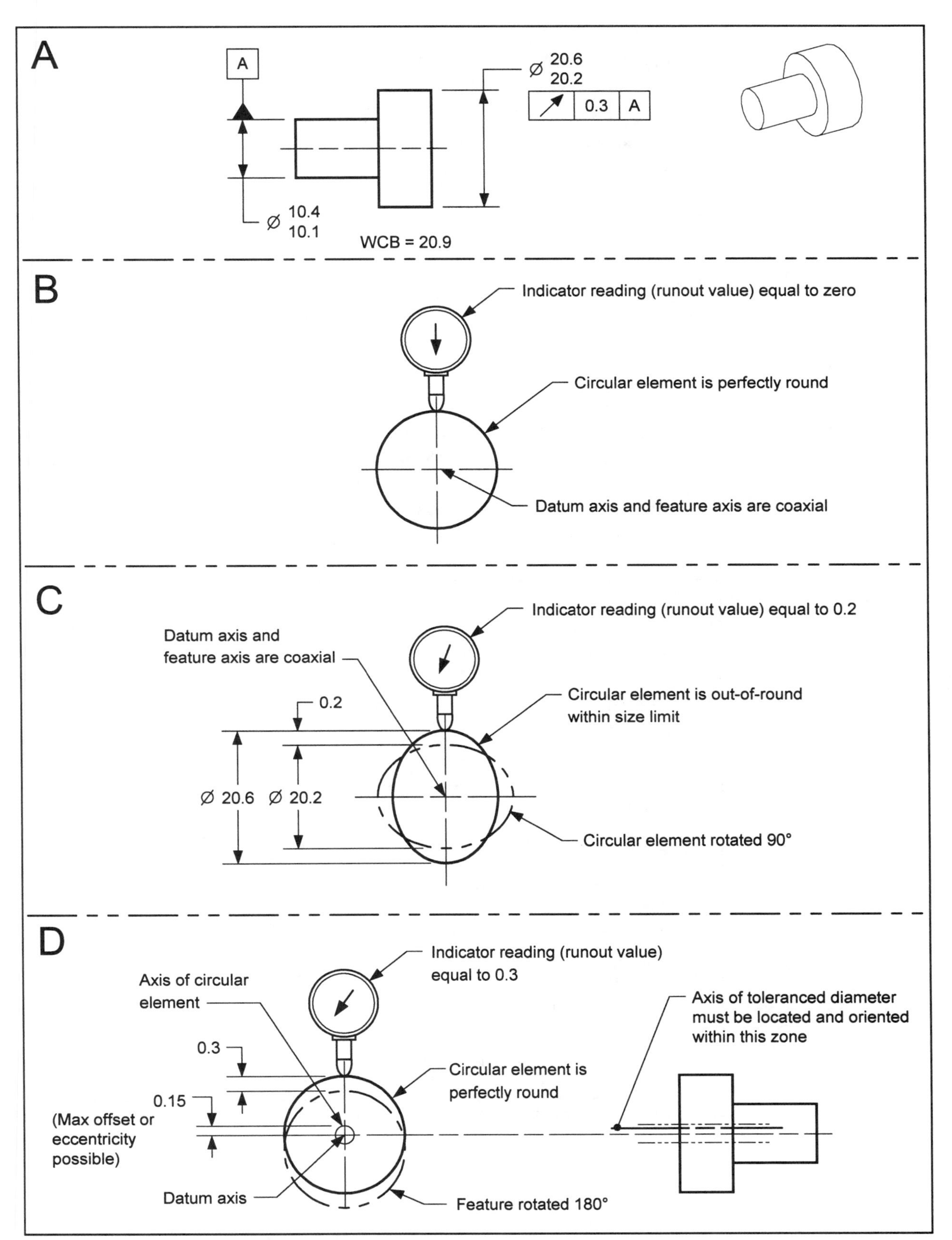

FIGURE 11-4 Circular Runout as a Composite Control

TECHNOTE 11-3 Runout Control Axis Offset

When a diameter is controlled by circular runout, its maximum possible axis offset from the datum axis is equal to one-half the runout tolerance value.

Circular Runout Applications

Figure 11-5 illustrates a circular runout application. In this application, the following conditions apply:

- The diameter must meet its size requirements.
- The worst-case boundary is affected (24.6 + 0.2 = 24.8).
- The runout control applies RFS.
- Runout applies at each circular element of the toleranced diameter.
- The runout tolerance zone is two coaxial circles 0.2 apart.
- The maximum possible axis offset is 0.1.

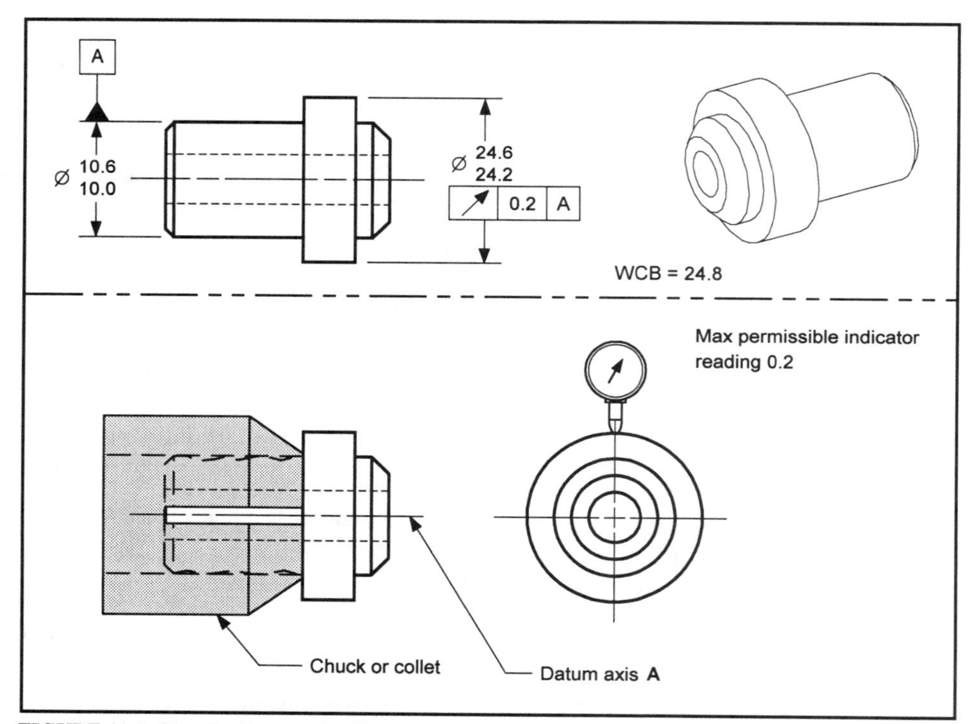

FIGURE 11-5 Circular Runout Applied to a Diameter

Figure 11-6 illustrates an application of circular runout applied to a surface that is perpendicular to the datum axis. This type of application is used to control the wobble of a surface. In this application, the following conditions apply:

- The runout control applies RFS.
- Runout applies at each circular element of the surface.
- The shape of the tolerance zone is two coaxial circles offset axially by 0.2 at the point of measurement.
- The circular runout control does not control the orientation of the surface.

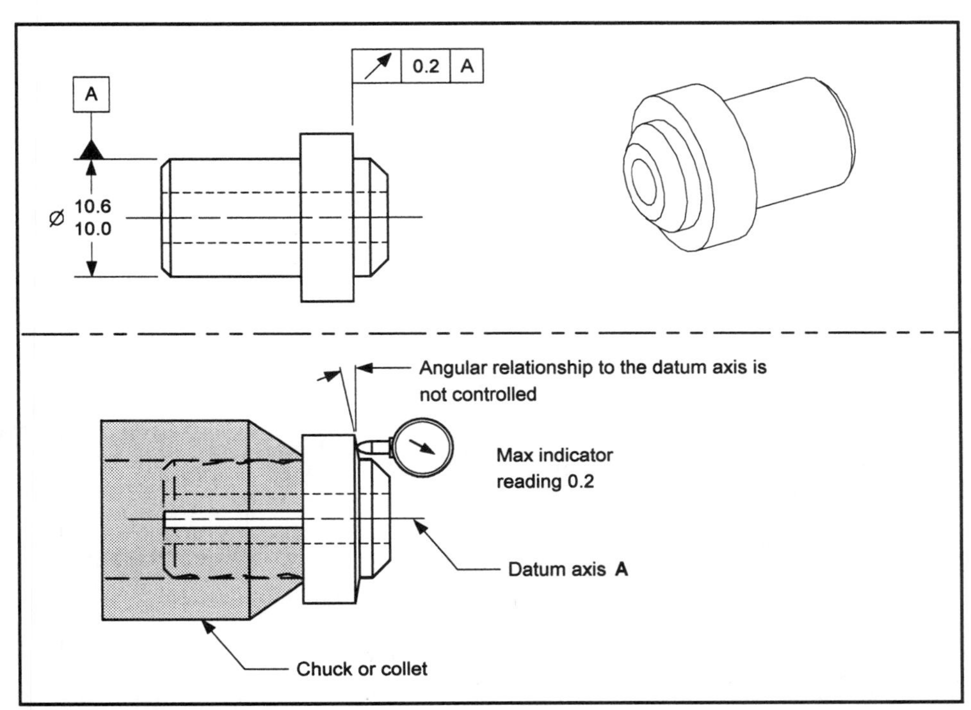

FIGURE 11-6 Circular Runout Applied to a Surface

Design Tip

It is preferred to use perpendicularity to control squareness of a surface to an axis.

Legal Specification Test for a Runout Control

For a runout control applied to a diameter to be a legal specification, it must satisfy the following conditions:

- A datum must be referenced in the feature control frame.
- Datums referenced must specify a proper datum axis.
- A runout control must be applied to a surface element that surrounds or is intersected by the datum axis.
- The runout control must be applied on an RFS basis.
- The runout control must not include any modifiers.

Figure 11-7 shows a legal specification flowchart for a runout control applied to a diameter.

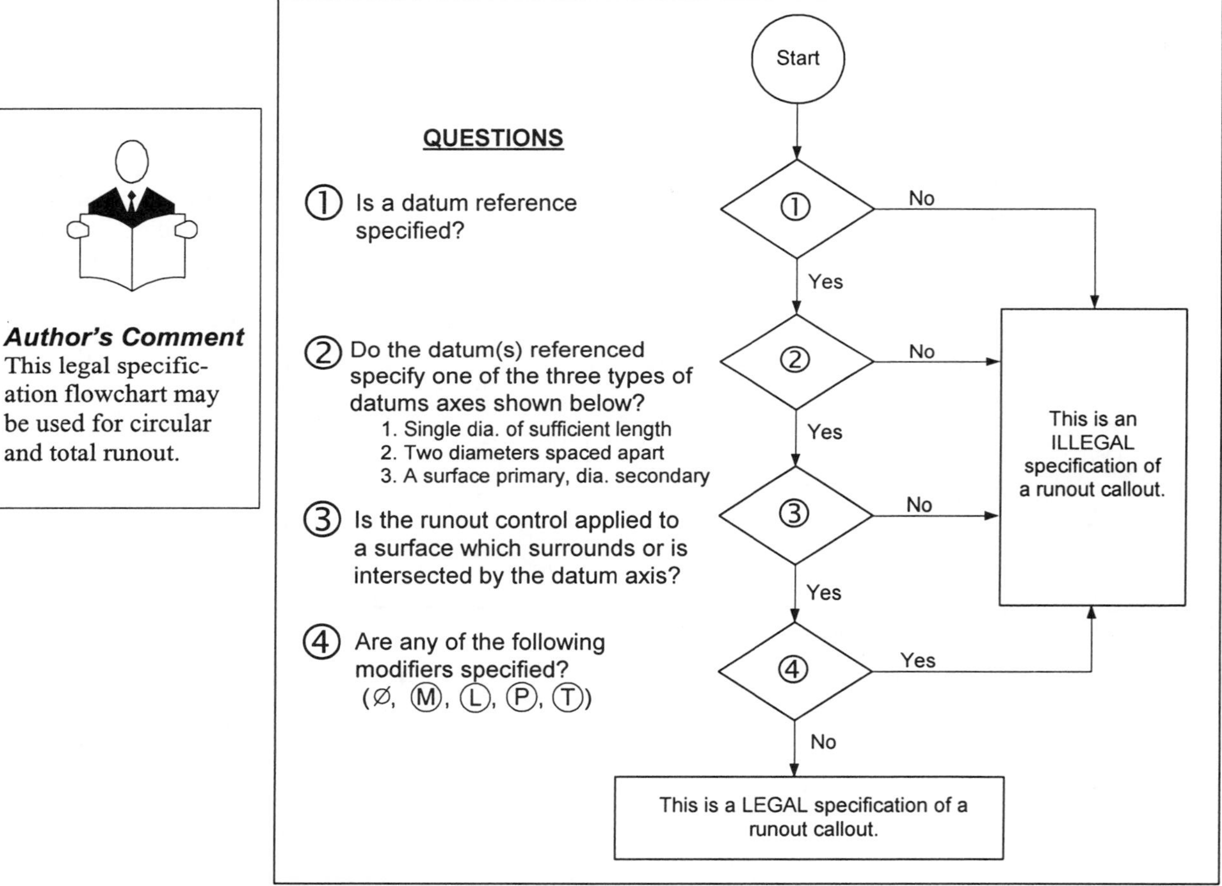

FIGURE 11-7 Legal Specification Flowchart for Runout

Verifying Circular Runout

In Figure 11-8, a circular runout control is applied to a diameter. When verifying this diameter, three separate checks should be made: the size of the diameter, the Rule #1 boundary, and the runout of the diameter. Chapter Two explains how to check the size and Rule #1 boundary. Now we will look at how to verify the runout requirement.

One way a circular runout control could be verified is shown in the lower half of Figure 11-8. First, the part is mounted in a chuck or collet to establish datum axis *A*, then a dial indicator is placed perpendicular to the surface of the diameter being inspected. The part is rotated 360°, and the total indicator reading (TIR) is the runout tolerance value for that circular element. The dial indicator is moved to another location on the diameter, and another indicator reading is obtained. The number of circular elements to be checked is usually left to the inspector's judgment.

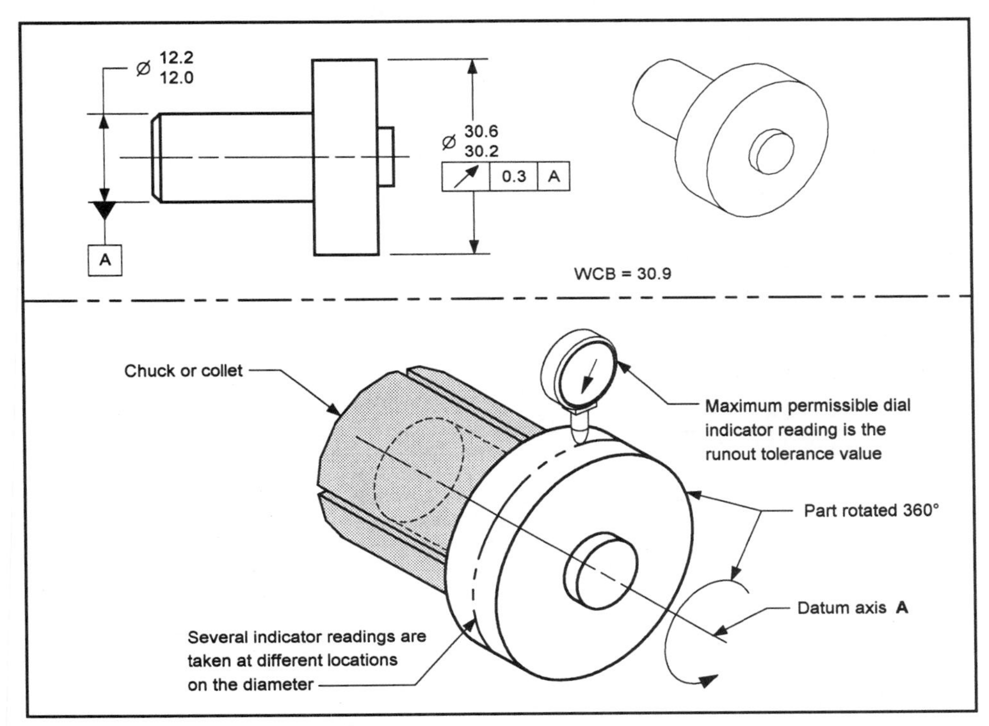

FIGURE 11-8 Verifying Circular Runout

TECHNOTE 11-4 Verifying Circular Runout

When verifying circular runout, the runout tolerance value is the maximum dial indicator reading for the circular element being checked.

TOTAL RUNOUT

Definition

For more info. . .
See Paragraph 6.7.1.2.2 of Y14.5.

Total runout is a composite control affecting the form, orientation, and location of all surface elements (simultaneously) of a diameter (or surface) relative to a datum axis. A ***total runout control*** is a geometric tolerance that limits the amount of total runout of a surface. It applies to the entire length of a diameter simultaneously. It is referred to as a composite control because it affects the form, orientation, and location of a part feature simultaneously. Total runout is frequently used to control the location of a diameter. When applied to a diameter, it controls the form (cylindricity), orientation, and location of the diameter relative to a datum axis.

When applied to a diameter, the tolerance zone shape for a total runout control is easily visualized; it is two coaxial cylinders whose centers are located on the datum axis. The radial distance between the cylinders is equal to the runout tolerance value. Figure 11-9 illustrates the tolerance zone for total runout.

The size of the larger cylinder is established by the radius to the surface element that is farthest from the datum axis. The second cylinder of the tolerance zone is radially smaller than the larger cylinder by the runout tolerance value. When verifying a diameter controlled with total runout, a dial indicator is placed on the surface being verified. The part is rotated 360°, and the dial indicator is moved along the surface of the diameter; it indicates the radial distance between the cylinders.

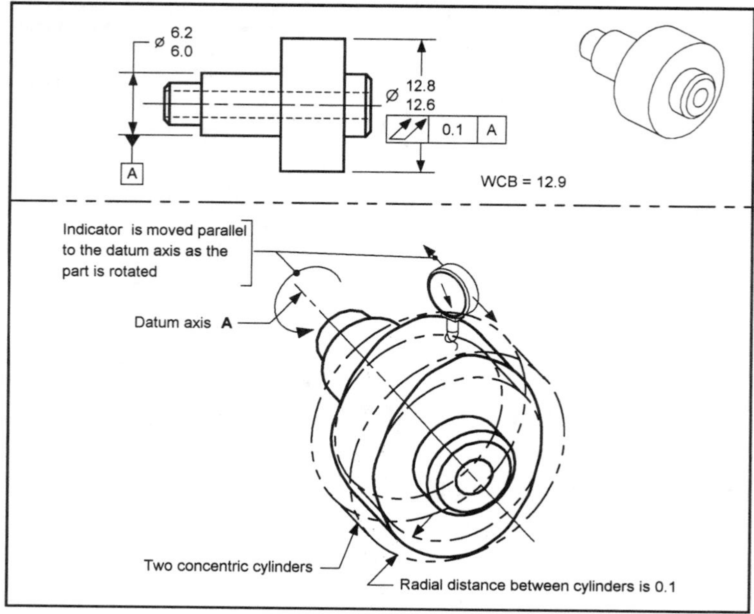

FIGURE 11-9 Total Runout

TECHNOTE 11-5 Total Runout Tolerance Zone

When applied to a diameter, the tolerance zone for a total runout control is two coaxial cylinders that are located on the datum axis.

Total Runout as a Composite Control

Total runout is a composite control; it limits the cylindricity, orientation, and axis offset of a diameter. When total runout is applied to a diameter, it affects the worst-case boundary of the diameter. When verifying total runout, the dial indicator reading includes all three types of part errors. Figure 11-10 illustrates how form and location errors are combined in a total runout verification. As a diameter is being verified with a dial indicator, an indicator reading would be produced by any error in feature roundness. Since the indicator is being moved along the axis as the part is rotated, the indicator would also be affected by the straightness and taper of the surface of the diameter. Even though there is no axis offset, the dial indicator detects the form errors.

A part could be produced such that the entire runout error results from the axis offset. An axis offset of 0.2 produces a runout gage reading of 0.4. The maximum possible axis offset for total runout is equal to one-half the runout tolerance value. Whenever any form error exists in the toleranced feature, the allowable axis offset will be reduced by the amount of the form error. However, in a total runout verification, the gage reading does not separate the form error from the axis offset.

Design Tip

The WCB of a diameter is the same whether circular or total runout is specified. Since circular runout is easier to manufacture, do not use total runout unless the additional control of the form (straightness or taper) is required for part function.

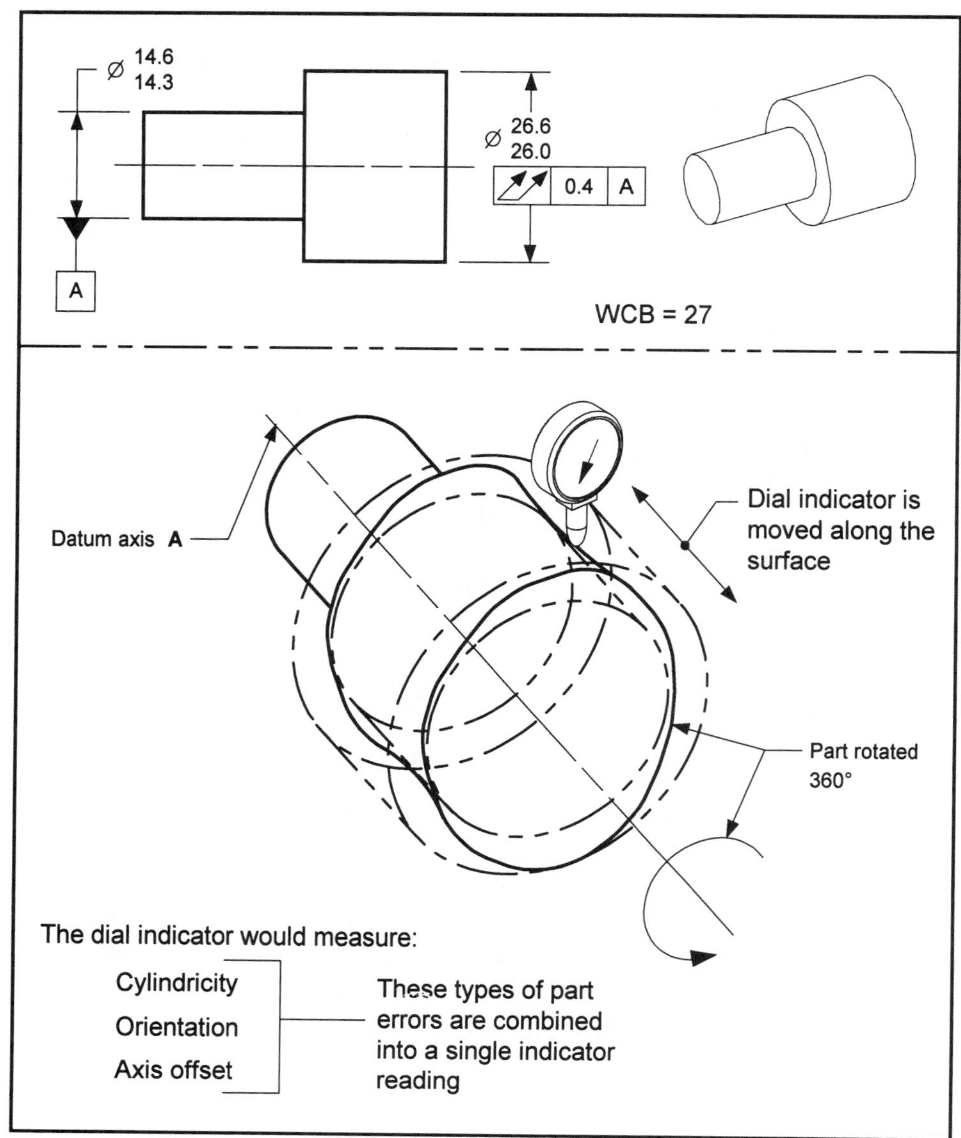

FIGURE 11-10 Total Runout as a Composite Control

TECHNOTE 11-6 Total Runout Control Axis Offset

When a diameter is controlled by total runout, its maximum possible axis offset from the datum axis is equal to one-half the runout tolerance value.

Total Runout Application

Figure 11-11 illustrates a total runout application. In this application, the following conditions apply:

- The diameter must meet its size requirements.
- The WCB is affected (24.6 + 0.2 = 24.8).
- The runout control applies RFS.
- The runout applies simultaneously to all elements of the diameter.
- The tolerance zone is two coaxial cylinders 0.2 apart.
- The maximum possible axis offset is 0.1.

Author's Comment

If the runout tolerance value is smaller than the size tolerance of the diameter, the roundness and straightness of the diameter will be limited by the runout control.

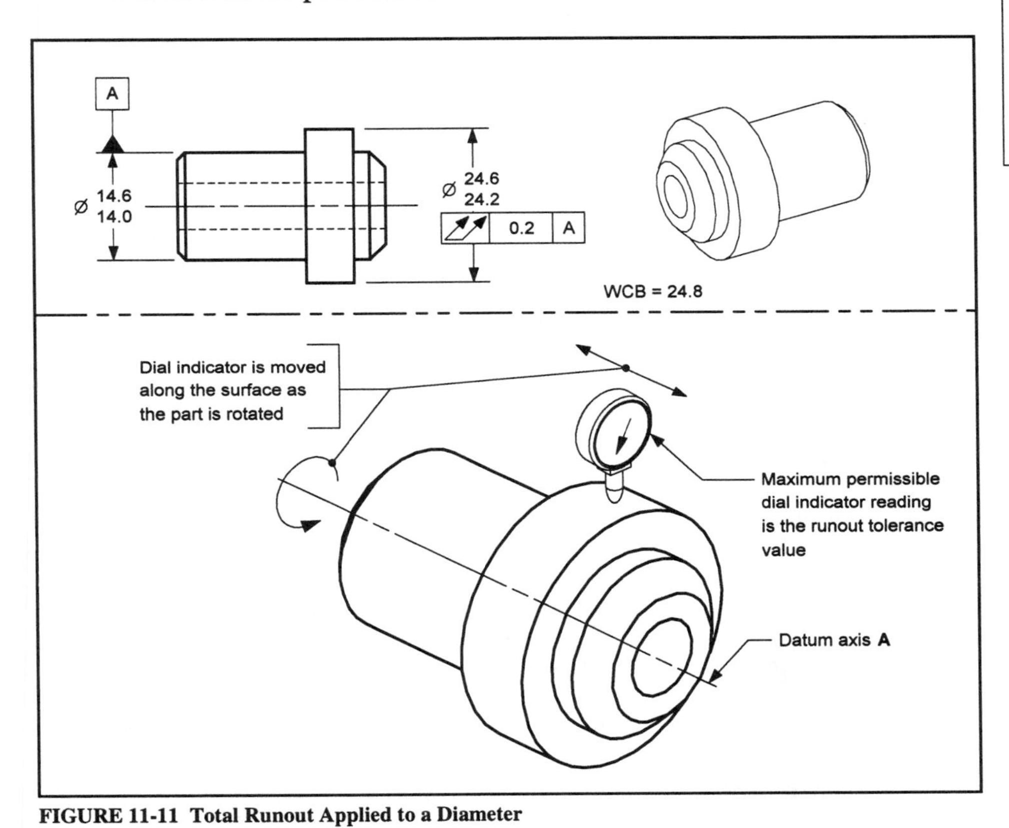

FIGURE 11-11 Total Runout Applied to a Diameter

Figure 11-12 illustrates an application of total runout applied to a surface that is perpendicular to the datum axis. This type of application is used to control the squareness of a surface to a datum axis. In this application, the following conditions apply:

- The runout control applies RFS.
- The runout applies to all elements of the surface simultaneously.
- The shape of the tolerance zone is two parallel planes perpendicular to the datum axis.
- The runout symbol controls the angular relationship (orientation) of the surface to the datum axis.
- The runout control also limits the flatness of the surface.

Design Tip
If the design intent is to control the squareness of the surface relative to the axis, it would be more straightforward to use a perpendicularity control.

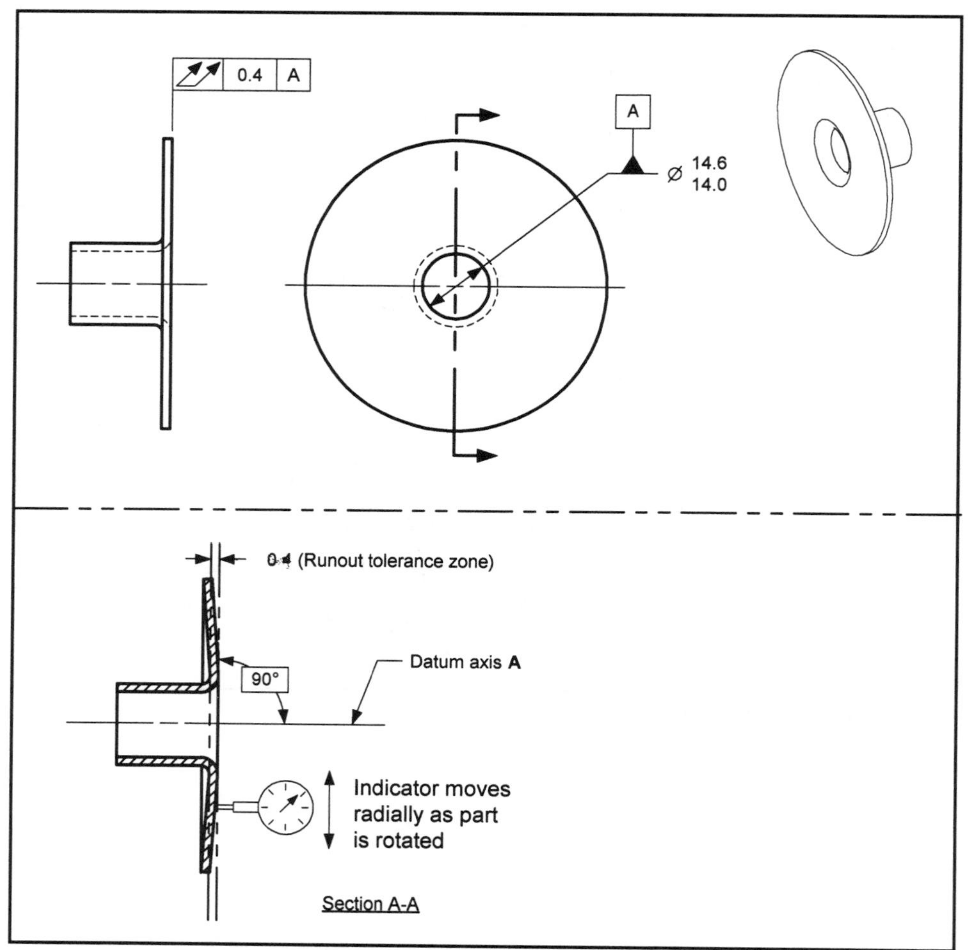

FIGURE 11-12 Total Runout Applied to a Surface

Legal Specification Test for Total Runout

The legal specification test for total runout is the same as the test for circular runout. The legal specification flowchart for circular runout is shown in Figure 11-7 on page 308.

Verifying Total Runout

In Figure 11-13, a total runout control is applied to a diameter. When verifying this diameter, three separate checks should be made: the size of the diameter, the Rule #1 boundary, and the runout of the diameter. Chapter Two explains how to check the size and Rule #1 boundary. Now we will look at how to inspect the runout requirement.

One way the total runout control could be verified is shown in Figure 11-13. First, the part is mounted in a chuck or collet to establish datum axis *A*. Then a dial indicator is placed perpendicular to the surface of the diameter being inspected. The part is rotated 360°, as the dial indicator is moved axially along the surface, and the total indicator reading (TIR) is the runout tolerance value for the surface.

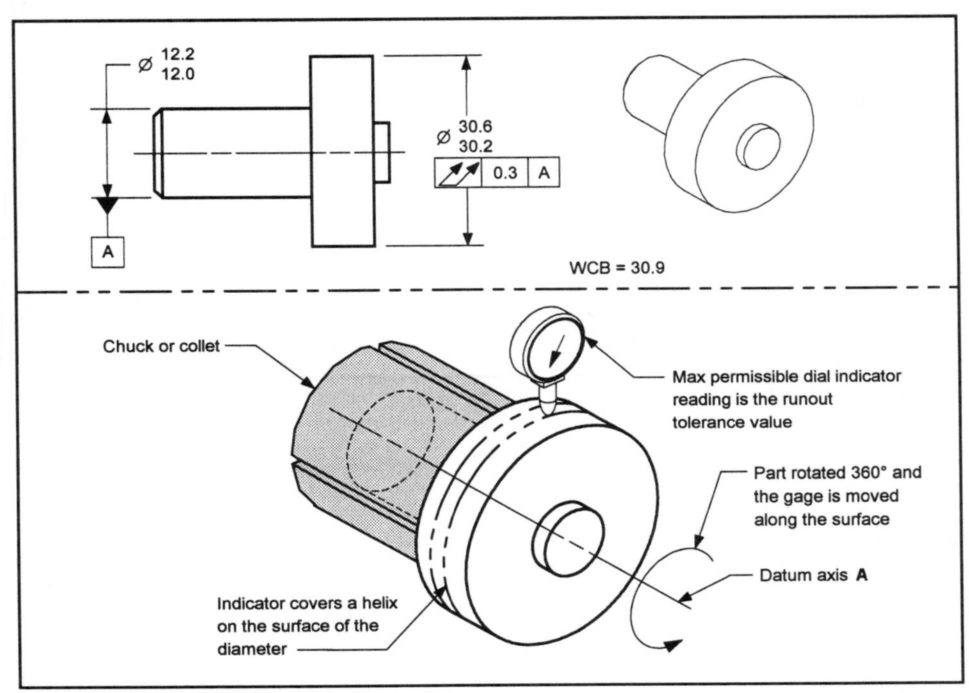

FIGURE 11-13 Verifying Total Runout

TECHNOTE 11-7 Verifying Total Runout

When verifying total runout, the dial indicator is moved axially along the surface, and the dial reading is the runout tolerance value for the diameter being checked.

In Figure 11-14, a runout control is applied to a surface. When inspecting this runout requirement, the part would be mounted in a chuck or collet to establish datum axis *A*. Then a dial indicator is placed perpendicular to the surface being inspected. The part is rotated as the dial indicator is moved radially, and the TIR is the runout tolerance amount for the surface.

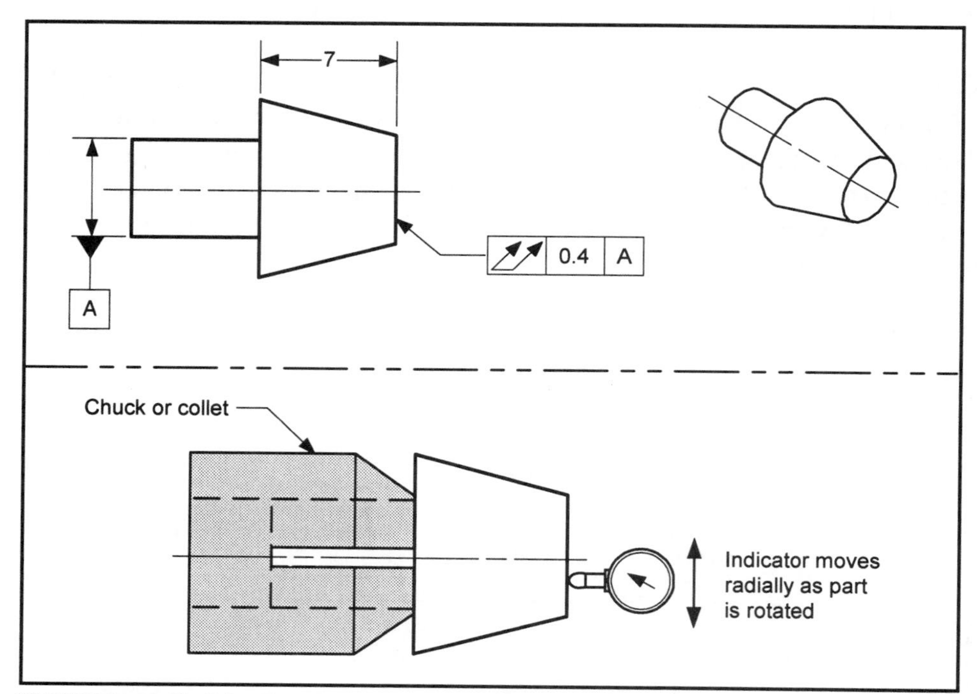

FIGURE 11-14 Verifying Total Runout

Comparison Between Circular and Total Runout

Circular and total runout controls are used to tolerance coaxial diameters. Both controls are similar, but some significant differences exist. The chart in Figure 11-15 compares these controls (when they are applied to a diameter).

Runout Applied to a Diameter		
Concept	**Circular Runout**	**Total Runout**
Tolerance zone	Two coaxial circles	Two coaxial cylinders
Relative cost to produce	$	$ $
Relative cost to inspect	$	$ $
Part characteristics being controlled	Location Orientation Circularity	Location Orientation Cylindricity

FIGURE 11-15 Comparison of Circular and Total Runout

A part could pass a circular runout verification and fail a total runout verification. Figure 11-16 shows two parts with zero circular runout error and 0.1 total runout error. In each case, the straightness or flatness error of the part surface would not be detected in a circular runout check, but would be detected in a total runout check.

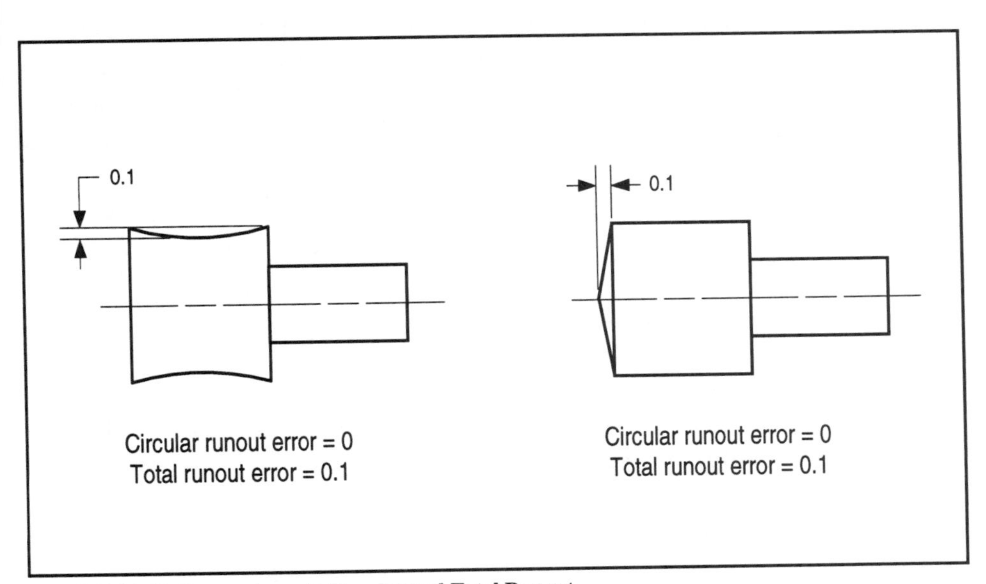

FIGURE 11-16 Comparison of Circular and Total Runout

RUNOUT CALCULATIONS

For more info. . .
This text provides a brief introduction to tolerance stacks. For more information on this topic see reference number six listed in the bibliography.

In industry, it is very common to make a calculation to find a max. or min. distance on a part. When calculating a max. or min. part distance, all of the part tolerances must be included. This section explains how to use runout controls in part calculations. The technique shown can be used for circular or total runout.

Tolerance Stacks Using Runout

Tolerance stacks on a part that uses runout is straightforward. In this section, a simple method of how to calculate the max. and min. distance between two diameters is explained. This calculation involves using the radius of the diameters and the runout tolerance between the diameters.

Figure 11-17 shows an example of a part calculation to find the max. and min. distance between two diameters of a part. When reading the steps below, refer to Figure 11-17.

1. Label the start and end points of the distance to be calculated. On the start point of the calculation, draw a double-ended arrow. Label the arrow that points towards the end point of the calculation as positive (+). Label the other arrow as negative (-). Each time a distance that is in the direction of the positive arrow is used in the calculation, the distance will be a positive value. When a distance is used in the negative direction, it will be a negative value.
2. Establish a loop of part dimensions or gage distances (as in Figure 1-17) from the start point to the end point of the calculation.
3. Calculate the answer.

When solving for a min. distance, half the runout tolerance value is subtracted from the calculation. When solving for a max. distance, half the runout tolerance value is added to the calculation.

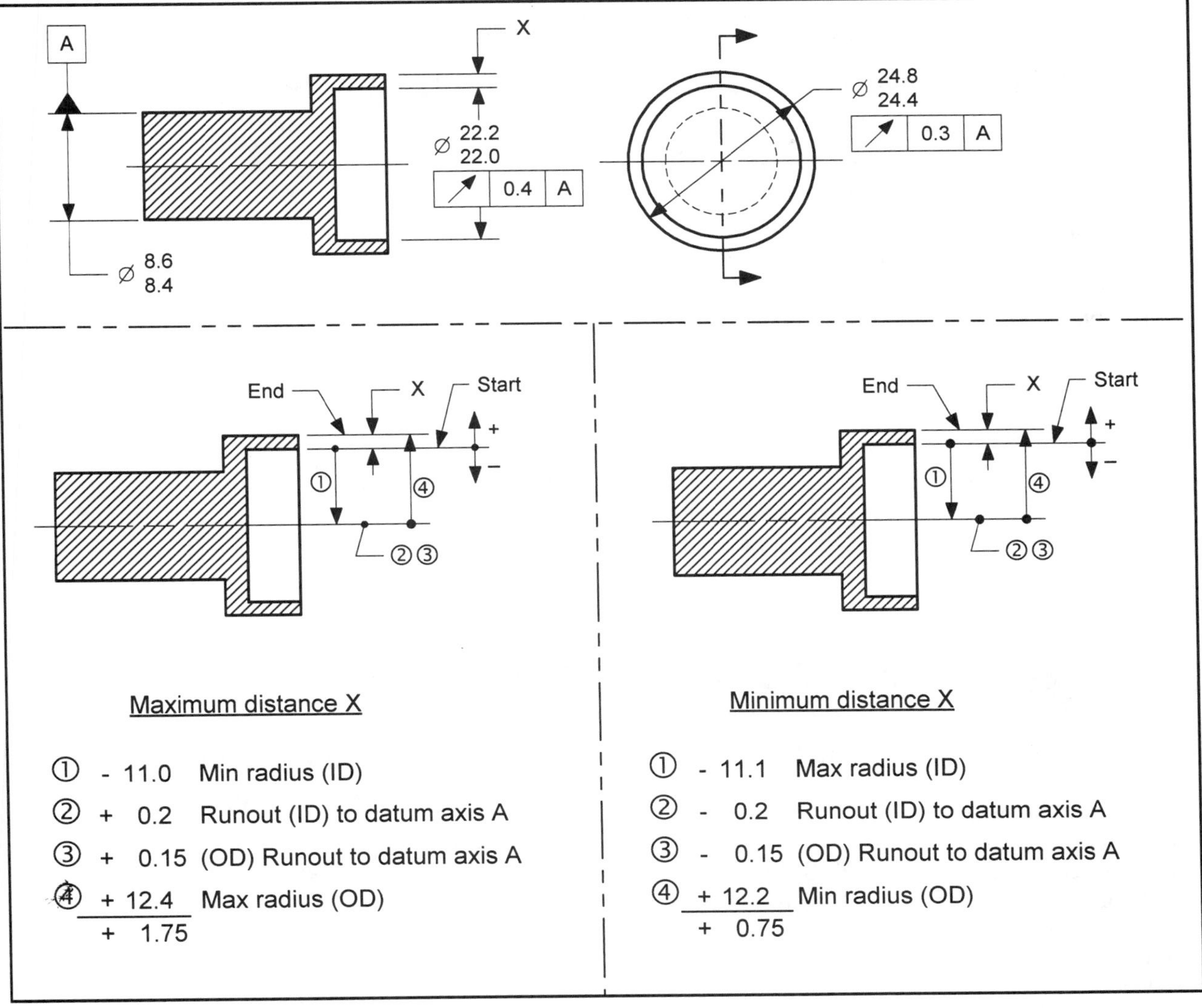

FIGURE 11-17 Tolerance Stacks Using Runout

Summary

A summarization of runout control information is shown in Figure 11-18.

Symbol	Datum reference required	Can be applied to a		Can affect WCB	Can use Ⓜ Ⓛ Ⓣ Ⓟ modifier	Can be applied at RFS	Can override Rule #1	Tolerance zone shape **
		Surface	FOS					
↗	Yes	Yes	Yes	Yes	No	Yes*	No	2 coaxial circles
⌰	Yes	Yes	Yes	Yes	No	Yes*	No	2 coaxial cylinders

* Is automatic per Rule #2 ** When applied to a diameter

FIGURE 11-18 Summarization of Runout Controls

Study Tip
Read each term. If you don't recall its meaning, look it up in the chapter.

VOCABULARY LIST

New Terms Introduced in this Chapter

Circular runout
Circular runout control
Composite control
Total runout
Total runout control

Author's Comment
These topics, plus advanced coverage of many of the topics introduced in this text, will be covered in my new book on advanced GD&T concepts.

ADDITIONAL RELATED TOPICS

Topic	ASME Y14.5M-1994 Reference
• Using runout to control datum features	Paragraph 6.7.1.3.4

QUESTIONS AND PROBLEMS

1. Describe what runout is.

____

____

2. List three ways a datum axis can be established for a runout application.

____

____

____

3. Describe the shape of the tolerance zone when circular runout is applied to a diameter.

____

4. Circular runout is considered a composite control. List three types of part errors a circular runout may affect.

____

____

____

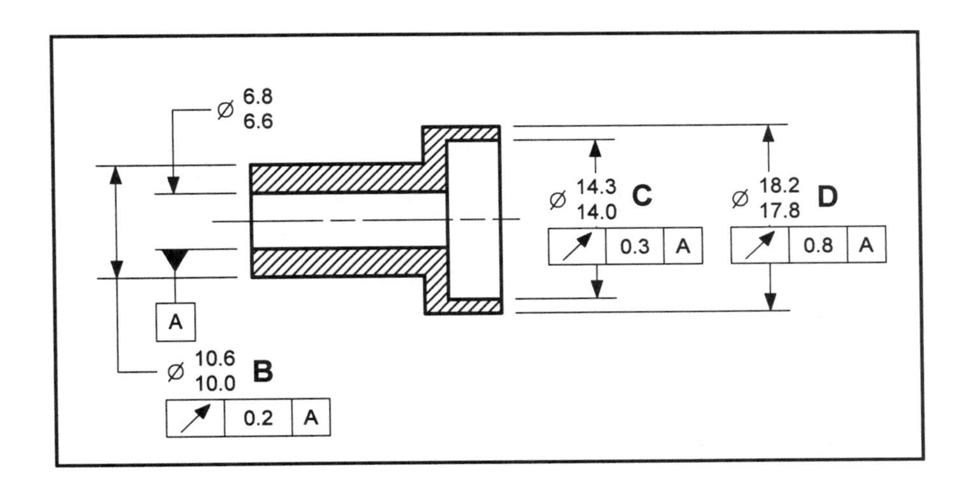

5. Use the drawing above to fill in the chart below.

Dia.	Max Possible Axis Offset from Datum Axis A
B	
C	
D	

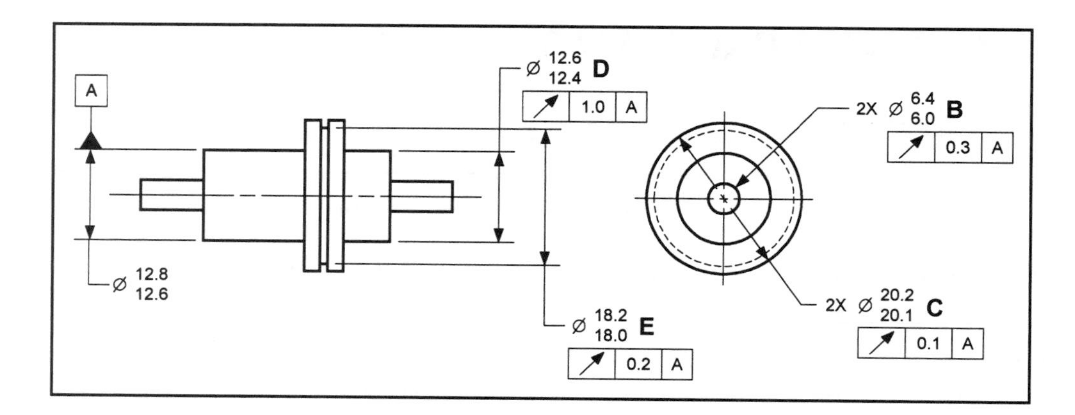

6. Use the drawing above to fill in the chart below.

QUESTION	APPLIES TO			
	DIA **B**	DIA **C**	DIA **D**	DIA **E**
The size of the diameter is limited to. . .				
The roundness of the diameter is limited to. . .				
The maximum offset between the diameter axis and datum axis A is. . .				
Describe the tolerance zone for the runout controls applied to the diameter.				
How many places should the runout control be checked on this diameter?				
What is the outer boundary (virtual condition) of this diameter?				

7. For each circular runout control shown below, indicate if it is a legal specification. If a control is illegal, explain why.

A. | ↗ | 0.2 | A | ____________________

B. | ↗ | 0.2 | B | A | ____________________

C. | ↗ | 0.2 Ⓜ | A | ____________________

D. | ↗ | ∅ 0.2 | A | ____________________

8. Describe what total runout is.

____

____

9. Describe the shape and size of the tolerance zone when total runout is applied to a diameter.

____

10. Total runout is considered a composite tolerance. List four types of part errors that a total runout control may affect.

____

____

____

____

11. For each total runout control shown below, indicate if it is a legal specification. If a control is illegal, explain why.

A. | ⌰ | Ø 0.2 | A | _______________________________

B. | ⌰ | 0.2 Ⓟ | A | _______________________________

C. | ⌰ | 0.2 Ⓢ | A | _______________________________

D. | ⌰ | 0.2 | B | A | _______________________________

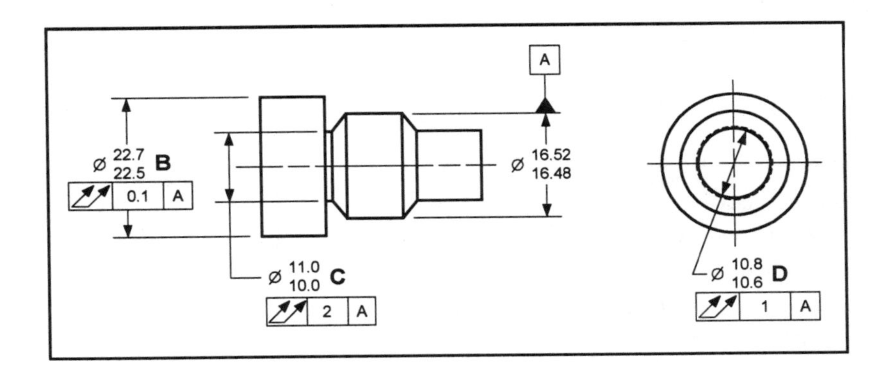

12. Use the drawing above to fill in the chart below.

Dia.	Max Possible Axis Offset from Datum Axis A
B	
C	
D	

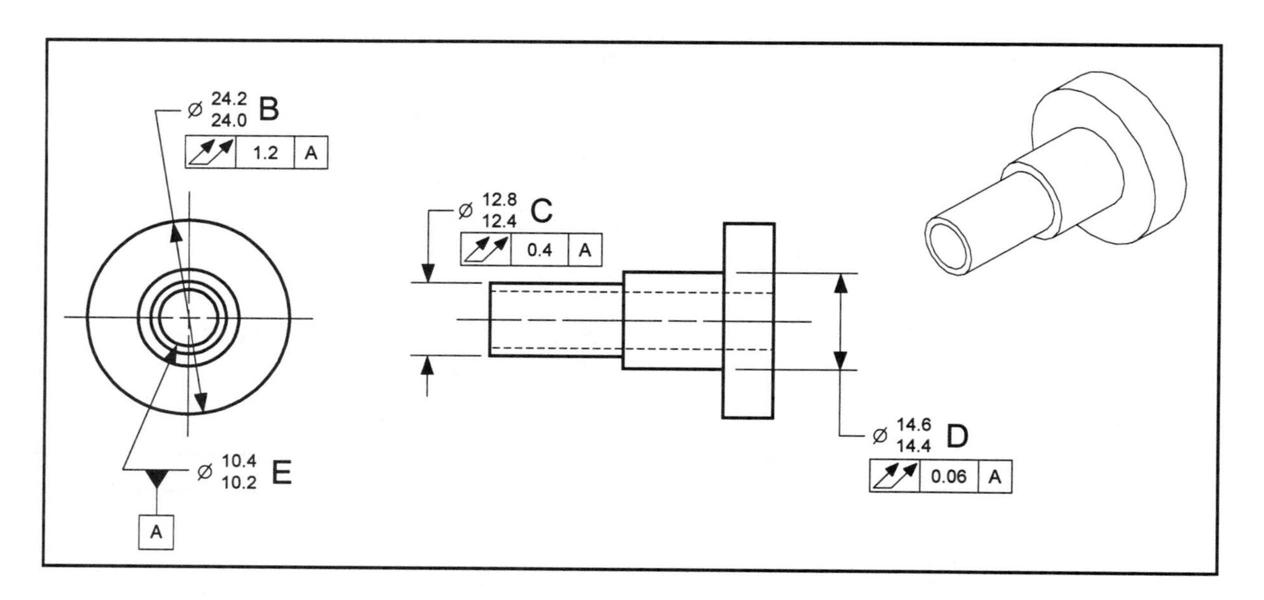

13. Use the drawing above fill in the chart below.

QUESTION	APPLIES TO			
	DIA **B**	DIA **C**	DIA **D**	DIA **E**
The size of the diameter is limited to. . .				
The roundness of the diameter is limited to. . .				
The maximum offset between the diameter axis and datum axis A is. . .				
Describe the tolerance zone for the runout controls applied to the diameter.				
What is the outer boundary (virtual condition) of this diameter?				

14. Circle the letters of the statements that apply when checking a total runout control applied to a diameter.
    A. The diameter must be within its size limits.
    B. The dial indicator reading is the runout tolerance value.
    C. The dial indicator reading is taken at several places along the diameter.
    D. The part is rotated about the datum axis.
    E. The dial indicator is moved along the datum axis as the part is rotated.

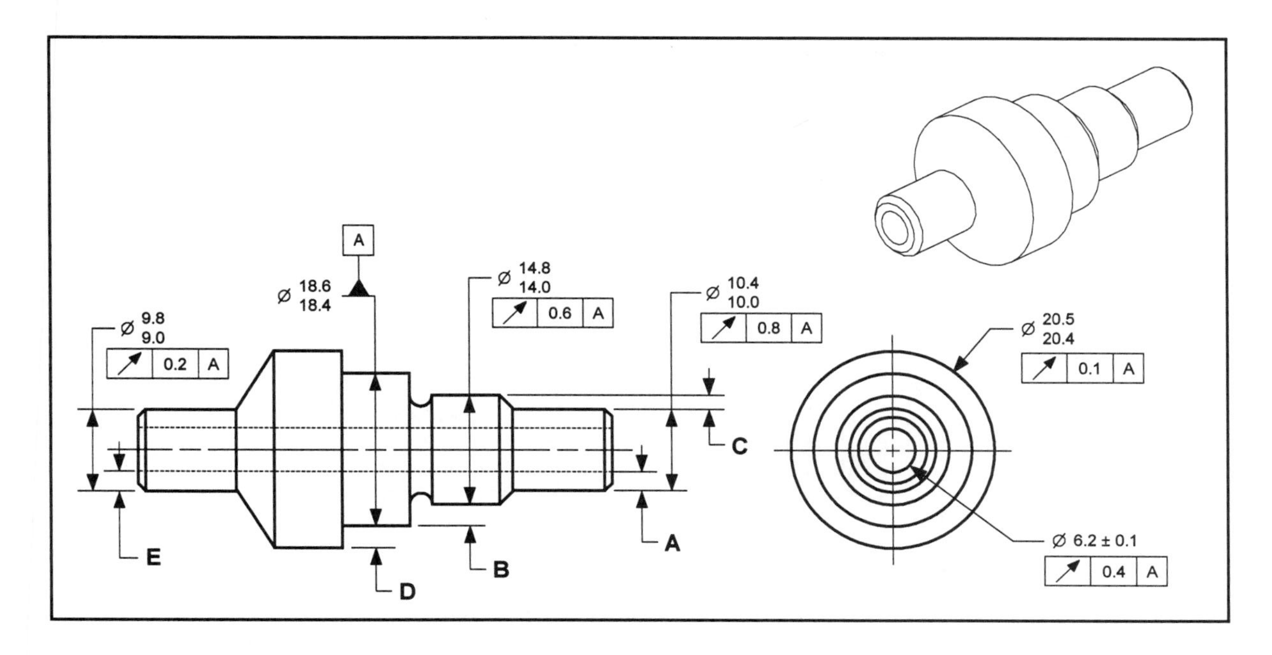

15. Use the drawing above to fill in the chart.

DISTANCE	MAX	MIN
A		
B		
C		
D		
E		

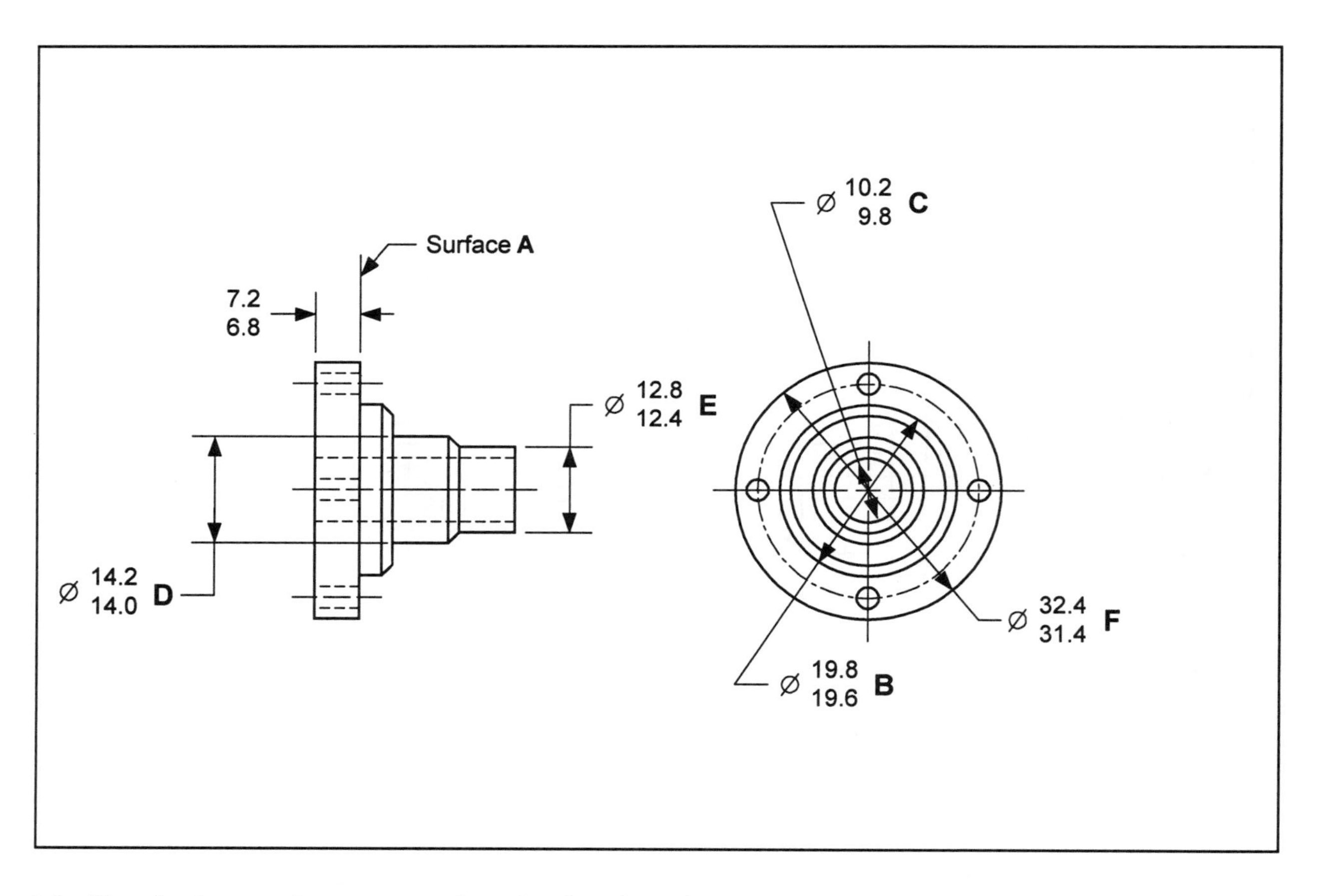

16. Use the instructions to complete the drawing above:
    a. Specify surface *A* as "datum *A*."

    b. Add a control to surface *A* that limits its flatness to 0.1 maximum.

    c. Add a control to diameter *D* that limits its perpendicularity to datum *A* within a 0.1 diameter cylindrical tolerance zone when diameter *D* is at MMC.

    d. Specify diameter *B* as datum *B*.

    e. Add a control to diameter *C* that limits the runout of its circular elements to within 0.2 relative to datums *A* and *B*.

    f. Specify diameter *C* as datum *C*.

    g. Add a control to diameter *D* that limits the runout of its circular elements to within 0.5 relative to datums *A* and *B*.

    h. Add a control to diameter *E* that limits the runout of its circular elements to within 0.8 relative to datum *C*.

    i. Add a control to diameter *F* that limits the runout of its circular elements within 1.0 relative to datums *A* and *B*.

Chapter 12

Profile Controls

INTRODUCTION

This chapter explains the concepts involved with using profile tolerancing. There are two types of profile controls: profile of a surface and profile of a line. The symbols for these profile controls are shown in Figure 12-1.

Profile of a surface is considered the most powerful control in the geometric tolerancing system. It can be used to control the size, location, orientation, and form of a part feature. Profile of a surface or line can be used to tolerance planar surfaces, cylinders, cones, curves, and irregular curves.

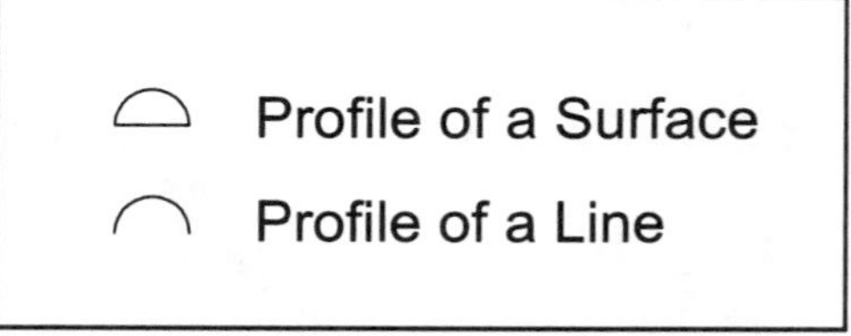

FIGURE 12-1 Profile Controls

CHAPTER GOALS AND OBJECTIVES

There are Three Goals in this Chapter:

12-1. Understand profile tolerancing.
12-2. Interpret the profile of a surface control.
12-3. Interpret profile of a line control.

Study Tip
Take a few minutes to fully understand these objectives. When reading this chapter, look for information to help you master these objectives.

Performance Objectives that Demonstrate Mastery of These Goals

Upon completion of this chapter, each student should be able to:

Goal 12-1 (pp. 329-334)

- Explain the effect of using profile with or without datum references.
- Describe the term, "true profile."
- List four part characteristics that a profile control can affect.
- Interpret a bilateral profile control tolerance zone.
- Interpret a unilateral profile control tolerance zone.
- Describe the extent to which a profile callout tolerance zone applies to a part.
- Describe the effect of the symbol for "between."
- Describe the effect of the symbol for "all around."
- List three advantages of using profile controls.

Goal 12-2 (pp. 335-343)
- Describe the shape and size of the tolerance zone for a profile of a surface application.
- Describe the shape and size of the tolerance zone for a profile of coplanar surfaces application.
- Explain when to use multiple single-segment profile controls.
- Describe the effect of each segment of a multiple single-segment profile of a surface application.
- Determine if a profile of a surface specification is legal.
- Describe how profile of a surface can be inspected.

Goal 12-3 (pp. 343-348)
- Describe a profile of a line control.
- Describe where a profile of a line control is most commonly used.
- Describe the shape and size of the tolerance zone in a profile of a line application.
- Describe the effect of profile of a line when used with coordinate tolerances.
- Determine if a profile of a line specification is legal.
- Describe how a profile of a line control can be inspected.
- Calculate distances on a part dimensioned with profile.

GENERAL INFORMATION ON PROFILE

Profile Terminology

A unique aspect of profile controls is that they can be specified with a datum reference (as a datum-related control) or without a datum (as a form control). When datums are referenced, the profile tolerance zone is related to the datum reference frame. When no datums are referenced, the profile tolerance zone applies where the part surface actually exists.

For more info. . .
See Paragraph 6.5.4 of Y14.5.

TECHNOTE 12-1 Profile as a Related Feature Control or as a Form Control

- When a profile control is specified with datum references, it is a related feature control.
- When a profile control is specified without datum references, it is a form control.

For more info. . .
See Paragraph 6.5 of Y14.5.

A ***profile*** is the outline of a part feature in a given plane. A ***true profile*** is the exact profile of a part feature as described by basic dimensions. A ***profile control*** is a geometric tolerance that specifies a uniform boundary along the true profile that the elements of the surface must lie within. A profile of a line control is a type of profile control that applies to line elements of the toleranced surface. Figure 12-2 illustrates these terms. Whenever a profile control is used, it is associated with a true profile (a surface defined with basic dimensions). The true profile may be located with basic or toleranced dimensions relative to the datums referenced in the profile control.

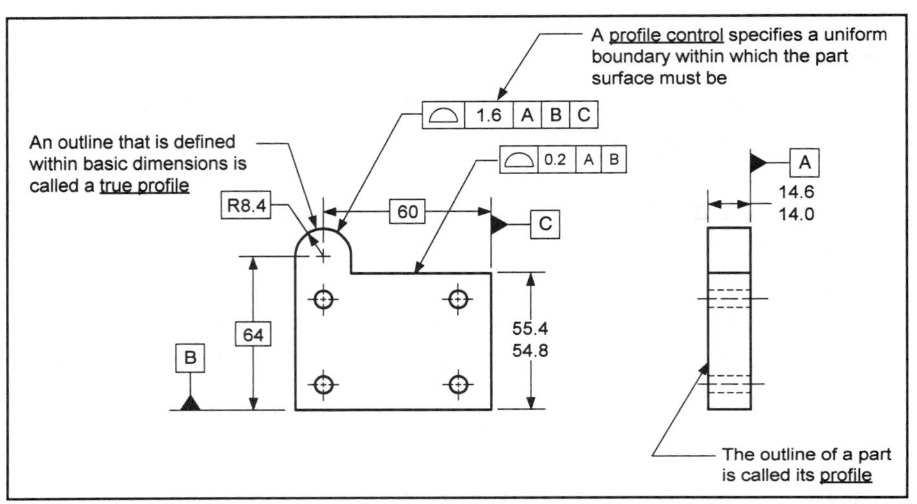

FIGURE 12-2 Profile Control

A profile control can be used to control four part characteristics: size, location, orientation, or form of a part feature. It can be applied to single or multiple features. Several profile controls may be applied to a single feature.

TECHNOTE 12-2 Profile

- A profile control must be applied to a true profile.
- A profile control can be used to control: size, location, orientation, or form

Profile Tolerance Zones

When a profile of a surface control is specified, the tolerance zone is a uniform boundary (a 3-D tolerance zone). It applies for the full length, width, and depth of the surface. When a profile of a line control is specified, the tolerance zone is two uniform lines (a 2-D tolerance zone). It applies for the full length of the surface.

Unless otherwise indicated, where a profile control (surface or line) is associated with a feature, the tolerance zone is a bilateral tolerance zone with equal distribution. An example is shown in the "default" section in Figure 12-3. This is the most common tolerance zone used with profile. However, when using profile controls (surface or line), three other distributions are permissible:

1. Bilateral tolerance zone (unequal distribution)
2. Unilateral tolerance zone (outside)
3. Unilateral tolerance zone (inside)

For more info. . .
See Paragraph 6.5.1 of Y14.5.

These tolerance zones are illustrated in the "optional" section in Figure 12-3. Note that for each tolerance zone (other than bilateral with equal distribution) phantom lines are added to the drawing. The phantom lines invoke the special tolerance zone types.

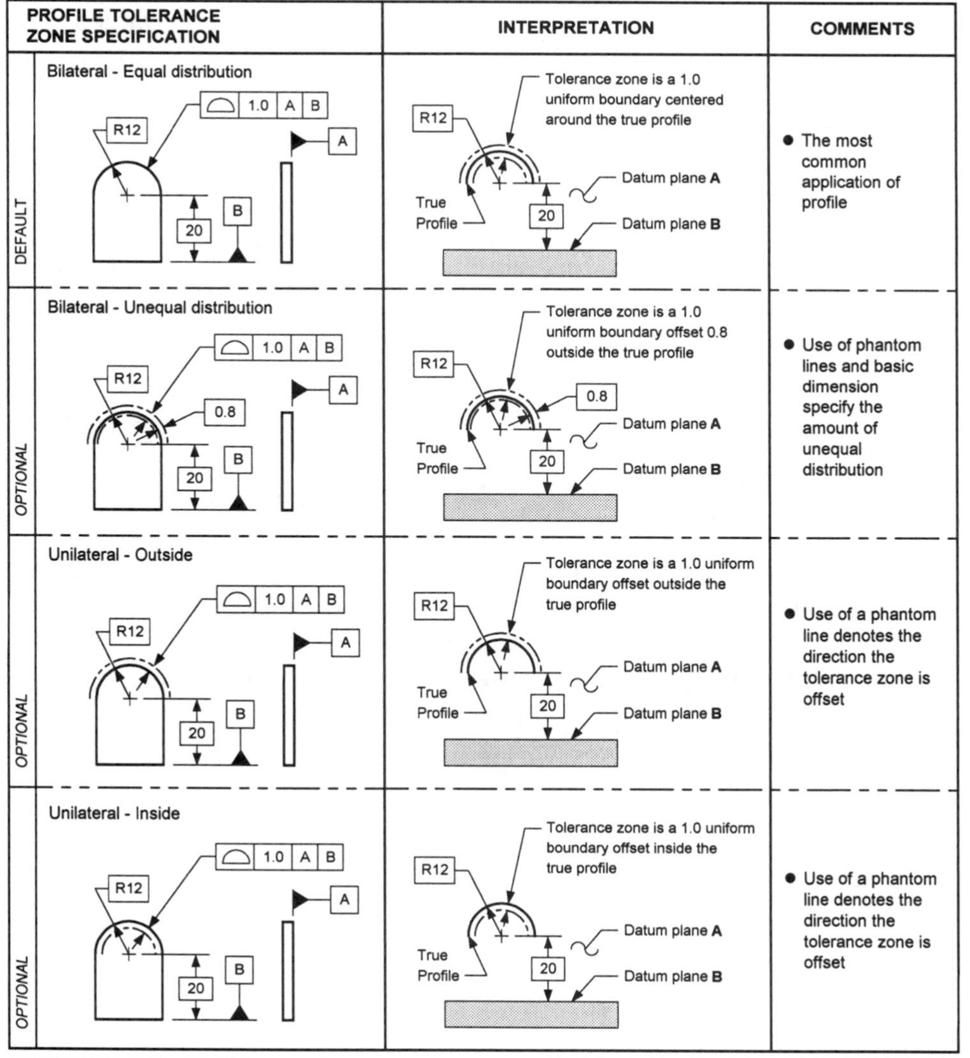

FIGURE 12-3 Profile Tolerance Zone

When a profile control (surface or line) points to a surface, its tolerance zone applies for the full length and width of the surface. If desired, the profile tolerance zone coverage can be extended to include additional surfaces or the entire profile. There are three ways to extend the coverage of a profile control:

1. A note associated with the profile control
2. The between symbol
3. The all around symbol

In Figure 12-4*A*, the profile control tolerance zone only applies to the surface to which it is pointing. In Figure 12-4*B*, a note has been added to the profile control to extend the coverage of the profile tolerance zone to include all the surfaces between points *A* & *B*.

In Figure 12-4*C*, a symbol has been used in place of the note. The symbol is the "between" symbol. The ***between symbol*** is a double ended arrow that indicates the tolerance zone extends to include multiple surfaces. This symbol indicates the profile tolerance zone is extended. The letters on each end of the arrow indicate the endpoints of the range surfaces included.

In Figure 12-4*D*, a circle has been placed on the bend of the profile control leader line. This symbol is called the "all around" symbol. The ***all around symbol*** is a circle placed on the bend of the leader line of a profile control. This symbol indicates that the profile tolerance zone applies all around in the view on which it is shown.

TECHNOTE 12-3 Profile Tolerance Zone Coverage

- A profile control tolerance zone (surface or line) applies only to the surface to which the control is directed, unless extended by one of the three methods shown below.
- The tolerance zone can be extended to cover additional surfaces by:
  1. the between symbol;
  2. the all around symbol; or
  3. a note

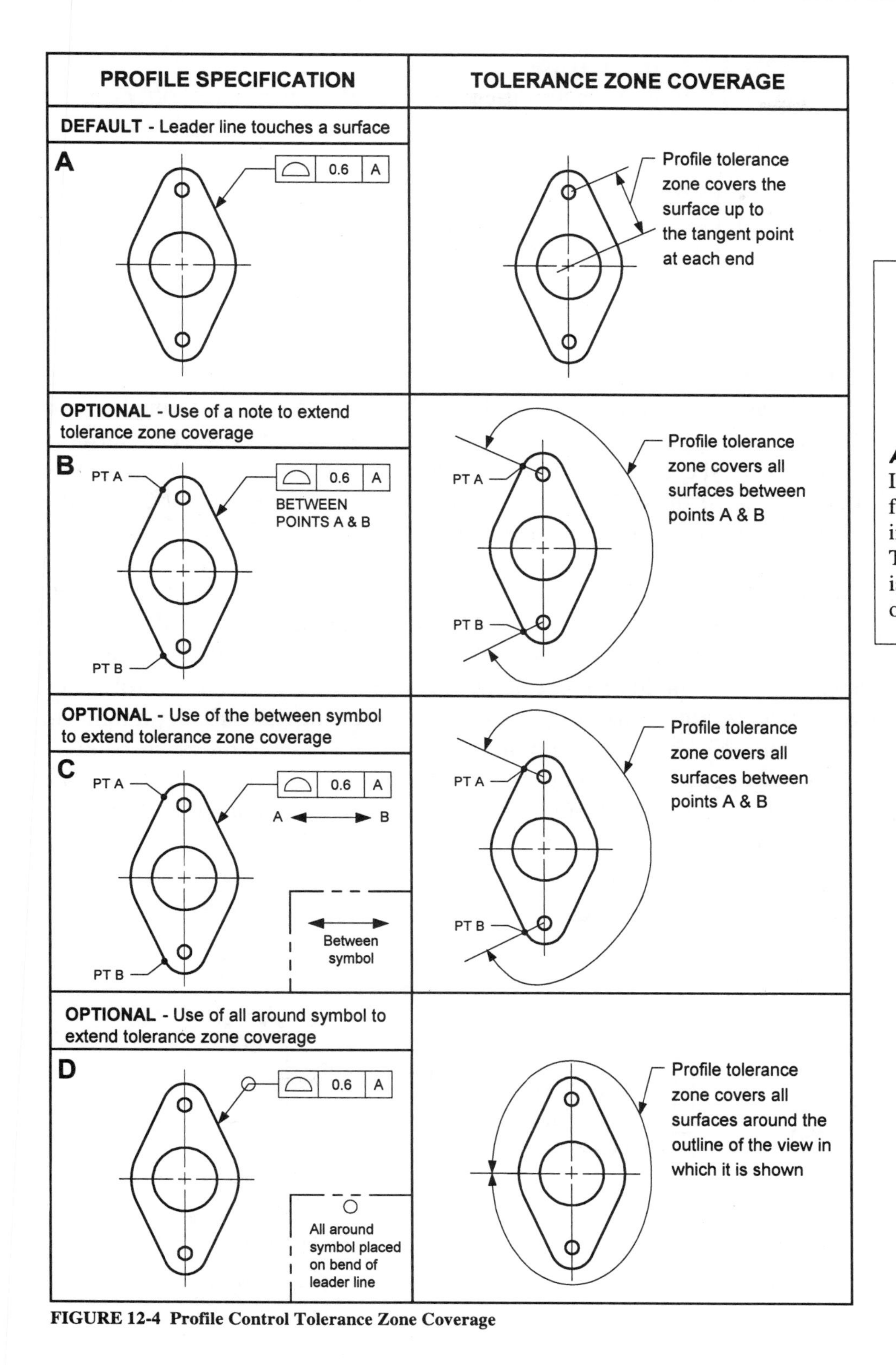

FIGURE 12-4 Profile Control Tolerance Zone Coverage

Author's Comment

In Y14.5, it is preferred to use symbols in place of notes. Therefore, option C is preferred over option B.

Advantages of Profile

In comparison to coordinate tolerancing, profile tolerancing offers many advantage. Three important advantages are:

1. It provides a clear definition of the tolerance zone.
2. It communicates the datums and datum sequence.
3. It eliminates accumulation of tolerances.

These advantages become obvious as profile tolerancing is compared to coordinate tolerancing. In Figure 12-5, a part is dimensioned with profile controls in panel *A* and with coordinate tolerancing in panel *B*. In panel *A*, the tolerance zone for the top surface of the part is clearly defined; it is a 0.4-wide uniform boundary centered about the true profile. The surfaces that contact the gage and the order in which they contact the gage are communicated through the feature control frames. The top surface of the part is affected by only one tolerance, the profile control.

Author's Comment
Profile tolerancing can be thought of as coordinate tolerancing with datums.

On the other hand, the coordinate tolerancing used in panel *B* doesn't communicate the design intent as well. The tolerance zone for the top surface of the part is not easily described. The radius has a tolerance, and the location of the radius has a tolerance. Together they produce a tolerance zone that is not uniform or logical. The drawing doesn't communicate which surface, and in which order, the part is to be held on for inspection. Also, the coordinate tolerancing uses toleranced dimensions to locate the radius. The tolerances from each dimension can accumulate and produce an unwanted tolerance accumulation

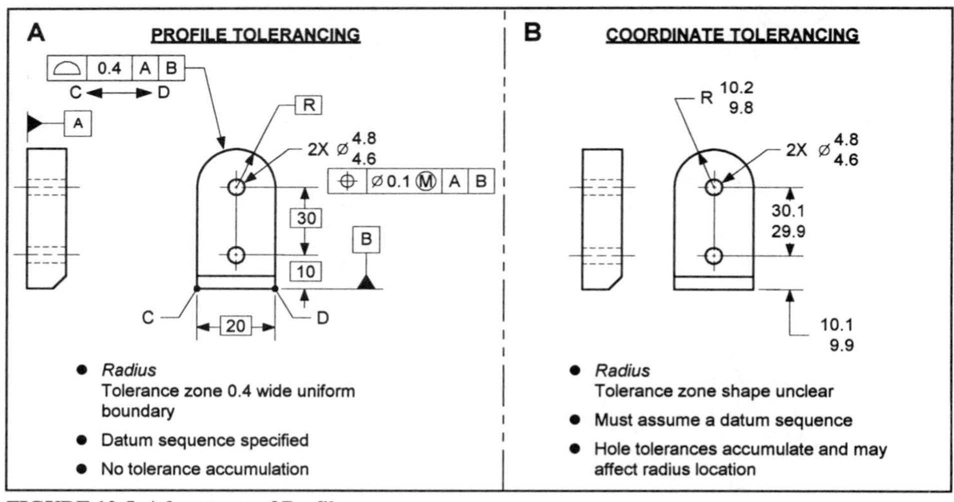

FIGURE 12-5 Advantages of Profile

Profile tolerancing communicates more clearly than coordinate tolerancing. In many organizations, profile tolerancing has been used to simply replace coordinate tolerancing.

PROFILE OF A SURFACE

A ***profile of a surface control*** is a geometric tolerance that limits the amount of error a surface can have relative to its true profile. Common applications for profile of a surface controls include controlling—either independently or in combination—the size, location, orientation and form of:

- Planar, curved, or irregular surfaces
- Polygons
- Cylinders, surfaces of revolution, or cones
- Coplanar surfaces

Design Tip
Profile is often used in this type of application to relate a dimensional measurement to a datum reference frame.

Profile Used to Tolerance a Surface Location

Figure 12-6 shows an example of profile controls used to tolerance the location, orientation, and form of a planar surface. This is the most common application of profile. In this application, profile is applied to a planar surface, and the following conditions apply:

- The profile callout is applied to a true profile.
- The true profile is related to the datums referenced with basic dimensions.
- The tol. zone is a uniform boundary centered around the true profile.
- All elements of the surface must be within the tolerance zone simultaneously.
- The tol. zone limits the location, orientation, and form of the surface.

For more info. . .
Figure 12-13 shows a method for inspecting this part.

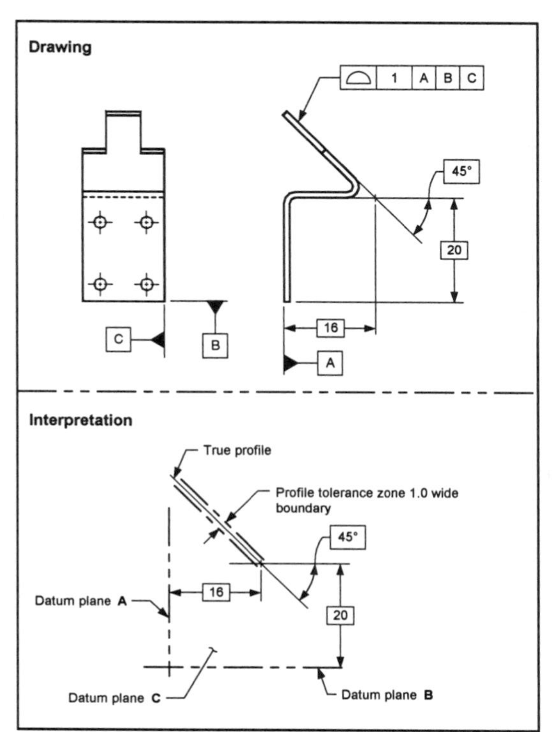

FIGURE 12-6 Profile Used to Tolerance a Surface Location

Profile Used to Tolerance a Polygon

Figure 12-7 shows an example of using profile controls to tolerance the size, location, orientation, and form of a polygon. In this application, the following conditions apply:

- The profile callout is applied to a true profile.
- The true profile is related to the datums referenced with basic dimensions.
- The all around symbol extends the profile tolerance zone to apply to all sides of the polygon.
- The tolerance zone is a uniform boundary centered around the true profile.
- All elements of the surfaces must be within the tolerance zone.
- The tolerance zone limits the size, location, orientation, and form of the polygon.
- A max. radius is used to limit the allowable radius on the corners of the polygon.

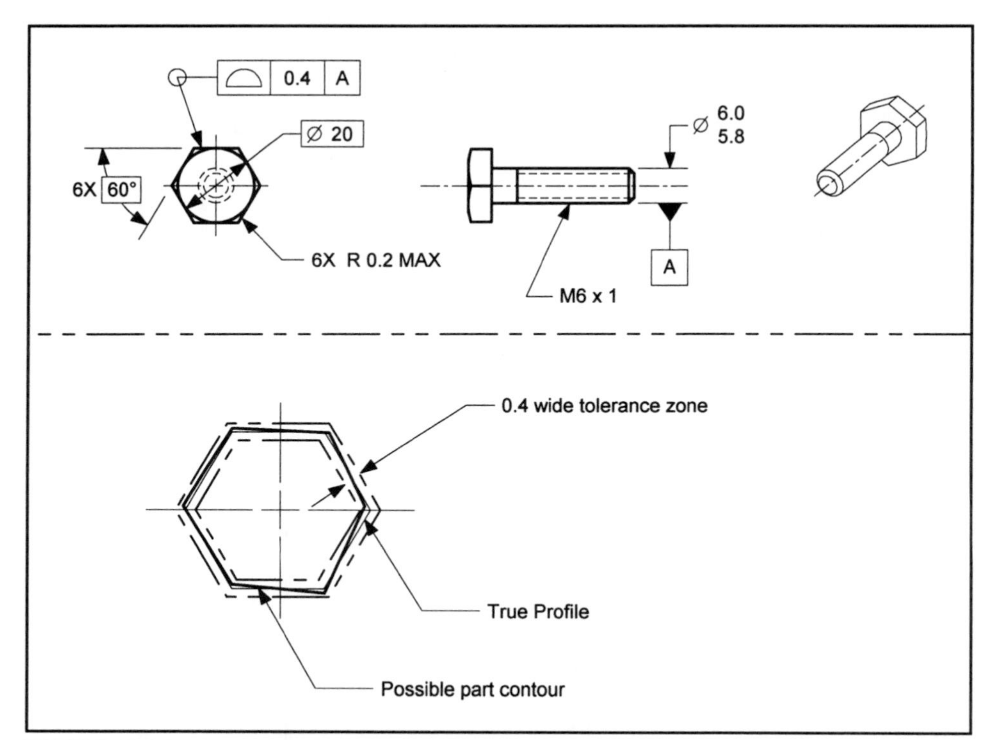

FIGURE 12-7 Profile Used to Tolerance a Polygon

Profile Used to Tolerance a Conical Feature

Profile of a surface can be applied to conical features. It may be applied as a form control (without datum references) or as a form and orientation control (with datum references). Figure 12-8 shows an example of profile applied to a conical feature.

For more info. . .
See Paragraph 6.5.8 of Y14.5.

In Figure 12-8, the profile controls the form and orientation of a conical feature. In this application, the following conditions apply:

- The profile callout is applied to a true profile. (The basic angle establishes the true profile.)
- The true profile is related to the datums referenced with basic dimensions. (An implied basic angle equal to half the specified angle exists.)
- The profile tolerance zone applies all around the cone. (When profile is applied to a surface of revolution, it automatically applies all around.)
- The tolerance zone is a uniform boundary centered around the true profile. (The equal bilateral default is in effect.)
- All elements of the surface must be within the profile tolerance zone. (The profile tolerance zone floats within the size tolerance zone.)
- The profile tolerance zone limits the orientation and form of the conical surface. (The size of the cone is limited by the toleranced dimension.)

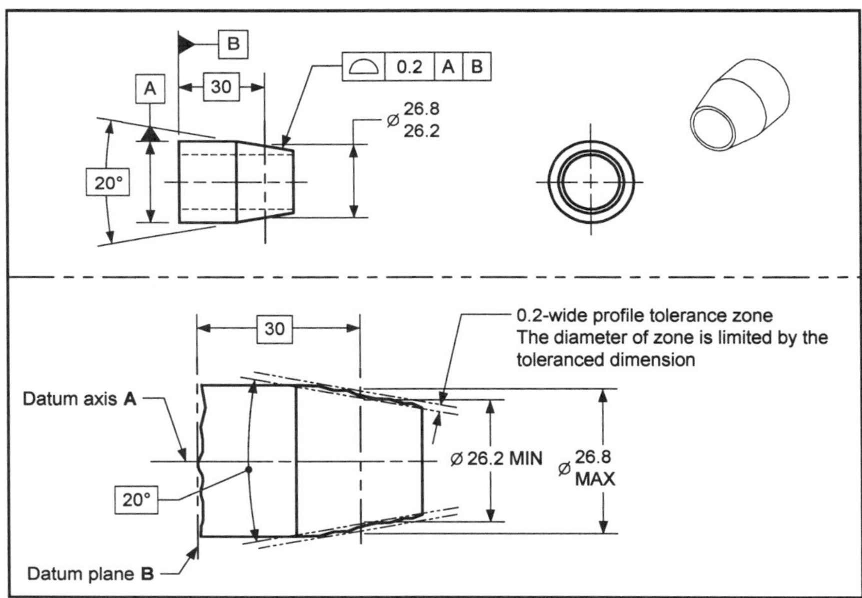

FIGURE 12-8 Profile Used to Tolerance a Conical Feature

In this application, a toleranced dimension is used to define the size of the cone. If a basic dimension is used to define the size of the cone, the profile tolerance zone would also limit the size of the cone. See Figure 12-14.

For more info. . .
See Paragraph 6.5.6.1 of Y14.5.

Profile Used to Tolerance Coplanar Surfaces

A profile control may be used when it is intended to treat two or more coplanar surfaces as a single surface. In this type of application, the profile control is a form control and does not use datum references; it simulates a flatness control. When profile is used as a form control, the tolerance zone is unilateral (away from the implied self-datum).

When profile of a surface (as a form control) is applied to coplanar surfaces, it controls the form of the surfaces as if they were a single surface. Figure 12-9 shows an example of profile applied to coplanar surfaces. In this application, the following conditions apply:

- The profile callout is applied to a true profile. (An implied basic zero between the surfaces establishes the true profile.)
- The number of surfaces being controlled is designated next to the profile callout.
- The tolerance zone is a unilateral boundary extending away from the implied datum. (The unilateral tolerance zone is automatic with an implied self-datum.)
- All elements of the surfaces to which the tolerance applies must be within the profile tolerance zone.
- The profile tolerance zone limits the form and coplanarity of the surfaces.

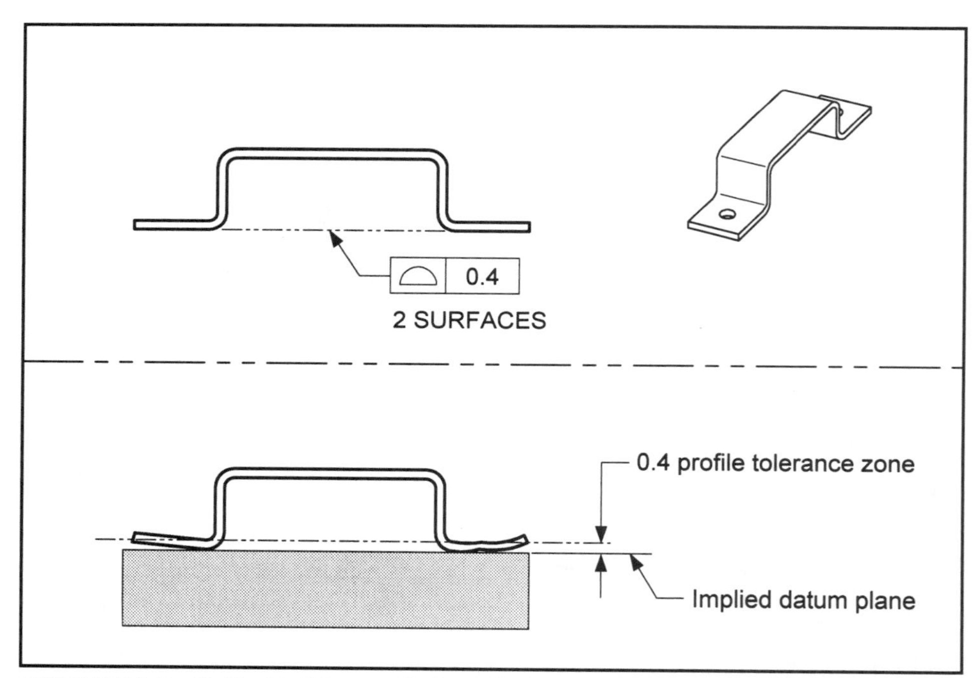

FIGURE 12-9 Profile Used to Tolerance Coplanar Surfaces

In an application where there are several coplanar surfaces, it may be desirable to use two surfaces to create a single datum plane and tolerance the remaining surfaces in reference to that datum plane. Figure 12-10 shows an example of an application of multiple coplanar surfaces toleranced with profile.

- The profile callout is applied to a true profile. (An implied basic zero between the surfaces establishes the true profile.)
- The number of surfaces being controlled is designated next to each profile callout.
- The tolerance zone for the profile control of the outer two surfaces is a unilateral boundary extending away from the implied datum. (The unilateral tolerance zone is automatic with a self implied datum.)
- The tolerance zone for the profile control of the inner two surfaces is a bilateral boundary equally centered about the true profile.
- All elements of the surfaces (which each profile control applies to) must be within the profile tolerance zone.
- The profile control on the outer surfaces limits the form and coplanarity of the surfaces.
- The profile control on the inner surfaces limits the location, orientation, and form of the surfaces.

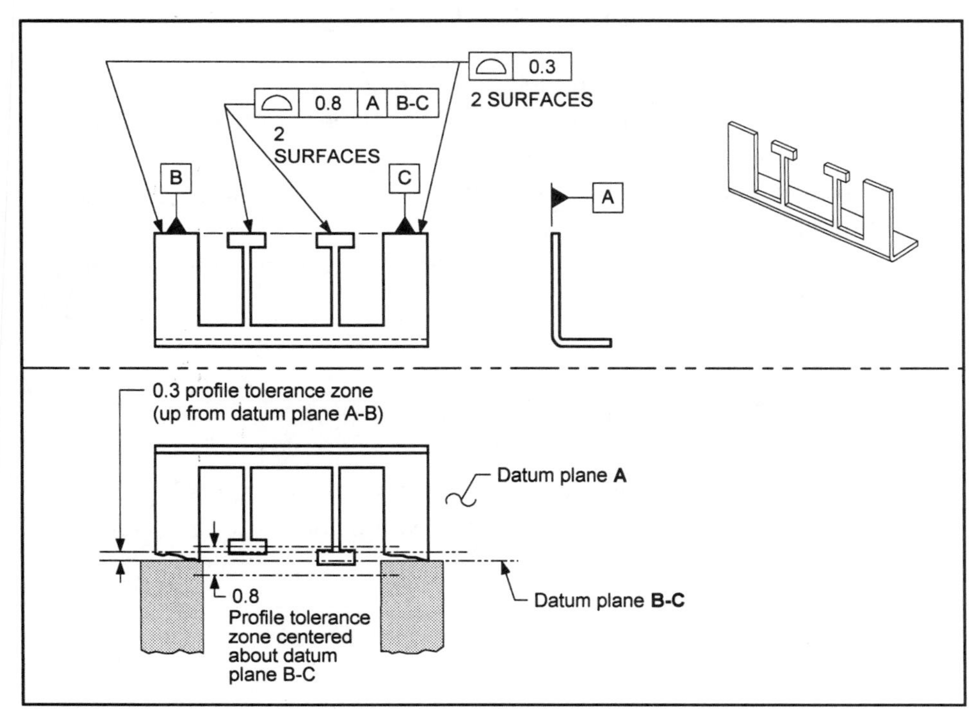

FIGURE 12-10 Profile Used to Tolerance Multiple Coplanar Surfaces

Profile Used in Multiple Single-Segment Controls

When profile is applied to a surface, it can control the size, location, orientation, and form of the surface. This is acceptable in many applications; however, often the functional requirements allow the size, location, orientation, and form tolerances to be at different values. This can be achieved by using multiple profile controls on a surface.

A ***multiple single-segment profile control*** is where two or more profile controls are tolerancing a surface relative to different datums. When using multiple single-segment profile controls, different levels of control can be achieved by using different tolerance values and adding or removing datum references. For example, if a profile control is applied to a surface, and it contains no datum references, it will only control the form of the surface. However, if a profile control that is applied to a surface contains three datum references, it will control the location, orientation and form of the surface.

For more info. . .
See Paragraph 6.5.9.1.2 of Y14.5.

Multiple single-segment profile controls can be used to specify a different amount of tolerance for different parameters of a surface. An example is shown in Figure 12-11.

In this application, the following conditions apply:

- The profile controls are applied to a true profile.
- The profile control with three datum references controls the location of the feature relative to datums *A, B,* and *C.*
- The profile control with one datum reference controls the orientation of the feature relative to datum *A.*
- The profile control with no datum references controls the size and form of the feature.

Author's Comment
The concept of using multiple profile controls on a part is only introduced in this text. Y14.5 contains a method called "Composite Profile Tolerancing" for dimensioning part surfaces with complex functional requirements.

TECHNOTE 12-4 Multiple Profile Callouts

In a multiple single-segment profile control application, each profile control should have a different tolerance value and more or fewer datum references.

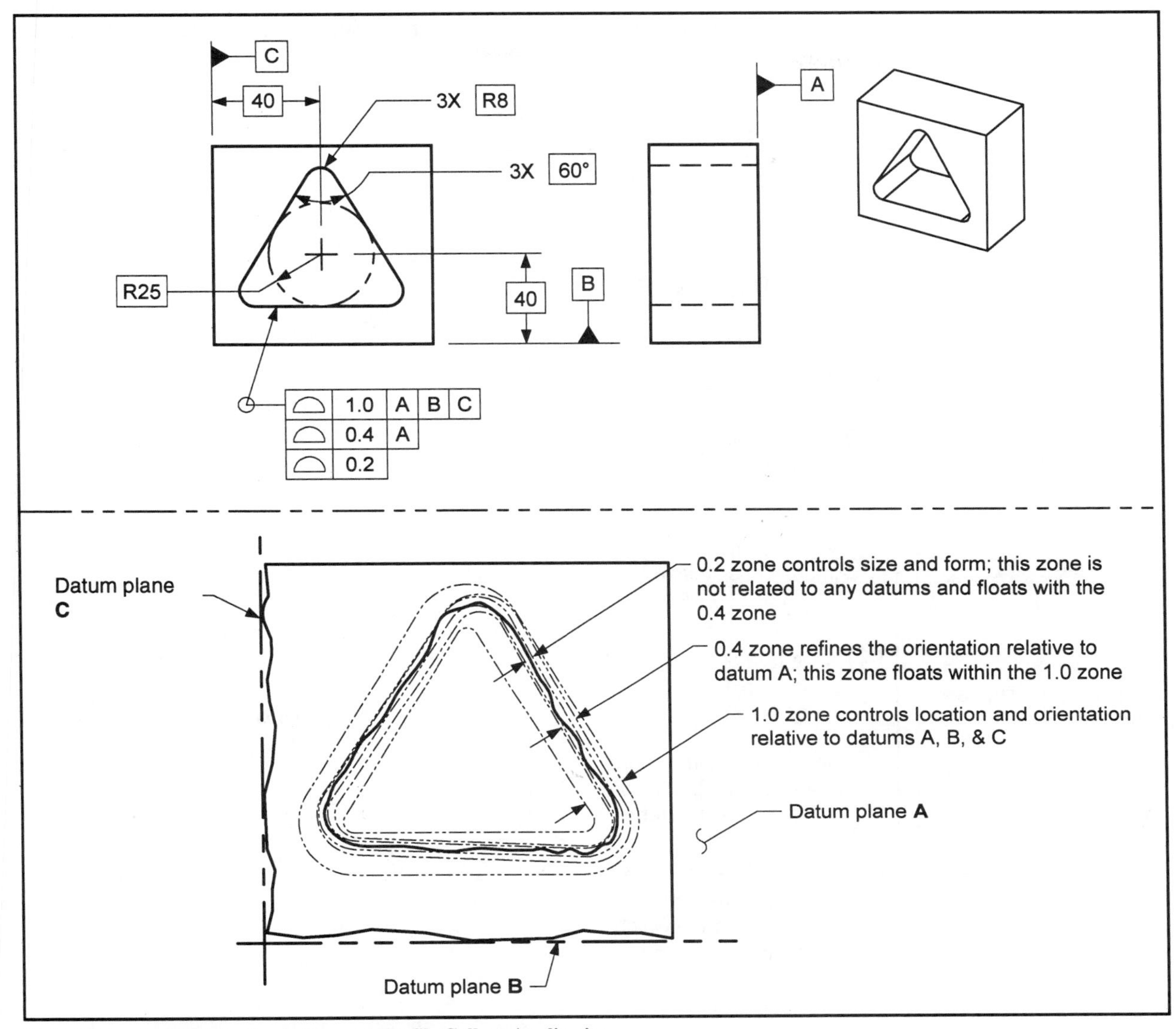

FIGURE 12-11 Multiple Single-Segment Profile Callout Application

Legal Specification Test for Profile of a Surface

For a profile of a surface control to be a legal specification, it must satisfy the following conditions:

- The true profile must be defined with basic dimensions or be a planar surface (or coplanar surfaces).
- No modifiers may be displayed in the tolerance portion of the feature control frame.
- If the true profile is located with toleranced dimensions, the profile tolerance must be a refinement of the tolerance value of the dimensions.

Figure 12-12 shows a legal specification flowchart for a profile of a surface control.

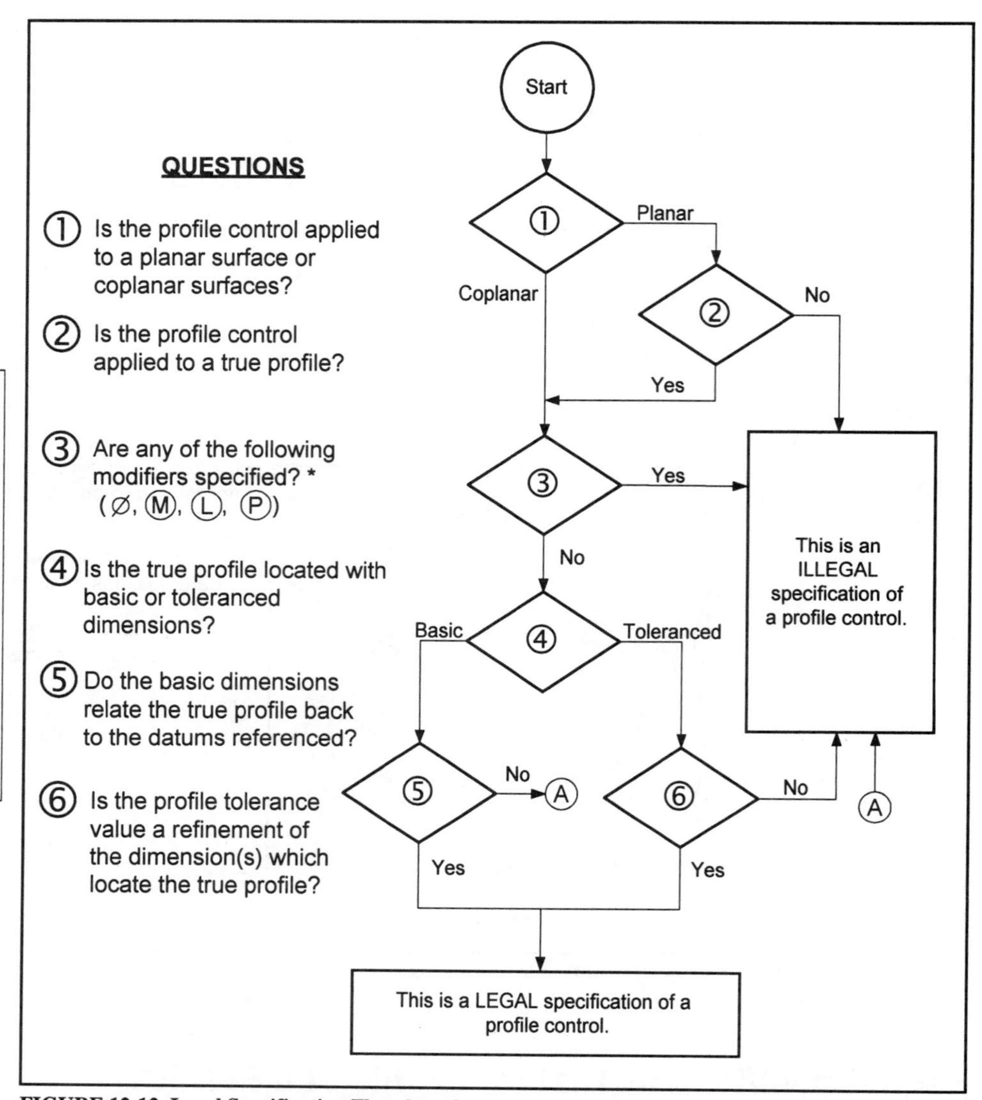

FIGURE 12-12 Legal Specification Flowchart for Profile of a Surface

Author's Comment

This legal specification flowchart may be used for profile of a surface and profile of a line.

* The MMC or LMC modifiers may be specified in the datum references of a profile control.

Inspecting Profile of a Surface

In Figure 12-6, a profile of a surface control is applied to a surface. There are many ways this surface could be inspected. One way is to use a special gage, as shown in Figure 12-13. First, the part is located in the datum reference frame. Then, a dial indicator is used to measure the distance from the toleranced surface to the true profile. Depending on the dial indicator reading of the part surface, the part surface will be determined to be in or out of the profile tolerance zone. The number of points to be checked is determined by the inspection plan.

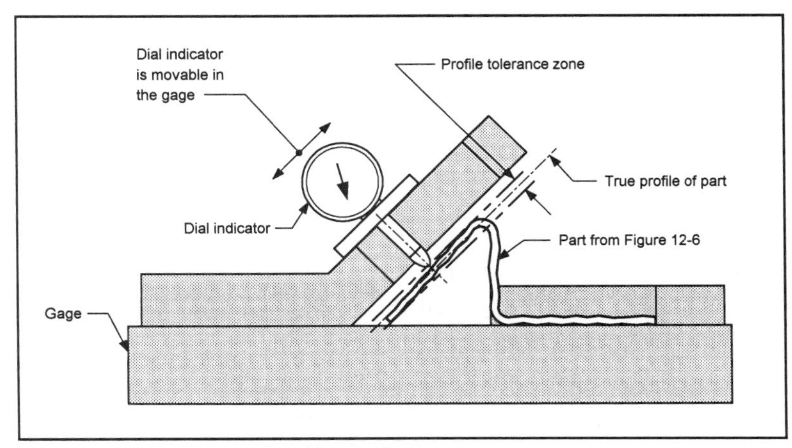

FIGURE 12-13 Inspecting Profile of a Surface

TECHNOTE 12-5 Inspecting Profile of a Surface

When inspecting profile of a surface, the inspector measures points along the surface to determine if the entire surface is in the tolerance zone.

PROFILE OF A LINE

The basic concepts of profile of a surface and profile of a line are similar. The primary difference is that the tolerance zone for profile of a surface is three-dimensional, and the tolerance zone for profile of a line is two-dimensional. A ***profile of a line control*** is a geometric tolerance that limits the amount of error for line elements relative to their true profile. The tolerance zone is established in the same manner as surface profile. The tolerance zone is two-dimensional; it is two uniform lines applied at any cross section of the surface. Profile of a line provides a control in one direction only. Therefore, profile of a line is often used as part of a multiple single-segment control for a surface. It is common to use both a profile of a surface and a profile of a line control on the same surface. Another common method of applying profile of a line is as a refinement of a coordinate tolerance. Examples of these applications are shown later in this section.

Profile of a line is often used in a multiple single-segment control as an orientation and/or form control. If datum references are specified, the line elements are oriented relative to the datums specified. If no datum references are specified, the line elements are being controlled for form only.

Profile of a Line Used to Control Form and Orientation in a Multiple Single-Segment Profile Control

Often, profile of a line is used to control the form and orientation of the line elements of a surface. Figure 12-14 shows an example where profile of a surface controls the location as well as the orientation of the surface (in one direction). Profile of a line controls the orientation of the line elements (in one direction) and the form of the line elements. In Figure 12-14, the following conditions apply:

- The profile callouts are applied to a true profile.
- The profile of a surface control limits the size, location, and orientation relative to the datums specified.
- The profile of a line control refines the form of the line elements.
- The profile of a line control refines the orientation of the line elements in one direction. (Note that the profile of a line tolerance value is less than the profile of a surface tolerance value.)

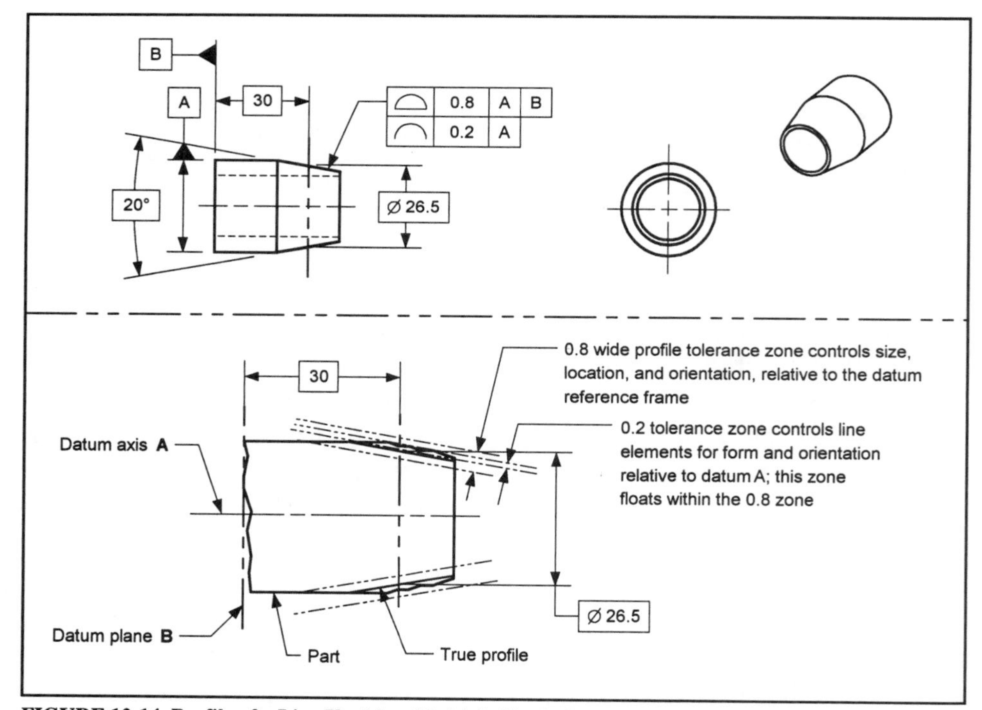

FIGURE 12-14 Profile of a Line Used in a Multiple Single-Segment Profile Control

Profile of a Line and a Coordinate Tolerance Used to Control Form and Location

In this example, profile of a line is used with a coordinate tolerance. The coordinate tolerance locates the surface, and the profile callout refines the form. The profile of a line control specifies two datum references; therefore, the profile of a line control affects the form and orientation of the line elements. Figure 12-15 shows an example of profile of a line and a coordinate tolerance used to control location, orientation, and form. In Figure 12-15, the following conditions apply:

- The profile callout is applied to a true profile.
- The coordinate tolerance locates the surface.
- The profile of a line control refines the form and orientation of the line elements in one direction. (Note that the profile of a line tolerance value is less than the coordinate tolerance value.)

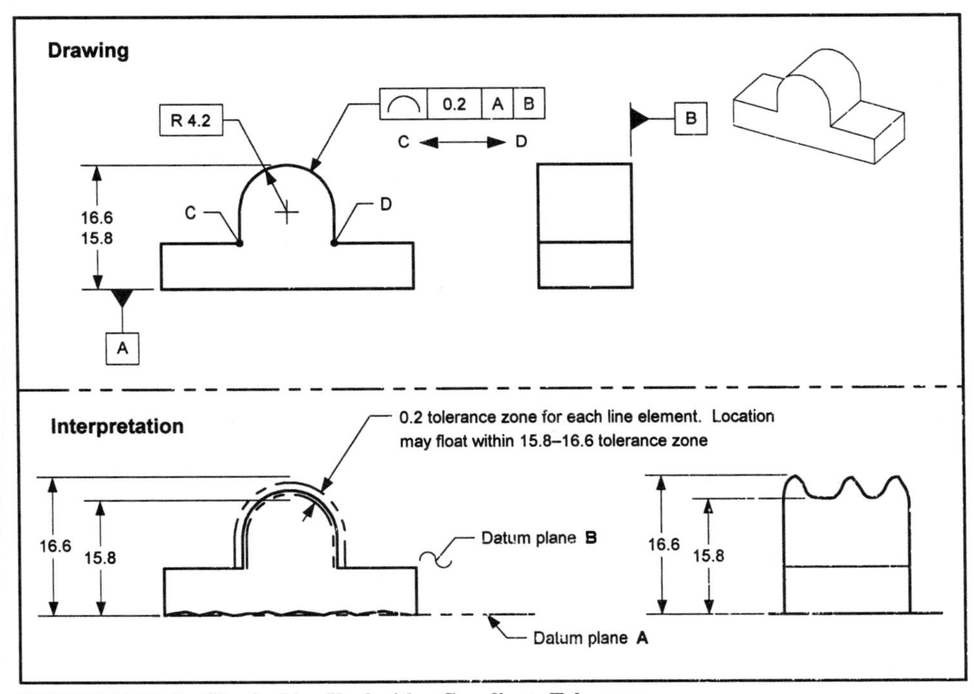

FIGURE 12-15 Profile of a Line Used with a Coordinate Tolerance

Legal Specification Test for Profile of a Line

For a profile of a line control to be a legal specification, it must satisfy the same conditions as profile of a surface. The legal specification flowchart for profile in Figure 12-12 applies to both profile of a surface and profile of a line.

Inspecting Profile of a Line

In Figure 12-15, a profile of a line control is used as part of a multiple control. The coordinate toleranced dimension would have to be inspected to ensure the location of the true profile of the surface. This would involve inspecting the height of the part. Now let's look at how to inspect the profile of a line control.

There are a number of ways the profile of a line control could be inspected. One way is to use a gage template as shown in Figure 12-16. First, the template is placed on the surface line element. Then, a gage pin with a diameter equal to the profile tolerance value is tried to check if it can be inserted into the gap between the part surface and the gage template. If the gage pin fits into the space between the gage template and part surface, the part surface is out of the profile tolerance zone.

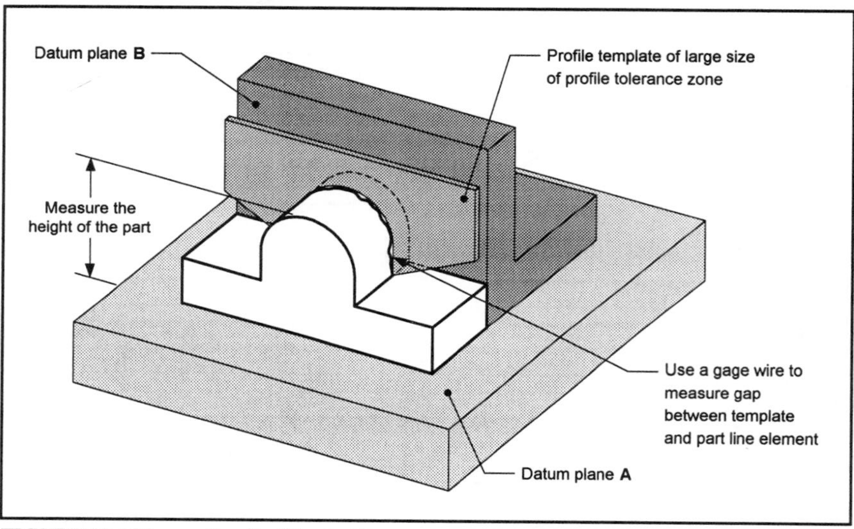

FIGURE 12-16 Inspecting Profile of a Line

TECHNOTE 12-6 Inspecting Profile of a Line

When inspecting profile of a line, the inspector measures points along the line element to determine if the entire line element is in the tolerance zone. Several line elements along the surface would be inspected. The number of points and line elements is frequently defined in the inspection plan.

PART CALCULATIONS

In industry, it is very common to make a calculation to find a max. or min. distance on a part. When calculating a max. or min. part distance, all of the part tolerances must be included. This section explains how to use profile tolerances in part calculations.

Tolerance Stacks Using Profile

Tolerance stacks on a part that uses profile (bilateral-equal distribution) is straightforward. In this section, a simple method of how to calculate the max. and min. distance on a part is explained. This calculation involves using the profile tolerance of the top surface and the basic dimensions that locate and define the radius. Figure 12-17 shows an example of a part calculation to find the max. and min. height of a part. When reading the steps below, refer to Figure 12-17.

For more info. . .
This text provides a brief introduction to tolerance stacks. For more information on this topic, see reference number six listed in the bibliography.

1. Label the start and end points of the distance to be calculated. On the start point of the calculation, draw a double-ended arrow. Label the arrow that points towards the end point of the calculation as positive (+). Label the other arrow as negative (-). Each time a distance that is in the direction of the positive arrow is used in the calculation, the distance will be a positive value. When a distance is used in the negative direction, it will be a negative value.
2. Establish a loop of part dimensions or gage distances from the start point to the end point of the calculation.
3. Calculate the answer.

When solving for a min. distance, half the profile tolerance value (for a bilateral-equal distribution control is subtracted from the calculation. When solving for the max. distance, half the profile tolerance value is added to the calculation.

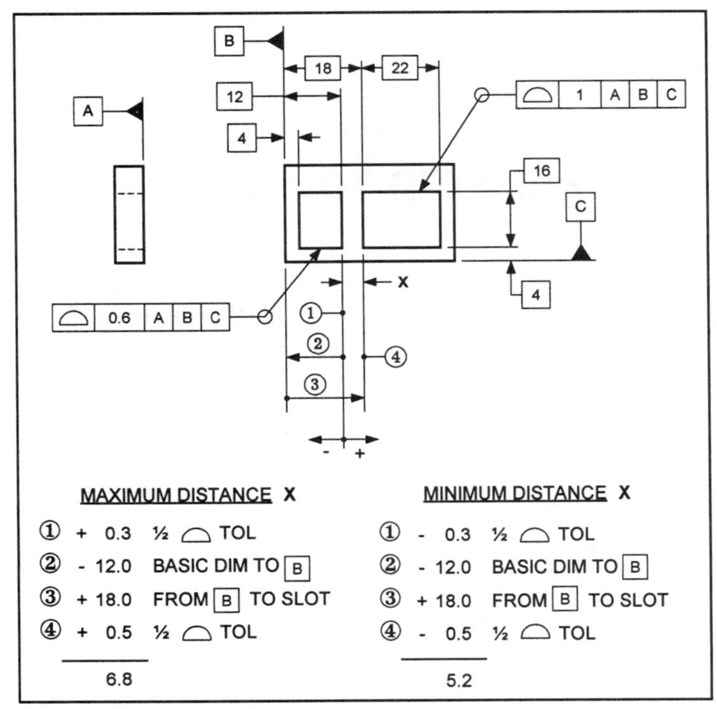

FIGURE 12-17 Part Calculations Involving Profile

Summary

A summarization of profile control information is shown in Figure 12-18.

Symbol	Datum reference required	Can be applied to a Surface	Can be applied to a FOS	Can use Ⓜ or Ⓛ modifier	Can be applied at RFS	Override Rule #1	Can use bonus tolerance concepts	Can use datum shift concepts
⌓	Yes*	Yes	No	No•	Yes**	No■	No	Yes
⌒	Yes*	Yes	No	No•	Yes**	No■	No	Yes

* Can be used with or without a datum reference
** Is automatic per Rule #2
■ Must be applied to a true profile therefore Rule #1doesn't apply
• These modifiers may be used in the datum portion of the feature control frame

FIGURE 12-18 Summarization of Profile Controls

VOCABULARY LIST

Study Tip
Read each term. If you don't recall the meaning of a term, look it up in the chapter.

New Terms Introduced in this Chapter

All around symbol
Between symbol
Multiple single-segment profile control
Profile
Profile control
Profile of a line control
Profile of a surface control
True profile

ADDITIONAL RELATED TOPICS

Author's Comment
These topics, plus advanced coverage of many of the topics introduced in this text, will be covered in my new book on advanced GD&T concepts.

Topic	ASME Y14.5M-1994 Reference
• Profile of a surface "all over"	Paragraph 6.5.2
• Boundary control for non-cylindrical feature	Paragraph 6.5.5.1
• Composite profile tolerancing	Paragraph 6.5.9.1

QUESTIONS AND PROBLEMS

1. When a profile control doesn't contain any datum reference, it is considered a ____________ control.

2. When a profile control contains datum references, it is considered a ____________________ control.

3. Describe the term "true profile." ____
____
____

4. List the four part characteristics that profile affects.
   1. ____
   2. ____
   3. ____
   4. ____

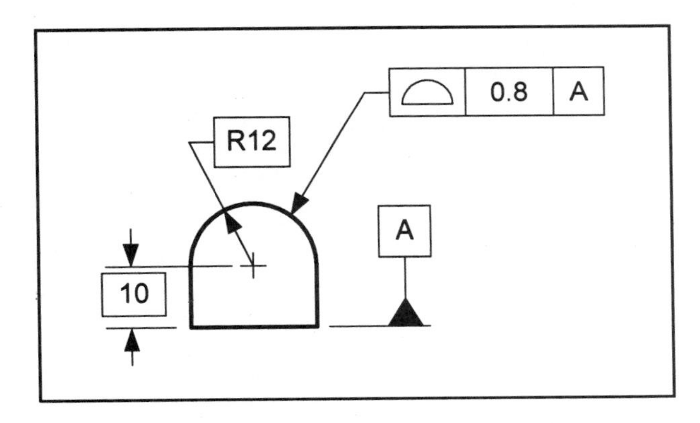

5. For the drawing above, describe the shape, size, and location of the tolerance zone for the top surface of the part.
____
____
____

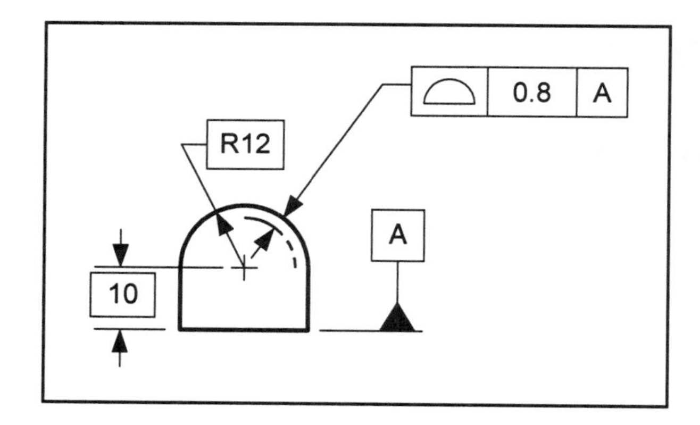

6. For the drawing above, describe the shape, size, and location of the tolerance zone for the surface of the part.

____

____

____

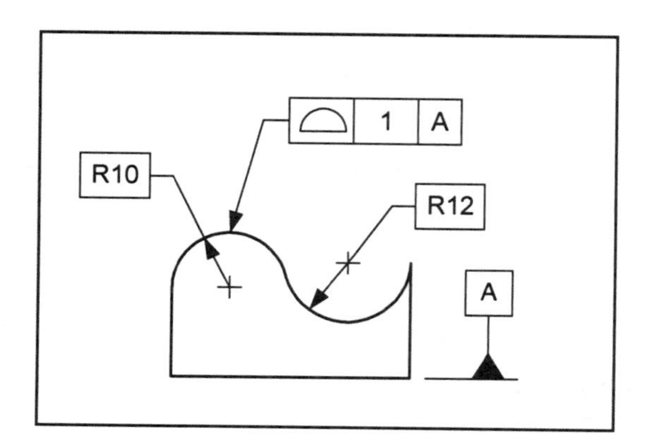

7. On the drawing above, indicate the area of the part to which the profile control applies.

8. List three ways to extend the coverage of a profile tolerance zone.
   1. ____
   2. ____
   3. ____

9. Three advantage of using profile controls are:
   1. ____
   2. ____
   3. ____

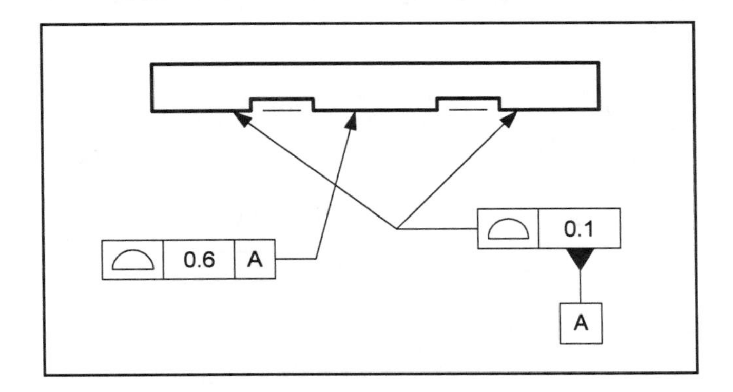

10. Draw datum plane *A* in the space below. Add the profile tolerance zones for the part shown in the figure above.

11. Use the drawing to fill in the chart.

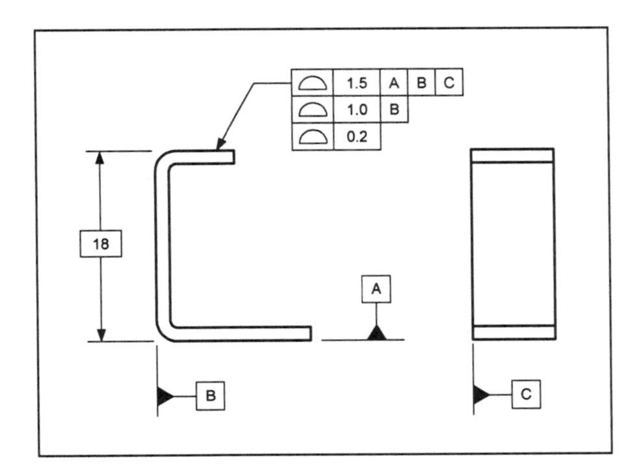

This profile callout	Controls the (size, location, orientation, form)	Within	Relative to
⌓ 1.5 A B C			
⌓ 1.0 B			
⌓ 0.2			

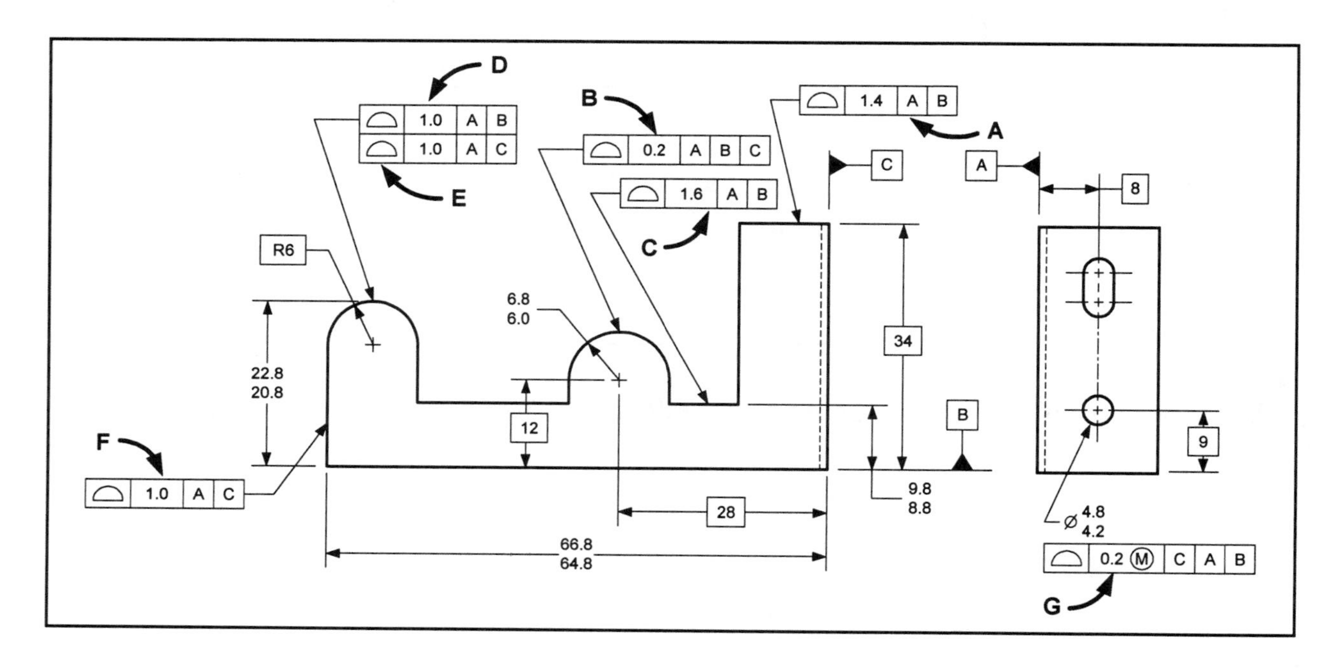

12. Using the drawing above, indicate if each of the profile specifications shown below is legal. If a specification is illegal, explain why.

A. | ⌓ | 1.4 | A | B | ____________________

B. | ⌓ | 0.2 | A | B | C | ____________________

C. | ⌓ | 1.6 | A | B | ____________________

D. | ⌓ | 1.0 | A | B | ____________________

E. | ⌓ | 1.0 | A | C | ____________________

F. | ⌓ | 1.0 | A | C | ____________________

G. | ⌓ | 0.2 Ⓜ | C | A | B | ____________________

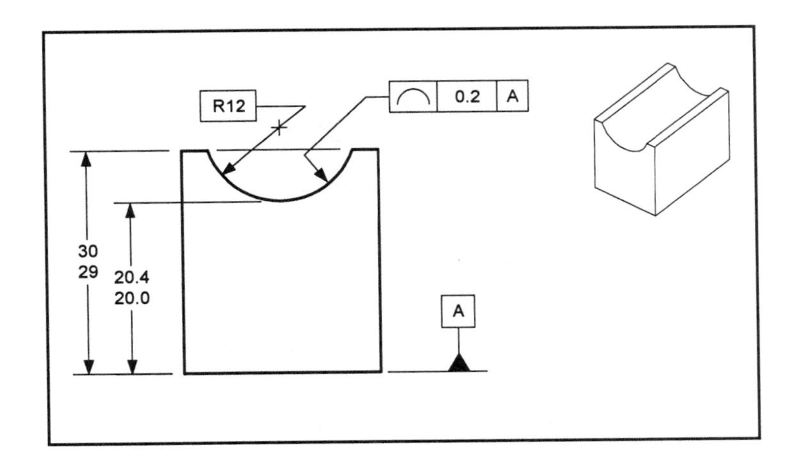

13. For the drawing above, describe the shape, size, and location of the tolerance zone for the profile callout.

____

____

____

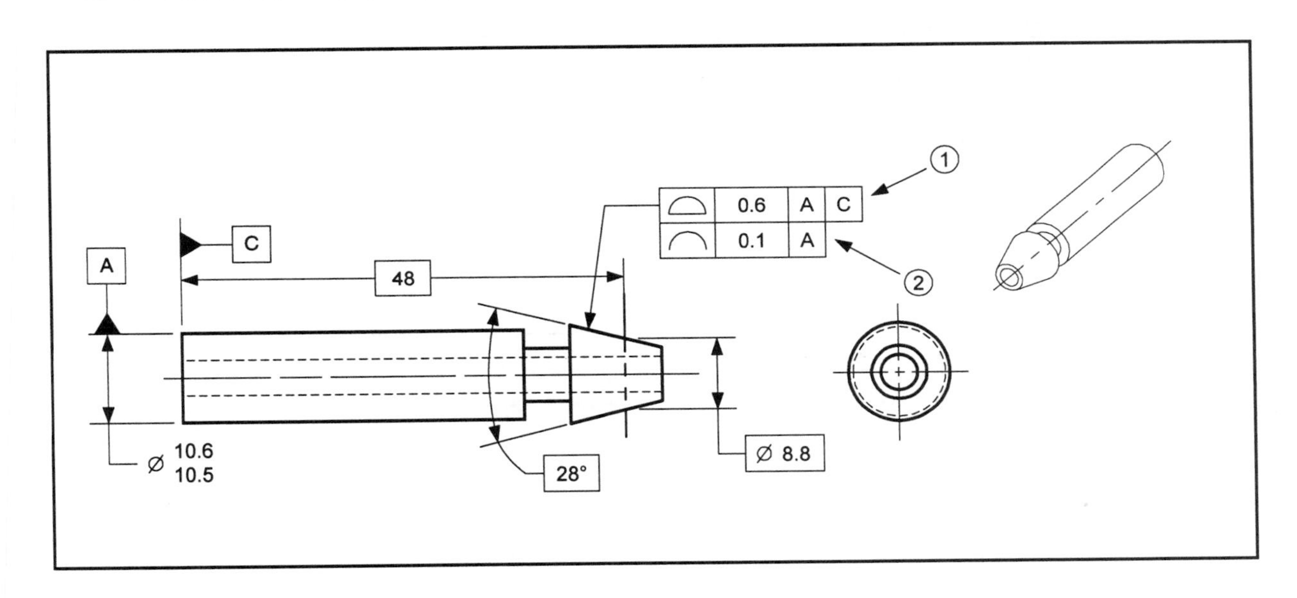

14. For the drawing above, describe the shape, size, and location of the tolerance zone for. . .

Callout ① ____

____

____

Callout ② ____

____

____

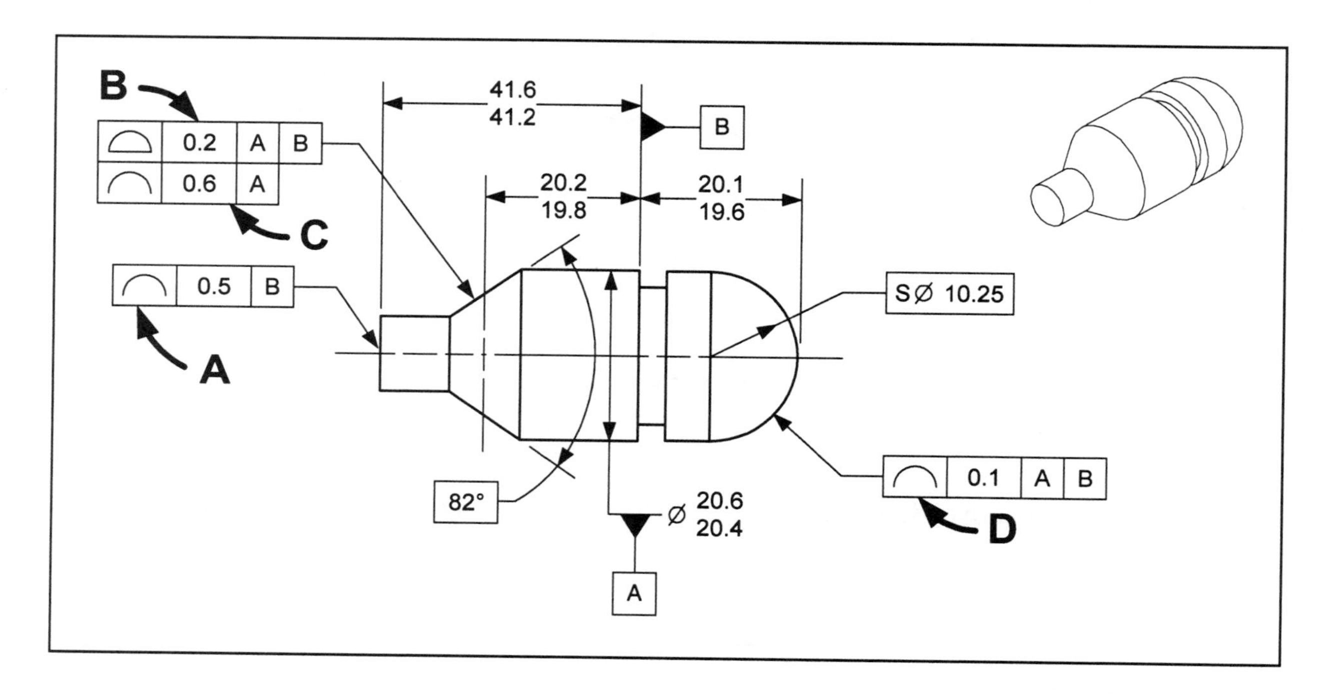

15. Using the drawing above, indicate if each of the profile specifications shown below is legal. If a specification is illegal, explain why.

A. | ⌒ | 0.5 | B | ________________________________

B. | ⌓ | 0.2 | A | B | ________________________________

C. | ⌒ | 0.6 | A | ________________________________

D. | ⌒ | 0.1 | A | B | ________________________________

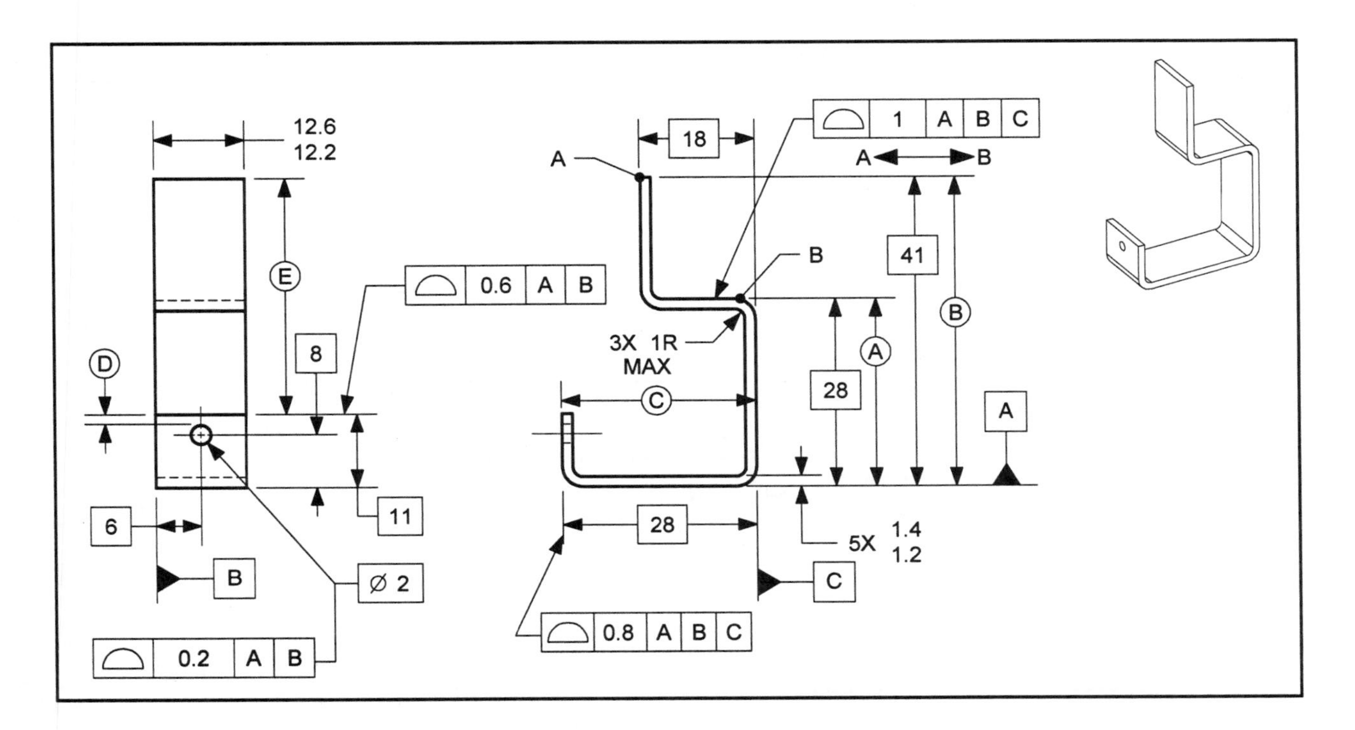

16. Using the drawing above, calculate the distances in the chart below.

Distance	Max	Min
A		
B		
C		
D		
E		

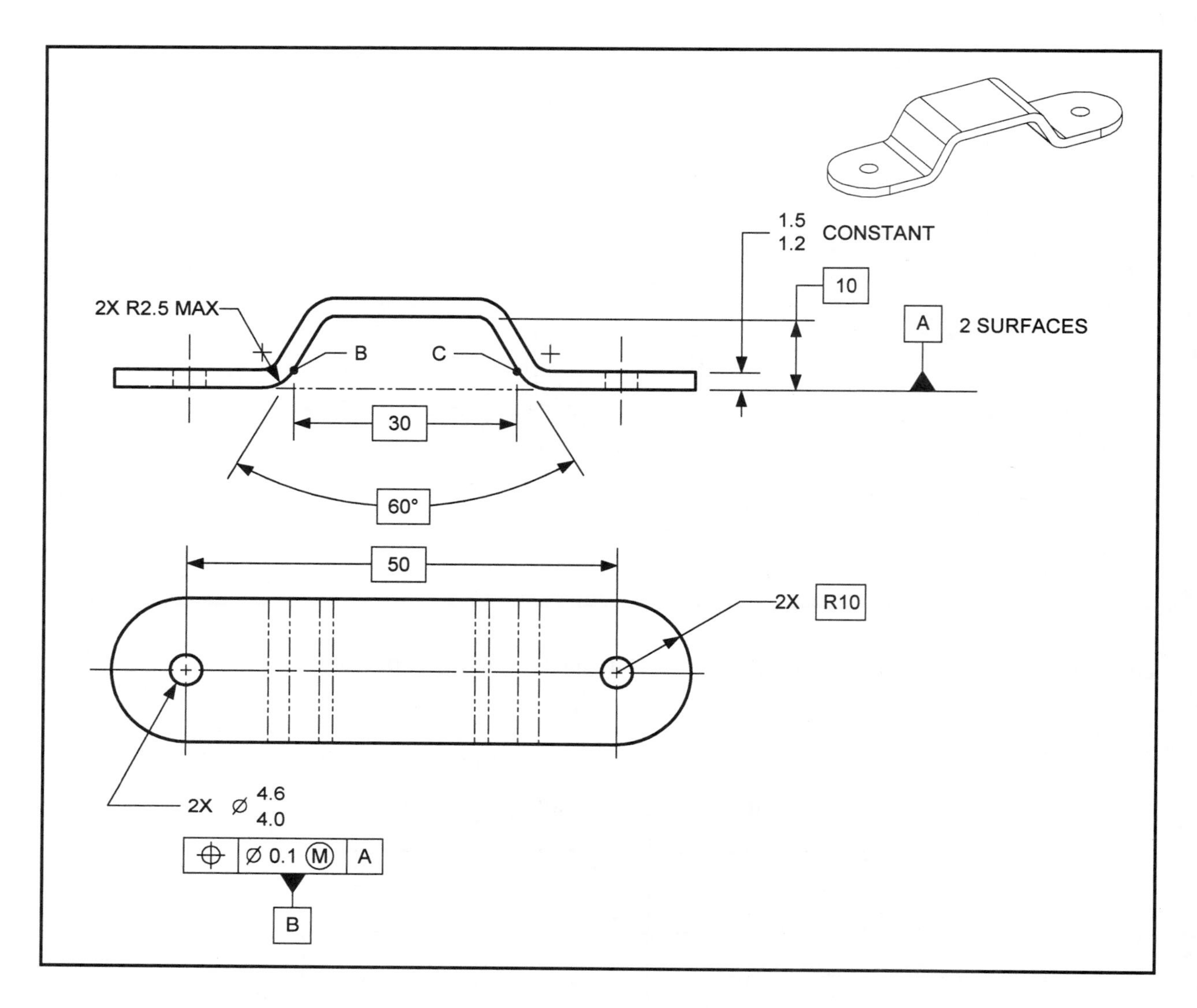

17. Use the instructions below to complete the drawing above.

   a. Add a profile of a surface control to control the coplanarity of datum feature *A*. The tolerance value should be 0.05.

   b. Add a profile of a surface control in the front view to control the profile of the part between points *B* and *C* relative to datum *A* primary and datum *B* at MMC secondary. It should be a bilateral control with a tolerance value of 0.2.

   c. Add a profile of a surface control in the bottom view to control the profile of the entire outline of the part relative to datum *A*. It should be a bilateral control with a tolerance value of 0.6.

Appendices

APPENDIX A

Answers to Selected Questions

CHAPTER ONE: QUESTIONS AND PROBLEMS

1. Limit
3. Plus-minus
5. Equal
7. Unequal
9. A zero precedes the decimal point.
11. ASME American Society of Mechanical Engineers
    Y14.5 The number of the standard
    M Metric
    1994 The year the standard was approved
13. a. Square tolerance zones
    b. Fixed-size tolerance zones
    c. Ambiguous instructions for inspection

CHAPTER TWO: QUESTIONS AND PROBLEMS

1.

Letter	Feature of size dimension	Non-feature of size dimension
A		X
B	X	
C		X
D		X
E		X
F		X
G		X
H	X	
I		X
J		X
K	X	

3. Actual local size is the value of any **individual** distance at any **cross section** of a feature.
5. There are two types of features of size: **internal** and **external**.
7. The **largest** perfect feature counterpart that can be inscribed about the feature
9. When a radius is specified, flats or reversals are **allowed**.
11. A **planar** FOS is a FOS that contains two parallel plane surfaces.
13. The five types of geometric characteristic symbols are Form, **orientation, profile, runout, and location**.

CHAPTER THREE: QUESTIONS AND PROBLEMS

1. Perfect form at MMC
3.

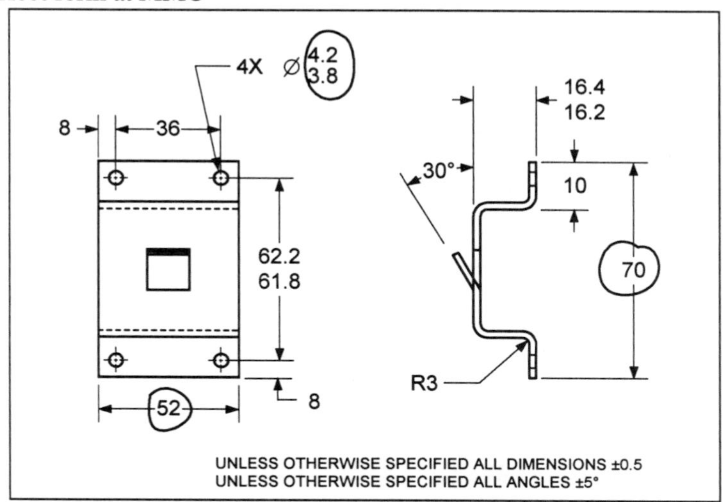

5. 9.8
7.

If dimension A was	The allowable form error on surface B is
12.8	0
12.7	0.1
12.6	0.2
12.5	0.3
12.4	0.4
12.3	0.5
12.2	0.6

9. RFS applies with respect to the individual tolerance, datum reference, or both, where no modifying symbol is specified.
11. a. To define theoretically exact part features
    b. To define datum targets
13. Bonus tolerance: an additional tolerance for a geometric control
15. Outer boundary: a worst-case boundary generated by the largest feature plus the stated geometric tolerance (and any additional tolerance, if applicable)
17.

Use N/A for not applicable				If a FOS dimension is identified,			If a feature control frame is identified,		
Letter	Letter identifies a...			Rule #1 Applies		VC, OB, or IB is...	It applies to a...		The amount of bonus tolerance permissible is...
	FOS Dimension	Non-FOS Dimension	Feature Control Frame	YES	NO		Feature	FOS	
A			✓				✓		0
B	✓			✓		63			
C			✓				✓		0
D							✓		0
E	✓			✓		4.0			
F			✓					✓	0.4
G		✓							
H			✓				✓		0
I	✓				✓	37			
J			✓					✓	0.6
K	✓			✓		29.1			
L			✓					✓	1.0

CHAPTER FOUR: QUESTIONS AND PROBLEMS

1. Flatness is the condition where a surface has all of its elements in one plane.
3. The high points of the toleranced surface locate the first plane of a flatness tolerance zone.
5. 0.4
7. Surface A= 0.1 Surface B= 0.4
9. No; it must be a refinement of the size tolerance.
11.

If the part was . . .	The flatness error of surface *B* would be limited to . . .	The flatness error of surface *A* would be limited to . . .
At MMC	0	0
At LMC	0.4	0.1
At 22.0	0.2	0.1

13. By contacting the toleranced surface against a surface plate and measuring the gap between the surface, the plate, and the part surface
15. Two parallel lines 0.05 apart
17. Rule #1 and the size dimension
19. a. Legal
    b. Illegal; MMC modifier not allowed on a surface
    c. Illegal; datum reference not allowed
    d. Illegal; ∅ modifier not allowed
21. Rule #1 and the size dimension
23. B; E; G; I
25. Circularity is a condition where all the points of a surface of revolution are equidistant from the axis.
27. Rule #1 and the size dimension
29. A; B; F
31.

Diameter	WCB	Max circularity error possible	Max straightness of axis error possible	Max straightness of line element error possible	Rule #1 applies (YES/NO)
A	9.8	0.8	0.8	0.8	YES
B	14.9	0.1	0.9	0.9	NO
C	10.4	0.04	0.4	0.04	YES
D	20.7	0.05	0.3	0.3	NO
E	5.8	0.2	0.5	0.5	NO
F	18.6	0.02	0.4	0.4	YES

33. Cylindricity is the condition of a surface of revolution in which all points of the surface are equidistant from a common axis.
35. Rule #1 and the size dimension
37. A; C; F
39. A. Illegal; no datum reference allowed
    B. Legal
    C. Illegal; cannot use the LMC modifier
    D. Illegal; cannot use the diameter modifier

41.

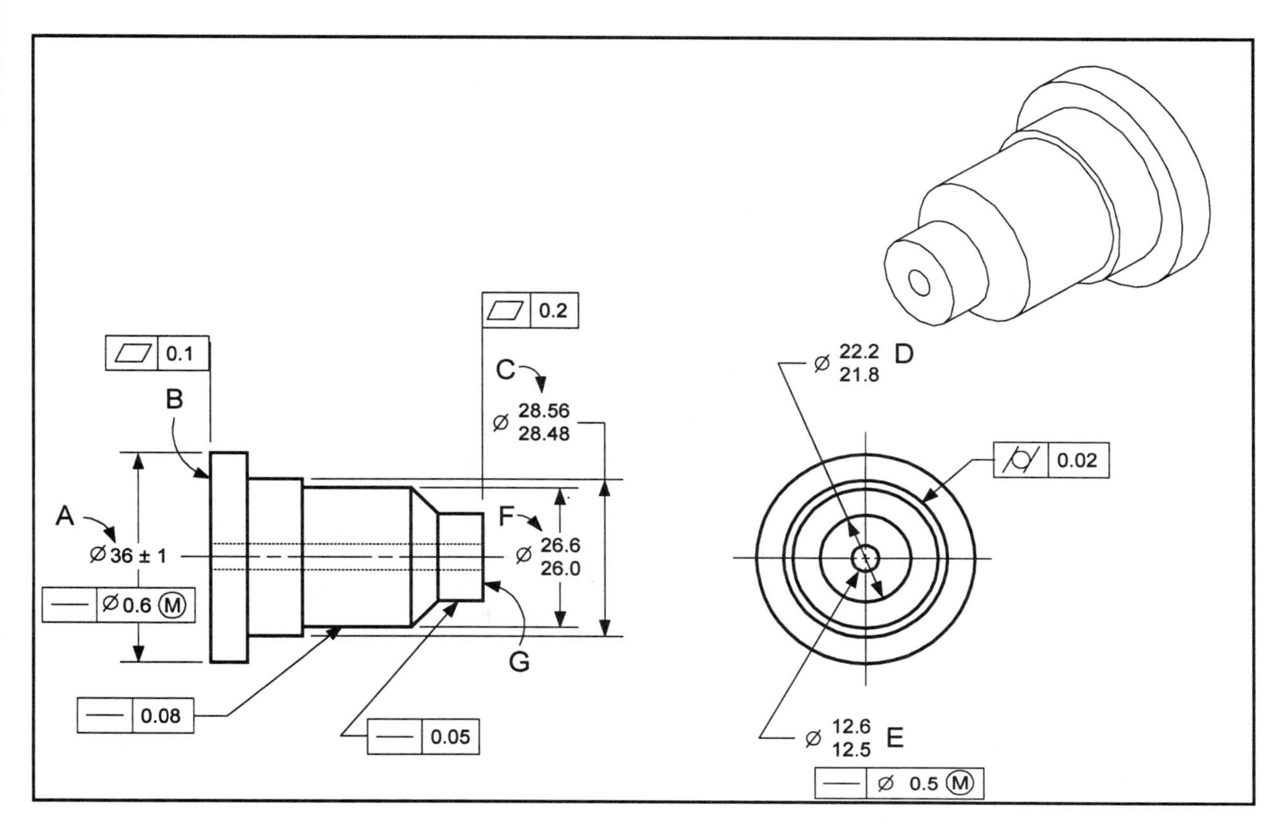

CHAPTER FIVE: QUESTIONS AND PROBLEMS

1. A system of symbols and rules that communicate to the drawing user how measurements are to be made
3. An assumed plane, axis, or point from which a measurement is made
5. a. Good parts rejected
   b. Bad parts accepted
7. A part feature that contacts a datum
9. [A]

11. How the part is mounted and located in its assembly
13. It is not shown; the general tolerance for angles either from the titleblock tolerances or a general note
15. a. Datum Plane E - 3
    b. Datum Plane B - 2
    c. Datum Plane A - 1
17. a. Rotation around the Z axis
    b. Movement along the Y axis

19.

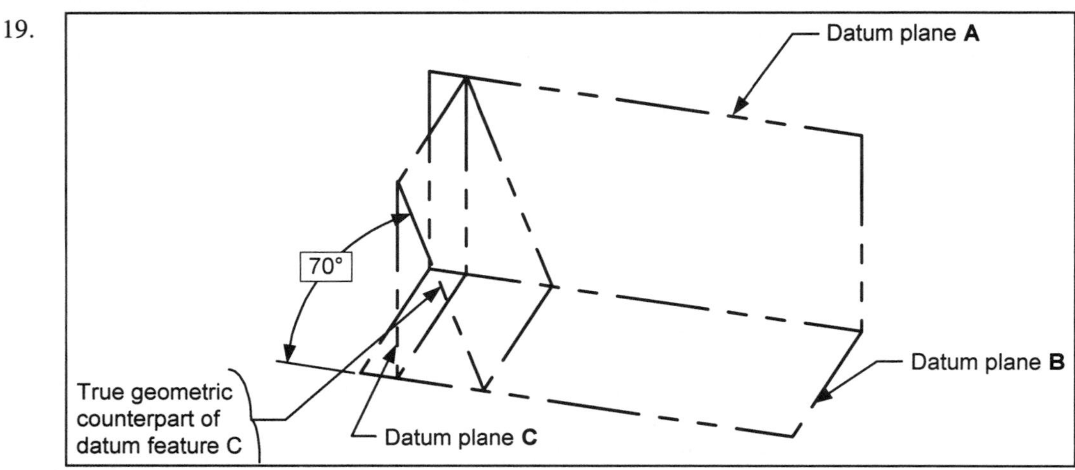

21. Symbols that describe the shape, size, and location of gage elements that are used to establish datum planes
23. To ensure there will be minimum variation between gages
25. 1. Basic dimensions should be used to define and locate the datum targets.
    2. The datum reference frame must restrain the part in all six degrees of freedom.
    3. The part dimensioning must ensure that the part will rest in the gage in only one orientation/location.
27.

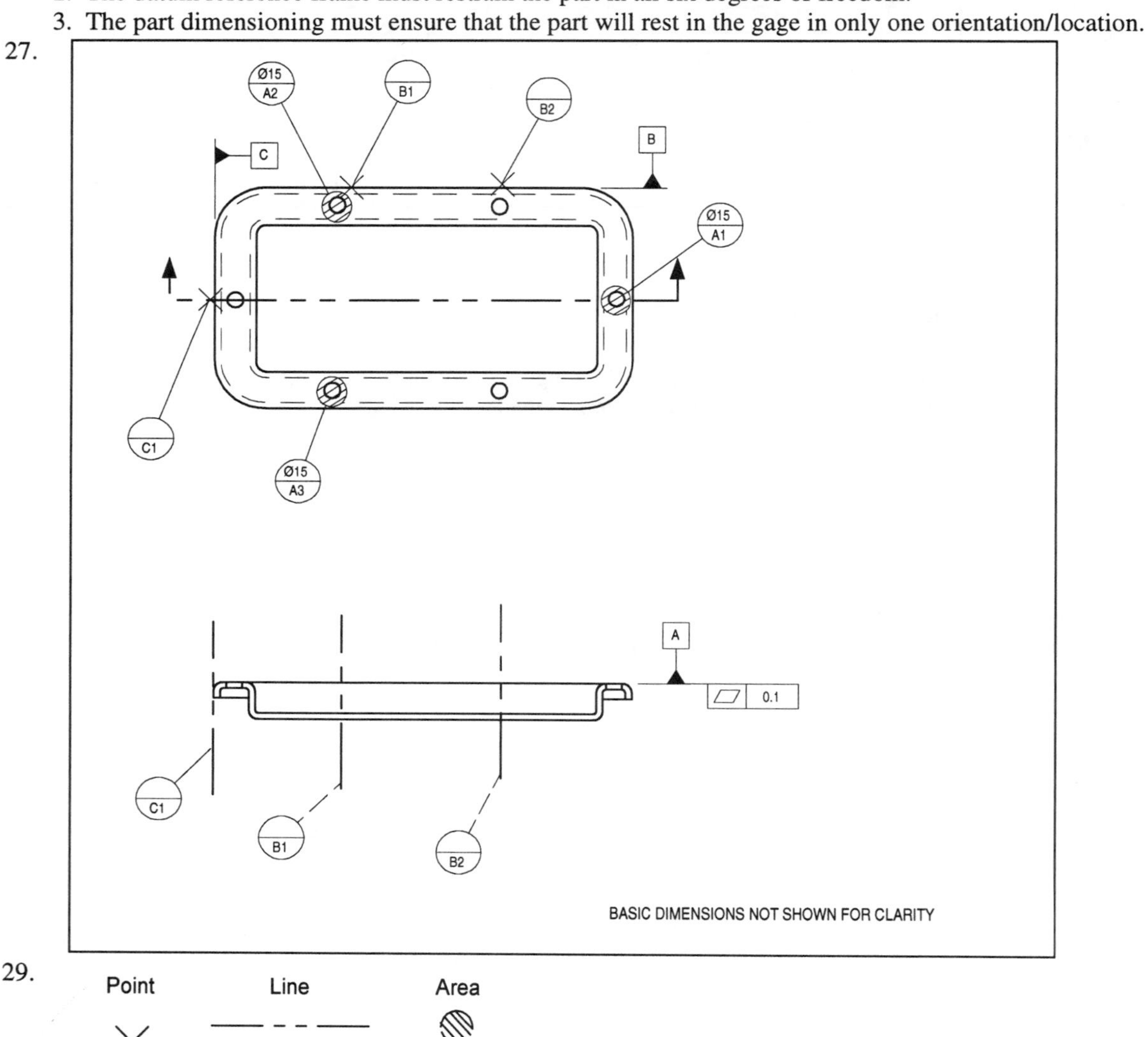

29.

Point	Line	Area
×	—— - - ——	(hatched circle)

31.

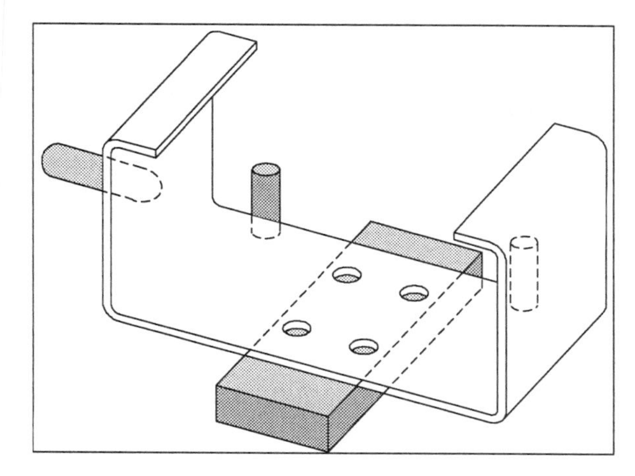

CHAPTER SIX: QUESTIONS AND PROBLEMS

1. Diameter; planar
3. Rule #2; modifier
5. Dimension; extension
7.

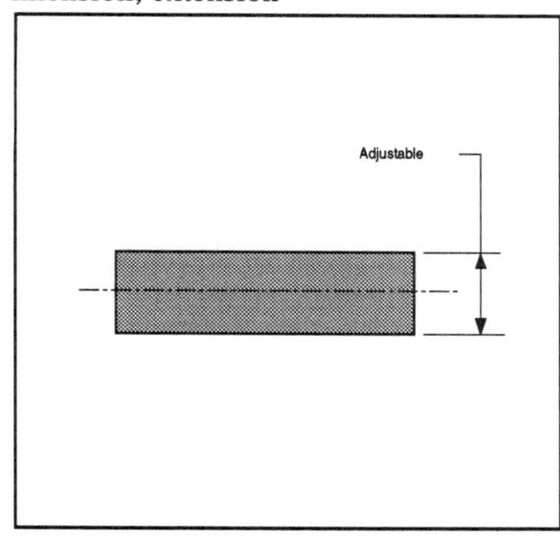

9.

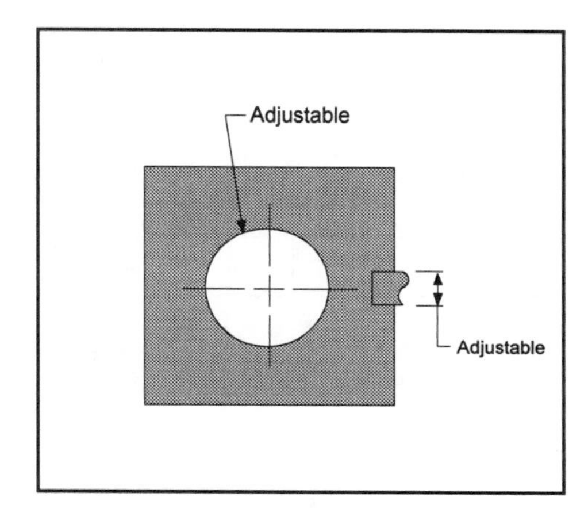

11. The datum feature simulator (gage) is a fixed size.
13. a. Where a straightness control is applied to a datum feature
    b. Where a secondary or tertiary datum feature of size in the same datum reference frame are controlled by a location or orientation control with respect to each other

15.

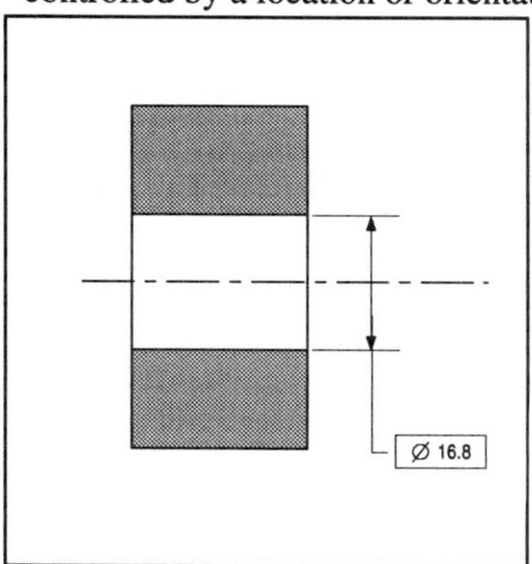

17.

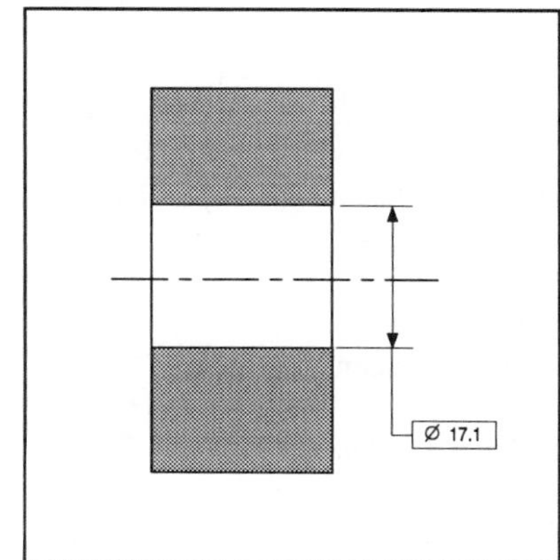

19.

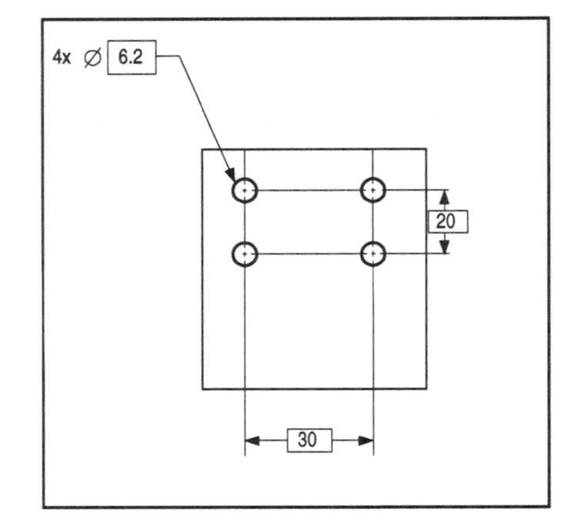

21.

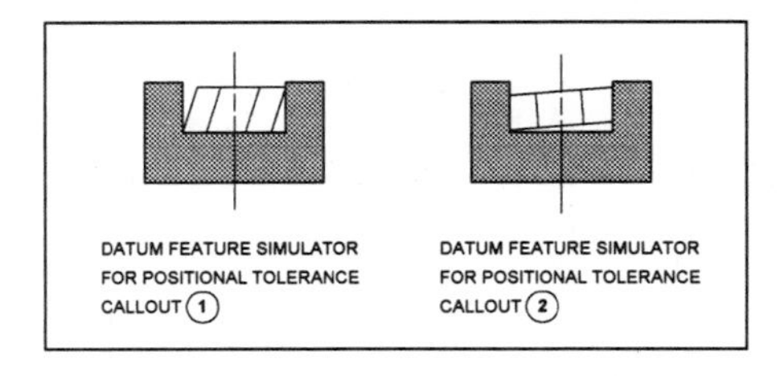

CHAPTER SEVEN: QUESTIONS AND PROBLEMS

1. Two parallel planes 0.2 apart
3. A general note
5. The tolerance zone would be oriented to both datum plane A and datum plane B.
7. a. The tolerance zone is two parallel planes.
   b. The tolerance value is the distance between the planes.
   c. All elements of the surface must be within the tol. zone.
   d. The flatness of the surface is also controlled.
9.

If the actual size of dia. B is...	The bonus tolerance possible is...	The perpendicularity tolerance zone diameter would be...
52.0	0	0.1
51.9	0.1	0.2
51.8	0.2	0.3

11. a. A bonus tolerance is permissible
    b. A fixed gage may be used.
    c. The axis or centerplane must be within the tolerance zone.
13. a. Two parallel planes
    b. A cylinder
15. Two parallel planes 0.1 apart.
17.

The flatness of surface. . .	Is limited to. . .
B	0.2
C	1.0

19. Yes
21. a. Tolerance zone is usually a cylinder
    b. Basic angle orients the tolerance zone in one direction
    c. Implied basic angle applies in the other direction
23. Two parallel planes 0.1 apart
25. The dimension between the surfaces
27. 0.1
29. 2 parallel planes 0.2 apart
31. a. Illegal; the tolerance value is too large.
    b. Illegal; the toleranced feature is perpendicular.
    c. Legal
    d. Legal
    e. Illegal; the toleranced feature is perpendicular.
    f. Illegal; cannot be parallel to itself

CHAPTER EIGHT: QUESTIONS AND PROBLEMS

1. A geometric tolerance that defines the location tolerance of a feature of size from its true position
3. a. Implied basic 90° angles
   b. Implied basic zero dimension
5. a. The distance between the features of size
   b. The location of the features of size
   c. The coaxiality of features of size
   d. The symmetry of features of size
7. The axis or centerplane of a FOS must be within the tolerance zone.
9. Two parallel planes 0.1 apart
11. a. Virtual condition boundary tolerance zone
    b. Bonus tolerance is permissible.
    c. A functional gage may be used.
13.

For the TOP callout labeled...	The shape of the tolerance zone is...	The max permissible bonus is...	The max permissible datum shift is...
A	**2.8 dia. boundary**	**0.6**	**0.6**
B	**42.3 dia. cylinder**	**1.0**	**0.6**
C	**5.9 boundary**	**0.8**	**0.6**

15. A gage that verifies functional requirements of part features as defined by the geometric tolerances
17. A sketch of a functional gage
19.

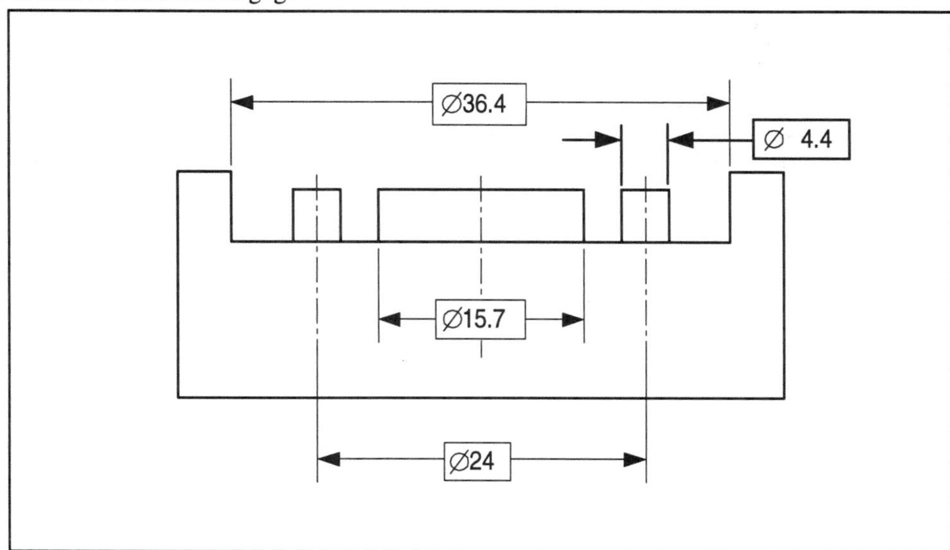

CHAPTER NINE: QUESTIONS AND PROBLEMS

1. cylindrical
3. 0.2 0.3 0.4 0.5 0.6
5.

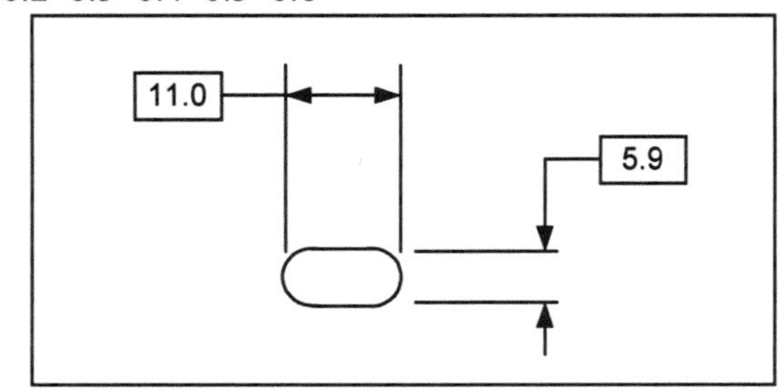

7. Projected tolerance zone
9.

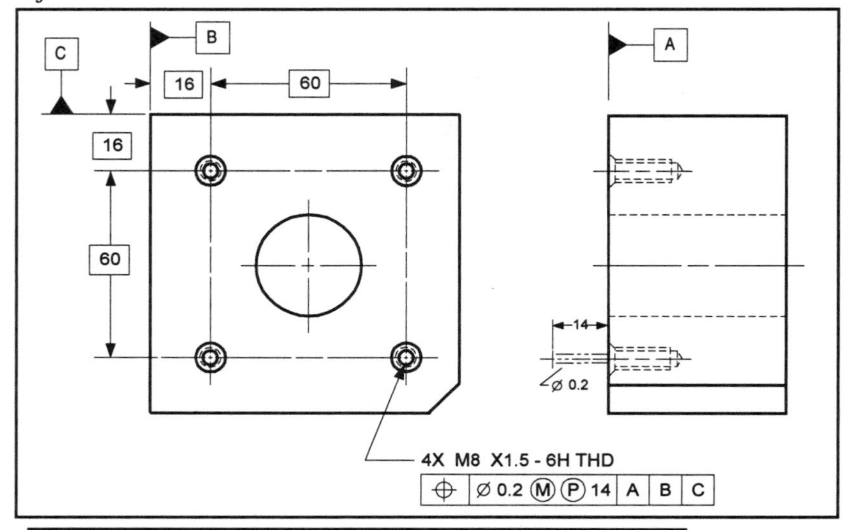

11.

Position Tolerance Zone Width at Centerline				
Toleranced Feature AME	Datum Feature AME 10.0	10.2	10.4	10.6
14.2	2.0	2.2	2.4	2.6
14.4	2.2	2.4	2.6	2.8
14.6	2.4	2.6	2.8	3.0

13.

Toleranced Hole AME	⌖ Tolerance Zone Diameter
4.6	0.2
4.4	0.4
4.2	0.6
4.0	0.8

15. 4.0
17. A tolerance stack is a calculation used to find the extreme max. or min. distance on a part.
19. Max. X = 16.6 Min. X = 15.4
21. 0.7 (for housing) 0.7 (for cover)
23. 0.3
    0.3

CHAPTER TEN: QUESTIONS AND PROBLEMS

1. A cylinder coaxial with the datum axis
3. The concentricity control
5. Yes
7. 0.01
9. a. Illegal; dia. symbol missing
   b. Illegal; cannot use projected tolerance zone modifier
   c. Illegal; cannot use RFS modifier
   d. Legal
   e. Illegal; cannot use MMC modifier
11. Two parallel planes 0.4 apart
13. Median points
15.

CONCEPT	SYMMETRY	POSITION (RFS)
Tolerance zone shape	Two parallel planes	Two parallel planes
What characteristic of the toleranced feature must be within the tolerance zone?	Median points of two point measurement	Centerplane of AME
Does Rule #1 apply to the toleranced feature?	Yes	Yes
What type of characteristics of the toleranced feature are being controlled?	Location; orientation	Location; orientation

17.

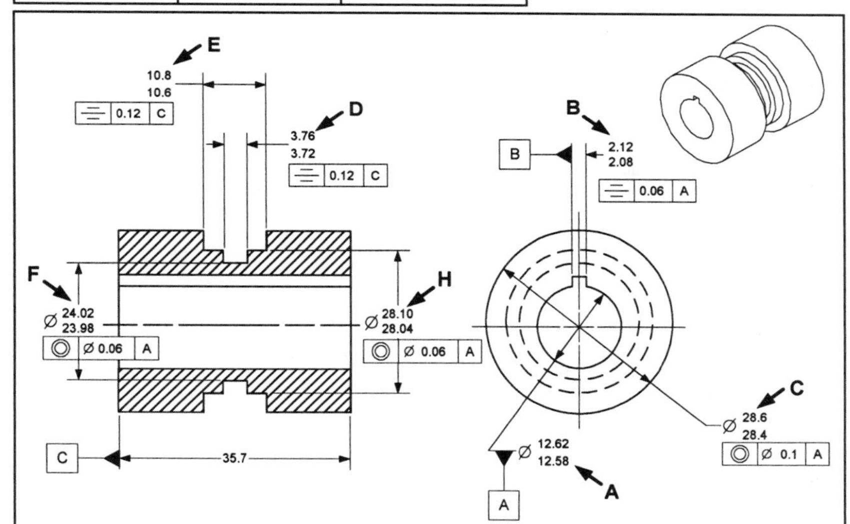

CHAPTER ELEVEN: QUESTIONS AND PROBLEMS

1. A composite tolerance that is to control the functional relationship (location, orientation, and form) of one or more features to a datum axis
3. Two coaxial circles
5.

DIA	MAX POSSIBLE AXIS OFFSET FROM DATUM AXIS A
B	0.1
C	0.15
D	0.4

7. a. Legal c. Illegal; cannot use MMC modifier
   b. Legal d. Illegal; cannot use diameter modifier
9. Two coaxial cylinders
11. a. Illegal; cannot use diameter modifier
    b. Illegal; cannot use projected tolerance zone modifier
    c. Illegal; cannot use RFS modifier
    d. Legal
13.

QUESTION	APPLIES TO			
	DIA B	DIA C	DIA D	DIA E
The size of the diameter is limited to?	0.2	0.4	0.2	0.2
The roundness of the diameter is limited to?	0.2	0.4	0.06	0.2
The maximum offset between the diameter axis and datum axis A is...	0.6	0.2	0.03	N/A
Describe the tolerance zone for the runout controls applied to the diameter.	2 coaxial cylinders with 1.2 radial seperation	2 coaxial cylinders with 0.4 radial seperation	2 coaxial cylinders with 0.06 radial seperation	N/A
What is the outer boundary (virtual condition) of this diameter?	25.4	13.2	14.66	10.2

15.

DISTANCE	MAX	MIN
A	2.75	1.25
B	2.60	1.50
C	3.10	1.10
D	1.10	0.85
E	2.15	1.05

CHAPTER TWELVE: QUESTIONS AND PROBLEMS

1. Form
3. The exact profile of a part as described by basic dimensions
5. A uniform boundary 0.8 wide, centered around the true profile
7.

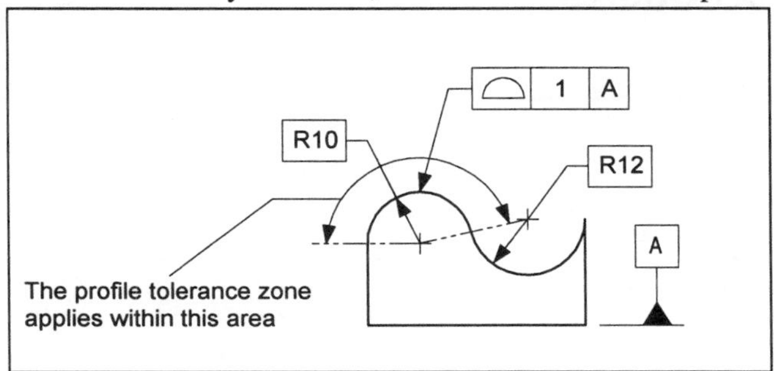

9. a. Clear definition of the tolerance zone
   b. Communicates datums and datum sequence
   c. Eliminates accumulation of tolerances
11. Location (or size) 1.5 A B C
    Orientation 1.0 B
    Form 0.2
13. Two uniform lines at any cross section of the surface
15. a. Illegal - in conflict with the location dimension
    b. Legal
    c. Illegal - tolerance value too large
    d. Legal
17.

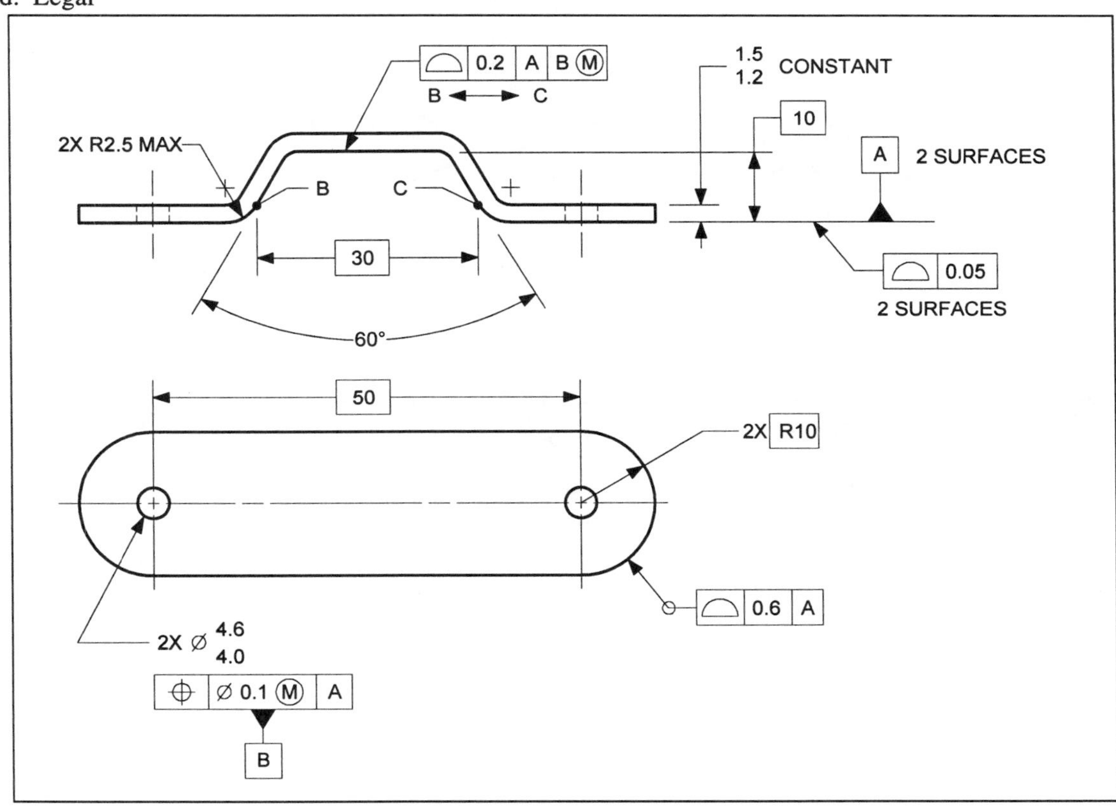

APPENDIX B

Feature Control Frame Proportion Charts

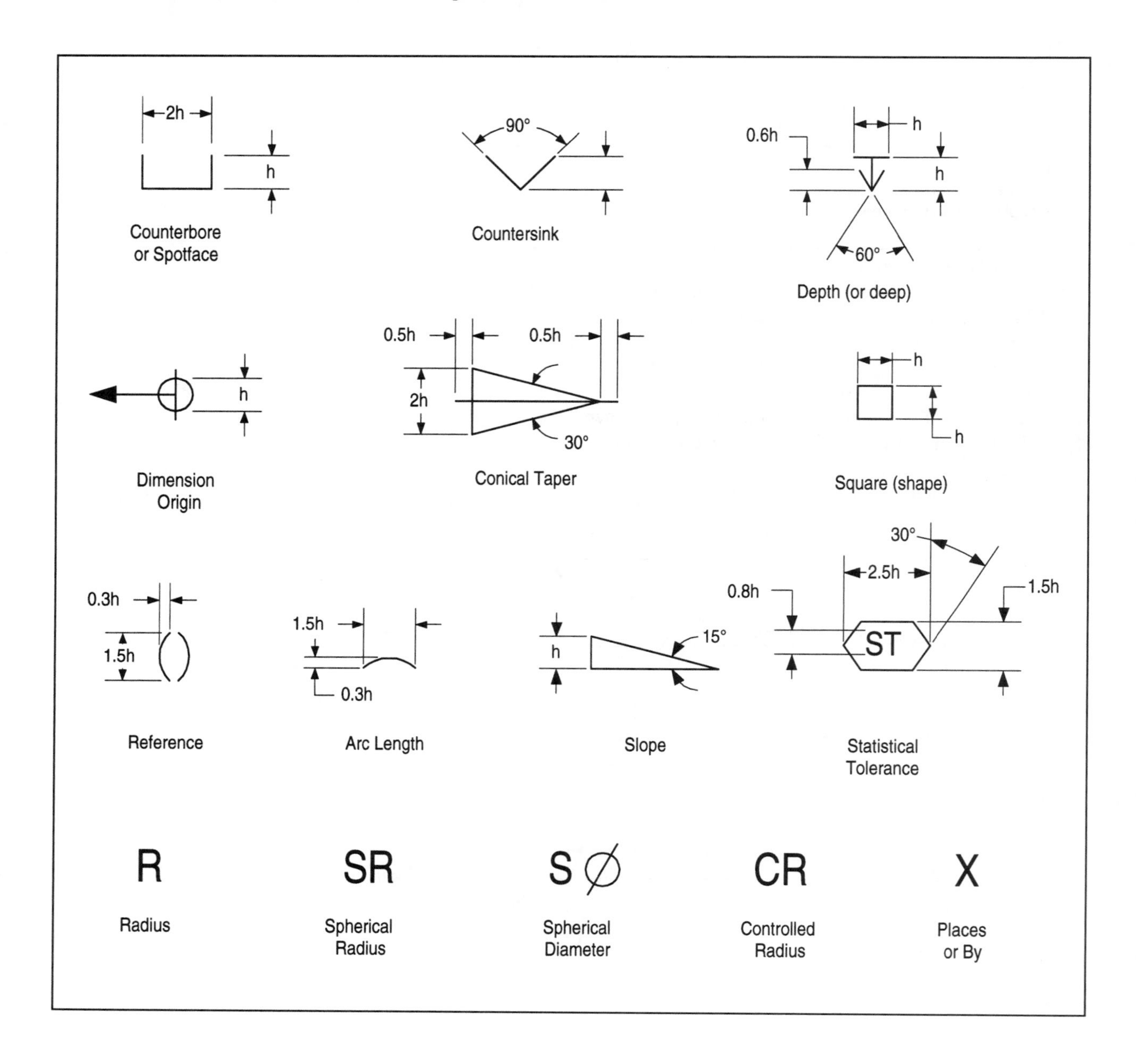

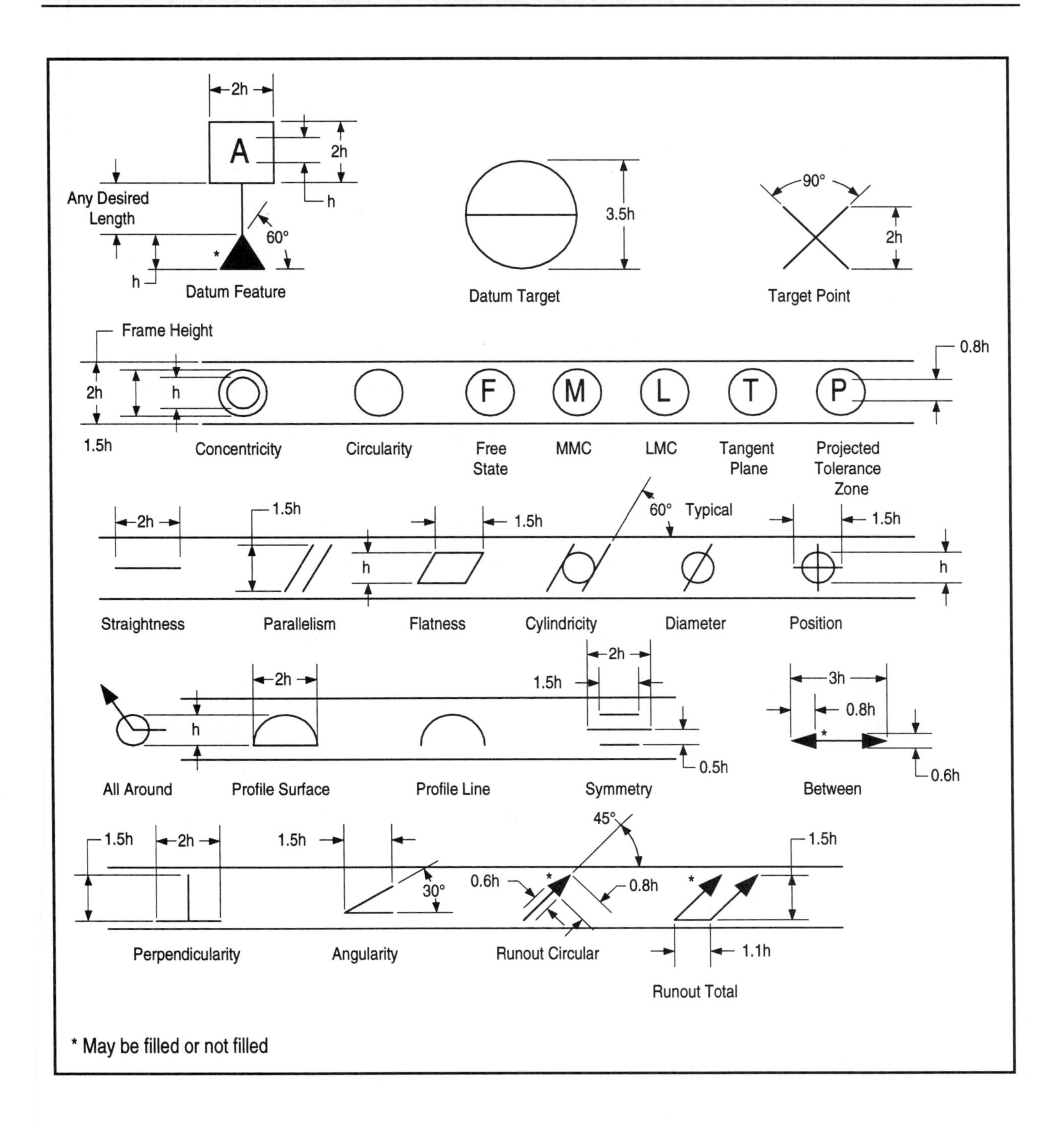
2h
A
2h
h
Any Desired Length
60°
h
Datum Feature
3.5h
Datum Target
90°
2h
Target Point
Frame Height
0.8h
2h
h
F
M
L
T
P
1.5h
Concentricity
Circularity
Free State
MMC
LMC
Tangent Plane
Projected Tolerance Zone
2h
1.5h
h
1.5h
60° Typical
1.5h
h
Straightness
Parallelism
Flatness
Cylindricity
Diameter
Position
2h
1.5h
2h
3h
0.8h
h
0.5h
0.6h
All Around
Profile Surface
Profile Line
Symmetry
Between
45°
1.5h
2h
1.5h
1.5h
30°
0.6h
0.8h
1.1h
Perpendicularity
Angularity
Runout Circular
Runout Total
* May be filled or not filled

APPENDIX C

ISO Dimensioning Standards

The following documents must be considered when adopting ISO/GD&T standards:

1.	ISO/1101-	Technical Drawings Geometrical Tolerancing
2.	ISO/5458-	Technical Drawings Positional Tolerancing
3.	ISO/5459-	Technical Drawings Datums and Datum Systems
4.	ISO/2692-	Technical Drawings Maximum Material Principle
5.	ISO/3040-	Technical Drawings Cones
6.	ISO/1660-	Technical Drawings Profiles
7.	ISO/129-	Technical Drawings General Principles
8.	ISO/406-	Technical Drawings Linear and Angular Dimensions
9.	ISO/10578	Technical Drawings Projected Tolerance Zones
10.	ISO/2692:1988/DAM 1	Technical Drawings Least Material Principle
11.	ISO/8015	Technical Drawings Fundamental Tolerance Principle
12.	ISO/7083	Technical Drawings Symbols Proportions
13.	ISO/10579	Technical Drawings Non-rigid Parts

Additional ISO standards involved:

1. ISO/1000- SI Units
2. ISO/286- Limits & Fits
3. ISO/TR5460- Technical Drawings - Verification Principles
4. ISO/2768-2 - General Geometrical Tolerances
5. ISO/1302- Surface Texture
6. ISO/2768-1 Tolerances for Linear and Angular Dimensions
7. Other peripheral standards on screw threads, gears, drills, welding, etc., may also be required for coverage beyond Y14.5 for product design.

APPENDIX D

ASME Y14.5/ISO Comparison Chart

Symbols	ASME Y14.5M	ISO
Feature Control Frame	⌖ | Ø0.03 Ⓜ | A | B | C	⌖ | Ø0.03 Ⓜ | A | B | C
Diameter	Ø	Ø
Spherical Diameter	SØ	SØ
Maximum Material Condition	Ⓜ	Ⓜ
Least Material Condition	Ⓛ	Ⓛ
Regardless of Feature Size	NONE	NONE
Projected Tolerance Zone	Ⓟ	Ⓟ
Free State	Ⓕ	Ⓕ
Tangent Plane	Ⓣ	Ⓣ (proposed)
Statistical Tolerance	⟨ST⟩	NONE
Radius	R	R
Controlled Radius	CR	NONE
Spherical Radius	SR	SR
Basic Dimension	50	50
Datum Feature	* B	* B or *
Datum Target	Ø8 A1, A1 8	Ø8 A1, A1 8
Target Point	×	×
Dimension Origin	⌀⟶	⌀⟶
Reference Dimension	(50)	(50)
Number of Places	8X	8X
Counterbore/Spotface	⌴	⌴ (proposed)
Countersink	⌵	⌵ (proposed)
Depth/Deep	↧	↧ (proposed)
Square	□	□
All Around	⟶○⟶	NONE
Dimension Not to Scale	150	150
Arc Length	⌒150	⌒150
Between	⟷	NONE
Slope	◺	◺
Conical Taper	▷	▷
Envelope Principle	NONE (implied)	Ⓔ

* May be filled or not filled

APPENDIX E

Bibliography

1. American Society of Mechanical Engineers. *Dimensioning and Tolerancing: ASME Y14.5M-1994 [Revision of ANSI Y14.5M-1982 (R1988)].* NY: ASME, 1995.

2. American Society of Mechanical Engineers. *Dimensioning and Tolerancing: ANSI Y14.5M-1982.* NY: ASME, 1983.

3. Henzold, G. *Handbook of Geometrical Tolerancing: Design, Manufacturing and Inspection.* England: Wiley & Sons, 1995.

4. Krulikowski, Alex. *Geometric Dimensioning and Tolerancing Self-Study Workbook.* Westland, MI: Effective Training, 1990.

5. Krulikowski, Alex. *Geometric Tolerancing Applications Workbook.* Westland, MI: Effective Training, 1994.

6. Krulikowski, Alex. *Tolerance Stacks: A Self-Study Course.* Westland, MI: Effective Training, 1992.

7. Krulikowski, Alex. *Tolerance Stacks Using GD&T.* Westland, MI: Effective Training, 1994.

GLOSSARY

Actual Local Size - The value of any individual distance at any cross section of a FOS.

Actual Mating Envelope (AME) of an External Feature of Size - A similar perfect feature counterpart of the smallest size that can be circumscribed about the feature so that it just contacts the surfaces at the highest points.

Actual Mating Envelope (AME) of an Internal Feature of Size - A similar perfect feature counterpart of the largest size that can be inscribed within the feature so that it just contacts the surfaces at their highest points.

All-Around Symbol - A circle placed on the bend of the leader line of a profile control.

Angularity - The condition of a surface, centerplane or axis being exactly at a specified angle

Angularity Control - A geometric tolerance that limits the amount a surface, axis, or centerplane is permitted to vary from its specified angle.

ASME Y14.5M-1994 - The national standard for dimensioning and tolerancing in the United States. ASME stands for American Society of Mechanical Engineers. The Y14.5 is the standard number. "M" is to indicate the standard is metric, and 1994 is the date the standard was officially approved.

Axis Theory - The axis (or centerplane) of a FOS must be within the tolerance zone.

Basic Dimension - A numerical value used to describe the theoretically exact size, true profile, orientation, or location of a feature or datum target.

Between Symbol - A double ended arrow that indicates the tolerance zone extends to include multiple surfaces.

Bi-directional Control- Where the location of a hole is controlled to a different tolerance value in two directions.

Bilateral Tolerance - A tolerance that allows the dimension to vary in both the plus and minus directions.

Bonus Tolerance - An additional tolerance for a geometric control. Whenever a geometric tolerance is applied to a FOS, and it contains an MMC (or LMC) modifier in the tolerance portion of the feature control frame, a bonus tolerance is permissible.

Boundary - The word "BOUNDARY" is placed beneath the feature control frames to invoke a boundary control.

Cartoon Gage - A sketch of a functional gage. A cartoon gage defines the same part limits that a functional gage would, but it does not represent the actual gage construction of a functional gage.

Circularity - A condition where all points of a surface of revolution, at any section perpendicular to a common axis, are equidistant from that axis.

Circularity Control - A geometric tolerance that limits the amount of circularity on a part surface.

Circular Runout - A composite control that affects the form, orientation, and location of circular elements of a part feature relative to a datum axis.

Circular Runout Control - A geometric tolerance that limits the amount of circular runout of a part surface.

Coaxial Datum Features - When coaxial diameters are used to establish a datum axis.

Coaxial Diameters - Two (or more) diameters that are shown on the drawing as being on the same centerline (axis).

Composite Control - Controls the form, location, and orientation of a part feature simultaneously (in a single gage reading).

Concentricity - The condition where the median points of all diametrically opposed elements of a cylinder (or a surface of revolution) are congruent with the axis of a datum feature.

Concentricity Control - A geometric tolerance that limits the concentricity error of a part feature.

Controlled Radius - A radius with no flats or reversals allowed. The symbol for a controlled radius is "CR."

Coordinate Tolerancing - A dimensioning system where a part feature is located (or defined) by means of rectangular dimensions with given tolerances.

Coplanar Datum Features - Two or more datum features that are on the same plane.

Coplanar Surfaces - Two or more surfaces that are on the same plane.

Cylindrical FOS - Contains one feature: the cylindrical surface.

Cylindricity - A condition of a surface of revolution in which all points of the surface are equidistant from a common axis.

Cylindricity Control - A geometric tolerance that limits the amount of cylindricity error permitted on a part surface.

Datum - A theoretically exact plane, point or axis from which a dimensional measurement is made.

Datum Feature - A part feature that contacts a datum.

Datum Feature Simulator - The inspection equipment (or gage surfaces) used to establish a datum.

Datum Reference Frame - A set of three mutually perpendicular datum planes.

Datum Shift - The allowable movement, or looseness, between the part datum feature and the gage.

Datum System - A set of symbols and rules that communicate to the drawing user how dimensional measurements are to be made.

Datum Target - A symbol that describes the shape, size, and location of gage elements that are used to establish datum planes or axes.

Dimension - A numerical value expressed in appropriate units of measure and used to define the size, location, orientation, form, or other geometric characteristics of a part.

Engineering Drawing - A document that communicates a precise description of a part. This description consists of pictures, words, numbers, and symbols.

Equal Bilateral Tolerance - A tolerance where the allowable variation from the nominal value is the same in both directions.

Feature - A general term applied to a physical portion of a part, such as a surface, hole, or slot.

Feature Control Frame - A rectangular box that is divided into compartments within which the geometric characteristic symbol, tolerance value, modifiers, and datum references are placed.

Feature of Size (FOS) - One cylindrical or spherical surface, or a set of two opposed elements or opposed parallel surfaces, associated with a size dimension.

Feature of Size Dimension - A dimension that is associated with a feature of size.

Fixed Fastener Assembly - Where the fastener is held in place (restrained) into one of the components of the assembly.

Fixed Fastener Formula- $H = F + 2T$ or $T = \frac{H - F}{2}$

Where:

T = position tolerance diameter

H = MMC of the clearance hole

F = MMC of the fastener

Flatness - The condition of a surface having all of its elements in one plane.

Flatness Control - A geometric tolerance that limits the amount of flatness error a surface is allowed.

Floating Fastener Assembly - Where two (or more) components are held together with fasteners (such as bolts and nuts), and both components have clearance holes for the fasteners.

Floating Fastener Formula - T = H - F
Where:
T = position tolerance diameter (for each part)
H = MMC of the clearance hole
F= MMC of the fastener

Functional Dimensioning - A dimensioning philosophy that defines a part based on how it functions in the final product.

Functional Gage - A gage that verifies functional requirements of part features as defined by the geometric tolerances.

Fundamental Dimensioning Rules - A set of general rules defined by ASME for dimensioning and interpreting drawings.

Geometric Characteristic Symbols (14) - The symbols are divided into five categories: form, profile, orientation, location, and runout.

Geometric Dimensioning and Tolerancing **(GD&T)** - A set of fourteen symbols used in the language of GD&T. It consists of well-defined of symbols, rules, definitions and conventions, used on engineering drawings to accurately describe a part. GD&T is a precise mathematical language that can be used to describe the size, form, orientation, and location of part features. GD&T is also a design philosophy on how to design and dimension parts.

Go Gage - A gage that is intended to fit into (for an internal FOS) or fit over (for an external FOS) the FOS.

Great Myth of GD&T - The misconception that geometric tolerancing raises product costs.

Implied Basic 90° Angles--A 90° basic angle applies where centerlines of features in a pattern (or surfaces shown at right angles on a drawing) are located and defined by basic dimensions, and no angle is specified.

Implied Basic Zero Dimension - Where a centerline or centerplane of a FOS is shown in line with a datum axis or centerplane, the distance between the centerlines or centerplanes is an implied basic zero.

Implied Datum - An assumed plane, axis, or point from which a dimensional measurement is made.

Inclined Datum Feature - A datum feature that is at an angle other than 90°, relative to the other datum features.

Inner Boundary (IB) - A worst-case boundary generated by the smallest feature of size minus the stated geometric tolerance (and any additional tolerance, if applicable).

International Standards Organization (ISO) - The organization that published an associated series of standards on dimensioning and tolerancing.

Least Material Condition - The condition in which a feature of size contains the least amount of material everywhere within the stated limits of size.

Limit Tolerance - When a dimension has its high and low limits stated. In a limit tolerance, the high value is placed on top, and the low value is placed on the bottom.

Maximum Material Condition - The condition in which a feature of size contains the maximum amount of material everywhere within the stated limits of size

Median Point - The mid-point of a two-point measurement.

Modifiers (8) - Communicate additional information about the drawing or tolerancing of a part.

Multiple Single-Segment Profile Control - When two or more profile controls are tolerancing a surface relative to different datums.

Multiple Single-Segment TOP Control - When two (or more) single segment TOP callouts are used to define the location, spacing, and orientation of a pattern of FOS.

No-Go Gage - A gage that is not intended to fit into or over a FOS. A No-Go gage is made to the LMC limit of the FOS.

Non-Feature of Size Dimension - A dimension that is not associated with a FOS.

Outer Boundary (OB) - A worst-case boundary generated by the largest feature of size plus the stated geometric tolerance (and any additional tolerance, if applicable).

Parallelism - The condition that results when a surface, axis or centerplane is exactly parallel to a datum.

Parallelism Control - A geometric tolerance that limits the amount a surface, axis, or centerplane is permitted to vary from being parallel to the datum.

Perpendicularity - The condition that results when a surface, axis, or centerplane is exactly 90° to a datum.

Perpendicularity Control - A geometric tolerance that limits the amount a surface, axis, or centerplane is permitted to vary from being perpendicular to the datum.

Planar Datum - The true geometric counterpart of a planar datum feature.

Planar FOS - A FOS that contains two features: the two parallel plane surfaces.

Plus-minus Tolerance - The nominal or target value of the dimension is given first, followed by a plus-minus expression of a tolerance.

Primary Datum - The first datum plane that the part contacts in a dimensional measurement.

Profile - The outline of a part feature in a given plane.

Profile Control - A geometric tolerance that specifies a uniform boundary along the true profile that the elements of the surface must lie within.

Profile of a Line Control - A geometric tolerance that limits the amount of error for line elements relative to their true profile.

Profile of a Surface Control - A geometric tolerance that limits the amount of error a surface can have relative to its true profile.

Projected Tolerance Zone - A tolerance zone that is projected above the part surface.

Radius - A straight line extending from the center of an arc or circle to its surface.

Regardless of Feature Size - The term that indicates a geometric tolerance applies at any increment of size of the feature, within its size tolerance.

Rule #1 - Where only a tolerance of size is specified, the limits of size of an individual feature prescribe the extent to which variations in its form--as well as in its size--are allowed.

Rule #2 - RFS applies, with respect to the individual tolerance, datum reference, or both, where no modifying symbol is specified.

Secondary Datum - The second datum plane that the part contacts in a dimensional measurement

Simulated Datum - The plane (or axis) established by the datum feature simulator.

Simultaneous Engineering - A process where design is a result of input from marketing, engineering, manufacturing, inspection, assembly, and service.

Special-Case FOS Datum - When a FOS datum feature is referenced at MMC, but simulated in the gage at a boundary other than MMC.

Straightness (Axis or Centerplane) - The condition where an axis is a straight line (or, in the case of a centerplane, each line element is a straight line).

Straightness of a Line Element - The condition where each line element (or axis or centerplane) is a straight line.

Straightness Control (FOS) - A geometric tolerance that, when applied to a FOS, limits the amount of straightness error allowed in the axis or centerplane.

Straightness Control (Surface) - A geometric tolerance that, when directed to a surface, limits the amount of straightness error allowed in each surface line element.

Symmetry - The condition where the median points of all opposed elements of two or more feature surfaces are congruent with the axis or centerplane of a datum feature.

Symmetry Control - A geometric tolerance that limits the symmetry error of a part feature.

Tertiary Datum - The third datum plane that the part contacts in a dimensional measurement.

3-2-1 Rule - Defines the minimum number of points of contact required for a part datum feature with its primary, secondary, and tertiary datum planes.

Tolerance - The total amount that features of the part are permitted to vary from the specified dimension.

Tolerance Analysis Chart - A means of graphically displaying the limits of a part as defined by the print specifications

Tolerance of Position (TOP) Control - A geometric tolerance that defines the location tolerance of a FOS from its true position.

Tolerance Stack - A calculation used to find the extreme max. or min. distance on a part.

Total Runout - A composite control affecting the form, orientation, and location of all surface elements of a diameter (or surface) relative to a datum axis.

Total Runout Control - A geometric tolerance that limits the amount of total runout of a surface.

True Geometric Counterpart - The theoretical perfect boundary or best fit tangent plane of a specified datum feature.

True Position - The theoretically exact location of a FOS as defined by basic dimensions.

True Profile - The exact profile of a part feature as described by basic dimensions.

Unequal Bilateral Tolerance - A tolerance where the allowable variation is from the target value, and the variation is not the same in both directions.

Unilateral Tolerance - A tolerance where the allowable variation from the target value is all in one direction and zero in the other direction.

Variable Gage - A gage capable of providing a numerical reading of a part parameter.

Virtual Condition **(VC)** - A worst-case boundary generated by the collective effects of a feature of size specified at MMC or at LMC and the geometric tolerance for that material condition.

Virtual Condition Boundary Theory - A theoretical boundary limits the location of the surfaces of a FOS.

Worst-case Boundary **(WCB)** - A general term to refer to the extreme boundary of a FOS that is the worst-case for assembly. Depending upon the part dimensioning, a worst-case boundary can be a virtual condition, inner boundary, or outer boundary.

Zero Tolerance at MMC- A method of tolerancing part features that includes the tolerance geometric value with the FOS tolerance and states a zero at MMC in the feature control frame.

INDEX

E

F